PRINCIPLES OF REINFORCED CONCRETE CONSTRUCTION

MAXWELL PRESS

PRINCIPLES OF REINFORCED CONCRETE CONSTRUCTION

F. E. TURNEAURE

Dean of the college of Engineering university of wisconsin

E. R. MAURER

Professor of mechanics, university of wisconsin

MAXWELL PRESS

Chennai Trichy New Delhi

MAXWELL PRESS

This edition has been published in india by arrangement with Carson Books, UK

ISBN 978-93-5528-224-8

Maxwell Press
No. 44, Nallathambi Street,
Triplicane, Chennai 600 005

Printed and bound in India

MJP 1298

Publisher : C. Janarthanan

Publisher's Note

The legacy of a country is in its varied cultural heritage, historical literature, developments in the field of economy and science. The top nations in the world are competing in the field of science, economy and literature. This vast legacy has to be conserved and documented so that it can be bestowed to the future generation. The knowledge of this legacy is slowly getting perished in the present generation due to lack of documentation.

Keeping this in mind, the concern with retrospective acquiring of rare books has been accented recently by the burgeoning reprint industry. Maxwell Press is gratified to retrieve the rare collections with a view to bring back those books that were landmarks in their time.

In this effort, a series of rare books would be republished under the banner, "Maxwell Press". The books in the reprint series have been carefully selected for their contemporary usefulness as well as their historical importance within the intellectual. We reconstruct the book with slight enhancements made for better presentation, without affecting the contents of the original edition.

Most of the works selected for republishing covers a huge range of subjects, from history to anthropology. We believe this reprint edition will be a service to the numerous researchers and practitioners active in this fascinating field. We allow readers to experience the wonder of peering into a scholarly work of the highest order and seminal significance.

MAXWELL PRESS

PREFACE TO THE SECOND EDITION.

In the preparation of this edition numerous changes have been made and a large amount of material has been added. Results of recent important experiments have been included and the analytical treatment has been considerably extended in certain directions. New experimental data are noted in nearly every division of the subject, but of especial importance are the results on bond strength, the strength of beams in shear, and the strength of columns. The general subject of the deflection of beams has been treated at considerable length, both theoretically and experimentally. The treatment of T-beams has been extended and several diagrams have been added in Chapter V for use in designing this form of beam. The chapter pertaining to working stresses and general constructive details has been largely rewritten, especial attention being given to the subject of shear reinforcement. In Chapter VII the treatment of continuous beams has been considerably amplified. In the chapter on arches the methods of calculation have been more fully explained by means of an additional example, fully worked out, in which use is made of influence lines for fiber stress. A chapter has been added on chimneys, which includes a fairly complete analytical treatment of this subject. It is believed that the changes and additions which have been made will measurably enhance the usefulness of the work.

F. E. T.
E. R. M.

Madison, Wis.,
April, 1909.

PREFACE TO THE FIRST EDITION:

IN the present volume the authors have endeavored to cover, in a systematic manner, those principles of mechanics underlying the design of reinforced concrete, to present the results of all available tests that may aid in establishing coefficients and working stresses, and to give such illustrative material from actual designs as may be needed to make clear the principles involved.

The work is essentially divided into two parts: Chapters I to VI treat of the theory of the subject and the results of experiments, while the remaining chapters treat of the use of reinforced concrete in various forms of structures. In Chapter II the properties of plain concrete and of steel are considered to a sufficient extent to give accurate notions of their relation to the general subject in hand. The subjects of adhesion and of relative contraction and expansion are also discussed in this chapter. In Chapter III is given a full theoretical treatment of reinforced concrete, avoiding so far as possible empirical rules and methods; and in Chapter IV are presented the most important available tests on beams and columns, analyzed and correlated, so far as may be, with reference to theoretical principles. The subjects of working stresses and economical proportions are considered in Chapter V. In Chapter VI are brought together in convenient form all the formulas and diagrams needed for practical use. There are also included tables relating to reinforcing bars and a comprehensive table of the strength of floor slabs. This chapter is, for most purposes, complete in itself, so that the reader need not refer to any other portion of the work in order to use it in designing.

Following the theoretical portions are chapters on the application of reinforced concrete to building construction, arches, retaining walls, dams, and miscellaneous structures. In these chapters the analysis of various features is given, where the use of reinforced concrete involves problems new and unfamiliar. A complete general analysis of the solid arch rib is also given, which, the authors believe, offers advantages over the usual graphical method. It is primarily an analytical method, but may be shortened by obvious simple, graphical aids. Stresses in the concrete and steel are readily calculated by the use of diagrams in Chapter VI. In the chapters on the application of reinforced concrete it has not been the aim to cover practical construction in all its phases; for this the reader is referred to the more voluminous works on the subject. It is hoped, however, that as a treatment of the principles of design the work may prove of service to the student and the engineer.

F. E. TURNEAURE.
E. R. MAURER.

MADISON, WIS., Sept., 1907.

CONTENTS.

CHAPTER VI.

CHAPTER VII.

CHAPTER VIII.

CHAPTER IX.

CHAPTER X.

CHAPTER XI.

REINFORCED-CONCRETE CONSTRUCTION.

CHAPTER I.

INTRODUCTORY.

1. **Historical Sketch.**—The invention of reinforced concrete is usually credited to Joseph Monier, but his first constructions are antedated by those of Lambot, who in 1850 constructed a small boat of reinforced concrete and in 1855 exhibited the same at the Paris Exposition. In this latter year Lambot took out patents on this form of construction; it was regarded by him as especially well adapted to shipbuilding, reservoir work, etc.

In 1861, Monier, who was a Parisian gardener, constructed tubs and tanks of concrete surrounding a framework or skeleton of wire. In the same year Coignet announced his principles for reinforcing concrete, and proposed construction of beams, arches, pipes, etc. Both he and Monier executed some work in the new material at the Paris Exposition of 1867. In this year Monier took out patents on his reinforcement. It consists of two sets of parallel bars, one set at right angles to and lying upon the other, thus forming a mesh of bars. This system, and slight modifications of it, are extensively used at the present time, particularly for slab reinforcement. Though even the early Monier patents covered principles of wide application, still the early work in reinforced concrete was confined to a comparatively narrow field.

In 1884–5 the German and American rights of the Monier patents fell into the hands of German engineers. One of these, G. A. Wayss, and J. Bauschinger at once began an experimental investigation of the Monier system, and in 1887 they published their findings. The investigation proved reinforced concrete a valuable means of construction, and furnished some formulas and methods for design. From this time on, the use of reinforced concrete in Austria spread rapidly, and a few years ago the engineers of that country were credited with having done more to develop the new construction than those of any other country. Among these engineers should be mentioned Melan, who in the early 90's originated a system in which I or T beams are the principal element of strength, providing compressive as well as tensile strength. In Germany government regulations hindered the application of reinforced concrete for a time, but now it is widely used in that country. Over two hundred systems of reinforcement, it has been stated, have been developed in Germany alone.

In France the Monier system was never developed as in countries already mentioned. Here, as elsewhere, many other systems of reinforcement were invented from time to time, among which should be mentioned that of Hennebique, who was probably the first to use stirrups and "bent-up" bars. This system is in general use, and the elements of Hennebique's system are probably more widely used than those of any other.

In England and America the first use of iron or steel with concrete arose in the effort to fireproof the former by means of the latter. Attempting to utilize also the strength of concrete, Hyatt built beams of concrete reinforced with metal in various ways, and with Kirkaldy of London performed tests on such beams and published the results of the investigation in 1877. The first reinforced-concrete work in the United States was done in 1875 by W. E. Ward, who constructed a building in New York state in which walls, floor-beams, and roof were made of concrete reinforced with metal to provide tensile strength. But the Pacific Coast saw the actual early development of this form of construction. H. P. Jackson, G. W. Percy, and E. L. Ransome were the pioneer workers. Jackson has been credited with reinforced constructions dating as far back as 1877, but Ransome executed the most notable early examples. Among these are a warehouse (1884 or '85), a factory building a few years later, the building of the California Academy of Science (1888 or '89), and the museum building of Leland Stanford Junior University (1892). Percy was the architect of the last two. The museum building contains spans of 45 feet and is reinforced throughout. This and the Academy building withstood the recent earthquake remarkably well—the museum better than its two brick annexes.

Other pioneer constructors in reinforced concrete in this country were F. von Emperger and Edwin Thacher. The former introduced the Melan system (1894) and built the first reinforced arch bridges of considerable span. Thacher also was—and still is—a bridge-builder. His first large reinforced-concrete bridge was built in 1896 and was without precedent here or in Europe.

America is the home of the "patent bar". Both Ransome and Thacher invented bars known by their respective names, the patented feature of which is to furnish a "grip" between bar and concrete; besides these two there are several others on the market designed to give additional grip or bond. There are also patented bars for supplying "shear reinforcement". Some of these forms have been introduced into Europe.

Reinforced-concrete construction has had a remarkable development, particularly in the last decade, and is now regarded by engineers and architects generally as a safe form of construction with a wide field of economical application. Common practice has already established itself in some directions, and rational principles are available for much design work. Outstanding uncertainties are under investigation in many quarters, and the time is not far distant when "good practice" in reinforced concrete will have been established.

2. Use and Advantages of Reinforced Concrete.—A combination of steel and concrete constitutes a form of construction possessing to a large degree the advantages of both materials without their disadvantages. It will be desirable at the outset to consider briefly these advantages in order better to appreciate the field in which this type of construction is likely to be most successful.

Steel is a material especially well suited to resist tensile stresses, and for such purposes the most economical form—the solid compact bar—is well adapted. To resist compressive stresses steel must be made into more expensive forms, consisting of relatively thin parts widely spread, in order to provide the necessary lateral rigidity. A serious disadvantage in the use of steel in many locations is its lack of durability; and, again, a comparatively low degree of heat destroys its strength, thus rendering it necessary to add a protective covering where a fire-proof structure is demanded. Steel is a relatively expensive building material, and its cost tends to increase.

Concrete is characterized by low tensile strength, relatively high compressive strength, and great durability. It is a good fire-proof material, and therefore serves as a good fire-proof covering for steel. It is also found that steel well covered by concrete is thoroughly protected from corrosion. Concrete is also a comparatively cheap material and is readily available in almost any location.

In the design of structural members these qualities of steel and concrete will lead to the use of the two materials about as follows: For those structural members carrying purely tensile stresses steel must be employed, but it may be surrounded by concrete as a protection against corrosion and fire, or merely for the sake of appearance. For those members sustaining purely compressive stresses concrete is fundamentally the better and cheaper material. With concrete costing 30 cents per cubic foot, for example, and steel 4 cents per pound, or about \$20.00 per cubic foot, and with working stresses of 400 and 15,000 lbs/in^2, respectively, the relative cost of the two materials for carrying a given load is as $\frac{30}{400}$ is to $\frac{2000}{15,000}$, or as 45 is to 80. For large and compact compressive members plain concrete will therefore naturally be used, especially where durability is a factor. For more slender members, however, such as long columns, plain concrete is too brittle a material, and therefore too much affected by secondary and unknown stresses to be satisfactory; and for such members steel alone, or the two materials in combination, will preferably be used. Steel may be used with concrete in the form of small rods to reinforce the concrete; or it may be used in larger sections and simply surrounded and held rigidly in place by the concrete, most of the load being carried by the steel; or, finally, a steel column may be used and merely fireproofed by the concrete. As the cost of steel in the form of rods is much less than in the form of built members, and as compressive stresses can, in general, be carried more cheaply by concrete than by steel, economical construction will lead to the use of the maximum amount of concrete and the minimum amount of steel consistent with safety, although this principle will be modified by various practical considerations.

For those structural forms in which both tension and compression exist, that is to say, in all forms of beams, the combination of the two materials is particularly advantageous. Here the tensile stresses are carried by steel rods embedded in the concrete near the tension side of the beam. The steel is thus used in its cheapest form, it is thoroughly protected by the concrete, and the compressive stresses are carried by the concrete. Concrete alone cannot be used to any appreciable extent to carry bending stresses on account of its low and uncertain tenacity, but a concrete beam with steel rods embedded in it to carry the tensile stresses is a strong, economical, and very durable form of structure.

From these considerations it follows that reinforced-concrete construction is advantageous to varying degrees in different types of structures. Some of the most important of these types will here be noted, together with the advantages accompanying the use of reinforced concrete in their design.

3. *Buildings.*—This type of construction is especially useful for floor-slabs and to a somewhat less degree for beams, girders, and columns. It is also well adapted for footings in foundations, being more economical than I-beam footings embedded in concrete.

4. *Culverts and small Girder Bridges.*—Very satisfactory on account of its simplicity and economy as compared to masonry arches, and because of its durability as compared to steel bridges.

5. *Retaining-walls, Dams, and Abutments.*—Often economical for such structures as compared to ordinary masonry. Plain masonry structures of this kind are designed to resist lateral forces by their weight alone, the resulting compressive stresses, except in extremely large structures, being very small and much below safe values. By the use of reinforced concrete these structures can be designed of a more economical type and so arranged as to utilize the concrete in the form of beams, thus developing more nearly the full compressive strength of the material. The steel reinforcement is fully protected from corrosion, a factor which prevents the use of all-steel frames for structures of this class.

6. *Arch Bridges.*—In this form of structure reinforced concrete possesses less advantage over ordinary masonry than in those forms where the compressive stresses are less important. In an arch the stresses are principally compressive, and these do not require steel reinforcement; it is only to provide for the relatively small bending stresses due to moving loads, or as a precaution against undesirable cracks, that steel is serviceable. No large economy can be obtained through its use. By reason of greater simplicity and the less expensive abutments required, a flat-top culvert or beam bridge, with abutments of reinforced concrete, is more advantageous for short spans than the arch.

7. *Reservoir Walls, Floors, and Roofs.*—Very well adapted as a durable material and lending itself to lighter design than common masonry.

8. *Conduits and Pipe Lines.*—Reinforced concrete can often be used to great advantage in a water-conduit or large sewer. It is also sometimes used for pipe lines and tanks under pressure, the steel being relied upon to resist the tensile stresses, while the concrete serves as a protection and as a water-tight covering. The amount of steel may thus be determined by considerations of strength alone, where otherwise a much larger amount of metal would be needed and in a more expensive form.

9. *Elevated Tanks, Bins, etc.*—Advantageous because of its durability and its adaptability in the construction of heavy floors and walls subjected to lateral pressure. Of especial value for coal-bins, either for flooring and lining alone, or for the entire structure.

10. *Chimneys and Towers.*—Possesses advantages over brick or stone masonry in the fact that it forms a structure of monolithic character, resulting in greater certainty in the stresses and economy in design.

11. *Piles, Railroad Ties, etc.*—The use of a moderate amount of steel with concrete so as to give to this material a reliable tensile and bending resistance has opened the way for its use in a great variety of forms, not only as complete structures, or important members of structures, but also in many special individual forms. Concrete piles are valuable substitutes for piles of wood where the latter would be subject to deterioration. Reinforced-concrete ties offer some evident advantages over ties of wood or steel. This material is also well adapted to many other special uses, particularly where durability is an important factor.

CHAPTER II.

PROPERTIES OF THE MATERIALS.

12. In a design where two or more materials are combined in the same member the stresses in the different materials depend upon the elastic properties as well as upon the superimposed loads. Therefore in making such designs a knowledge of these elastic properties is quite as necessary as a knowledge of the strength of the materials.

CONCRETE.

13. **General Requirements.**—The conditions to be met in reinforced-concrete construction require the use, generally, of a concrete of relatively high grade. In this type of construction the strength of the material is of much greater importance than it is in many forms of plain concrete design, as the dimensions of the structures are more directly dependent upon strength and less upon weight. A comparatively strong concrete is therefore found to be economical.

It is especially important, also, that the concrete be of uniform quality and free from voids, as the sections are comparatively small and the stability of the structure, to a much greater extent than is the case with massive concrete, is dependent upon the integrity of every part. Thoroughly sound concrete is also required in order to insure good adhesion to the steel reinforcement and adequate protection of the steel from corrosion and from fire. These requirements call for great care in the preparation and placing of the material.

Concrete is subject to great variations in its properties, owing to the great variations in the character and proportions of its ingredients and in its preparation. It is therefore difficult to judge from results of tests made under certain conditions as to what may fairly be expected of a concrete prepared under other conditions; so that it is very important that regular and systematic tests of the material as actually used be made during the progress of the work.

14. Cement.—Portland cement only should be used; it should meet such standard specifications as those of the American Society of Civil Engineers. The rapidity of hardening of different cements varies considerably and may be an element requiring special attention where the structure is to receive its load very early or where such load is to be long deferred.

15. Sand.—The sand should be free from clay and preferably of coarse grain. A fine sand requires more cement than a coarse sand for equal strength, and more water for a like consistency. In the case of a very fine sand the difference may be very marked, so that unless care is taken and special tests made, the resulting concrete is likely to be porous and deficient in strength and adhesive power. Where the use of fine sand is contemplated, tests of strength may show that a considerable extra cost may be justified in securing a coarser material. The effect of size of sand is shown in Art. 19.

16. Broken Stone and Gravel.—Both materials are satisfactory, but they should be screened to remove the dust or sand and to remove particles larger than the maximum size desired. Beyond this, the screening of stone to size is undesirable unless an artificial mixture is to be made, as it tends to increase the proportion of voids. Gravel may be sufficiently uniform in quality so that the sand need not be removed, but it will usually require screening in order to insure a concrete of definite proportions.

The maximum desirable size of stone or gravel depends upon the size of the structural forms and the size and spacing of the reinforcement, it being desirable to use as large a size of aggregate as will admit of convenient working. Maximum sizes of stone of ¾ inch to 1¼ inches are common, but on heavy work, with rods widely spaced, there is no objection to still larger sizes.

The crushing strength of a gravel concrete is usually a little less than one of broken stone of the same proportion of voids, but the difference is unimportant. The difference in tensile strength is not well determined, but the few tests available indicate about the same relative difference as in compressive strength.

17. Proportions of Ingredients.—The proportions commonly used vary from about 1:1½:3 to 1:3:6 of cement, sand, and broken stone respectively; or the equivalent proportions if gravel be used. Richer mixtures than 1:2:4 are not common, nor poorer mixtures than 1:2½:5, although with well-graded material a very satisfactory concrete can be made of 1:3:6 proportions. Occasionally where the design is determined by other considerations than strength and cost, a very rich mixture or a poor one may be desirable, but where these elements determine the design, the most economical concrete will be a rich concrete of about the proportions above indicated. Customary proportions, such as 1:2:4, should not be blindly adopted. In any important work a careful study of the materials and of the best proportions to use for economy and strength will be well repaid. To secure sound and reliable work, with good adhesion and tensile strength, there must be no unfilled voids in the stone and little or none in the sand. The former is of more importance than the latter, and if cost and strength are to be reduced it should be done by using a poorer mortar to fill the voids in the stone. For equal amounts of cement, the denser the mixture (or the smaller the percentage of voids) the stronger the concrete.*

18. Consistency.—The tendency in all kinds of concrete construction is to use a wetter mixture than formerly. Relatively dry concrete thoroughly tamped will give slightly greater

* See valuable paper by Fuller and Thompson on "Proportioning of Concrete," Trans. Am. Soc. C. E., 1907, LIX, p. 67.

strength than a wet mixture; however, if not too wet the difference is not great, and considering the difficulty and expense of securing the necessary amount of tamping of the dry mixture, better results can usually be secured by using a plastic mixture. This is especially true with reference to obtaining a dense, homogeneous concrete. The usual practice now is to make the consistency such that the concrete will require only moderate tamping or puddling to bring the mass to a homogeneous condition. Such concrete, while somewhat weaker than the ideal compacted concrete, will, under actual conditions, be much more reliable and will be free from voids. In the case of reinforced-concrete work reliability is more important than maximum strength, and is promoted by using concrete of such consistency that it can readily be worked into place in the forms and around the reinforcing steel. In practice the consistency varies. Some use a concrete which requires considerable tamping and working, while others use a concrete which will practically flow into place. The dryer the concrete the closer the inspection required when the material is placed; on the other hand very wet concrete is not as strong and needs to be promptly poured to prevent segregation of the materials.

19. Compressive Strength.—The compressive strength of concrete is dependent upon many factors so that it is difficult and at the same time somewhat misleading to present "average values". Obviously, in any important work, the strength should be determined under the actual conditions under which the concrete is used. Uniformity is quite as important as average strength.

One of the best series of tests is that made at the Watertown Arsenal for Mr. George A. Kimball, Chief Engineer of the Boston Elevated Railway Company.* The concrete was made of five brands of Portland cement, coarse, sharp sand, and broken stone up to 2½-inch size. The concrete was well rammed into the molds, water barely flushing to the surface.

* Tests of Metals, 1899, p. 717.

The specimens were buried in wet ground after being taken from the molds. The average results were as follows:

TABLE No. 1.

COMPRESSIVE STRENGTH OF CONCRETE.

WATERTOWN ARSENAL, 1899.

Mixture.	Brand of Cement.	Strength, Pounds per Square Inch.			
		7 Days.	1 Month.	3 Months.	6 Months.
1:2:4	Saylor	1724	2238	2702	3510
	Atlas	1387	2428	2966	3953
	Alpha	904	2420	3123	4411
	Germania	2219	2642	3082	3643
	Alsen	1592	2269	2608	3612
	Average	1565	2399	2896	3826
1:3:6	Saylor	1625	2568	2882	3567
	Atlas	1050	1816	2538	3170
	Alpha	892	2150	2355	2750
	Germania	1550	2174	2486	2930
	Alsen	1438	2114	2349	3026
	Average	1311	2164	2522	3088

In a series of tests made at the Watertown Arsenal for Mr. George W. Rafter, the following average values were obtained on concrete about 20 months old.* The voids in the broken stone were practically filled. The mixture was of damp-earth consistency:

Cement.	Sand.	Strength.
1	1	4467 lbs/in^2
1	2	3731 "
1	3	2553 "

Results as high as indicated by the preceding values cannot be safely counted upon in practice. Wet concrete will show a lower strength than concrete as dry as that in the above tests, especially for the earlier periods, but the difference becomes less with lapse of time, and a fairly soft plastic con-

* Tests of Metals, 1898.

crete will acquire about the same strength as dry concrete within three or four months. A very wet concrete will, however, continue to be somewhat weaker than one containing less water, and while such a concrete may, on the whole, be desirable, its deficiency in strength as compared to maximum values should not be overlooked. Other variations in conditions, such as rapid drying out, or the use of very fine sand, for example, may give results materially below those here quoted.

The following average results of a large number of tests in the series made for Mr. Rafter, already referred to, show the relative strengths of dry, plastic, and wet concrete at the age of about twenty months. The dry mixtures were only a little more moist than damp earth and required much ramming; the plastic mixtures required a moderate amount of ramming to bring water to the surface; the wet mixtures quaked like liver under moderate ramming. Five brands of cement were used:

Consistency.	Mean Compressive Strength.
Dry	2348 lbs/in^2
Plastic	2203 "
Wet	2129 "

In actual practice results are very likely to be less favorable to dry mixtures on account of the great difficulty of securing adequate tamping.

The effect of size of sand has been thoroughly investigated by Feret. Fig. 1, from Johnson's "Materials of Construction", shows results obtained by Feret on 1:3 mortar after hardening one year in fresh water. The sand used consisted of mixtures of various proportions of fine (.0 to .5 mm.), medium (.5 to 2 mm.), and coarse (2 to 5 mm.) sand, and in the figure the result from any particular mortar is recorded in the triangle at such distances from the three base-lines as will represent the proportions of each size sand used. Lines of equal strength were then drawn in the diagram. Thus the strength of the mortar in which only fine sand was used was only 1400 lbs/in². The maximum strength of 3500 lbs/in² was obtained from a mixture containing about 85% of coarse sand and 15% of fine, with a very little sand of medium size. This diagram shows in a striking manner the effect of size of sand.

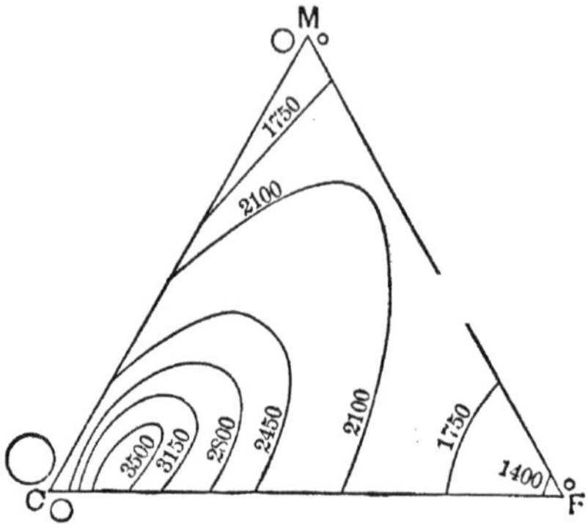

FIG. 1.—Effect of Size of Sand.

The tests quoted in the foregoing were made on specimens of cube form, which has generally been considered the standard form of compression specimen. More recently, however, the prismatic form, of a height of 2 to 3 diameters, has been commonly used because of its advantages in the measurement of distortions and in studying the results of tests on columns. In the prismatic form the strength will generally be from 10 to 30% less than in the cube form, as there is greater freedom for shearing action to take place. The results are, however, likely to be more uniform as they depend less upon the nature of the bedding, etc. For comparison with the strength of concrete in a beam the cube form is satisfactory.

The effect of varying conditions on the quality of concrete is well shown by results obtained at various times in the laboratories of the University of Wisconsin.* Tests made in 1906

* See Bulletins No. 1 and 2, Vol. 4, Engineering Series.

on thirty-six cylinders, 6×18 in. in size, of 1:2:4 concrete, 30 days old, hand-mixed, gave an average strength of 1750 lbs/in^2, with many results below 1500. Tests in 1907 on twenty-five cylinders, 10×24 in. in size, of machine-mixed concrete of the same kind, gave an average strength of 1940 lbs/in^2, individual results varying from 1500 to 2600 lbs/in^2. A fairly fine sand was used and the specimens were cured in air. More recent tests indicate that with good materials and careful manipulation an average strength of 2200 lbs/in^2 can readily be obtained on cylinders for 60-day tests.

Considering the various results noted it may be concluded that under reasonably good conditions as to character of material and workmanship an average strength of about 2000 lbs/in^2 may be expected of a 1:2.4 concrete in 30 to 60 days, on cylindrical specimens, the rate of hardening depending upon the consistency and the temperature, and for a 1.3:6 concrete a strength of about 1600 lbs/in^2.

It is important that the strength be determined by actual tests of the material proposed to be used, and if the results are too low the ingredients or proportions should be modified until a satisfactory result is obtained. Where the usual proportions give low results it will generally be advisable to increase the richness of the concrete rather than to reduce the working stresses.

20. **Tensile Strength.**—The tensile strength of concrete is quite as important as the compressive strength. In fact the most common type of failure of a reinforced concrete beam is closely related to the tensile strength of the concrete. The tensile strength is generally from one-tenth to one-twelfth of the compressive strength, but this ratio varies considerably. The character of the material and workmanship has probably a greater influence upon the tensile strength than upon the compressive.

Tests by Mr. M. O. Withey on 1:2:4 concrete, 28 days old, gave results averaging 189 lbs/in^2, varying from 142 to 160 lbs/in^2. The compressive strength was 1940 lbs/in^2.* Tests by Mr. W. H. Henby † gave results as follows:

Mixture.	Compressive Strength.	Tensile Strength.
1:2:4	3000 lbs/in^2	180 lbs/in^2
1:3:6	1800 "	115 "

Tests by Professor W. K. Hatt † gave the following results:

Kind of Concrete.	Age, days.	Compressive Strength, lbs/in^2.	Tensile Strength, lbs/in^2.
1:2:4 (broken stone)	30	——	311
1:2:5 "	90	2413	359
1:2:5 "	28	2290	237
1:5 (gravel)	90	2804	290
1:5 "	28	2400	253

Tests by Professor Ira H. Woolsen § on 1:2:4 mixtures 5 to 7 weeks old gave an average tensile strength of 161 lbs/in^2, compared to 1753 lbs/in^2 compressive strength.

Professor Talbot obtained values for 1:3:6 concrete from 50 to 84 days old of 178, 160, and 170 lbs/in^2.‖

21. Tensile Strength as Determined by Transverse Tests.—The transverse strength of plain concrete depends almost entirely upon its tensile strength, although the modulus of rupture is considerably greater than the strength in plain tension owing to the curved form of the stress-strain diagram. Feret ¶ found a very nearly constant ratio of 1.95 of modulus of rupture to tensile strength. The value of this ratio will ordinarily range from 1.8 to 2. Transverse tests of different concretes should therefore show about the same relative results as tensile tests. They are in fact quite as significant in this connection.

Some of the best tests on transverse strength are those made by William B. Fuller, and given in full in Taylor and

* Bulletin No. 2, Vol. 4, Univ. of Wis., 1908.

† Jour. Assn. Eng. Soc., Sept. 1900.

‡ Jour. West. Soc. Eng., Vol. IX, 1904, p. 234.

§ Eng. News, Vol. LIII, 1905, p. 561.

‖ Bulletin No. 1, Univ. of Ill., 1904.

¶ Etude Expérimentale du Ciment Armé. Paris, 1906.

Thompson's work on Concrete.* The following average results were obtained for 33–35-day tests.

Mixture by Volume.	Average Modulus of Rupture.
1:2.16:4.08	439 lbs/in²
1:2.16:5.1	380 "
1:3.24:5.1	285 "
1:3.24:6.12	226
1:3.24:7.14	239 "

Here we find the strength of the 1:3.24:6.12 mixture only about one-half that of the 1:2.16:4.08 mixture, indicating the relative weakness in tension of the lean mixture.

The results herein given, both of tensile and of transverse tests, indicate that the quality of the concrete has a greater relative effect on the tensile strength than on the compressive strength, the strength of a 1:3:6 mixture being not more than two-thirds that of a 1:2:4 mixture. Reasonable values for ultimate tensile strength would appear to be about as follows:

1:2:4 mixture......... 160–200 lbs/in²
1:3:6 " .... 100–125 "

22. Shearing Strength.—There is a lack of uniformity among writers as to just what is meant by the term "shearing strength", resulting in a wide variation in the suggested values for working stresses. In this work the authors will use the term as it is commonly thought of among American engineers, to denote the strength of the material against a sliding failure when tested as a rivet or bolt would be tested for shear; that is, when the maximum shearing stresses are confined to a single plane.

Tests made under the direction of Professor C. M. Spofford on cylinders 5 inches in diameter with ends securely clamped in cylindrical bearings gave results as follows:

Mixture.	Shearing Strength, lbs/in².	Compressive Strength, lbs/in².	Ratio of Shearing to Comp. Strength.
1:2:4	1480	2350	.63
1:3:5	1180	1330	.89
1:3:6	1150	1110	1.04

* Concrete, Plain and Reinforced. N. Y., 1906.

Tests made at the University of Illinois on rectangular specimens tested in a similar manner gave the following average results:

Mixture.	Shearing Strength, lbs/in².	Compressive Strength, lbs/in².	Ratio of Shearing to Comp. Strength.
1:2:4	1418	3210	.44
1:3:6	1250	2290	.57

Tests made by punching through plates gave shearing strengths varying from 37 to 90 per cent of the compressive, the value depending upon the form of test-piece.*

Tests by M. Feret on mortar prisms gave results for shearing strength equal to about one-half the crushing strength.

The ordinary crushing failure is really a failure by shearing, and under such conditions the crushing stress is, theoretically, twice the shearing stress, the angle of shear being 45°. Results of tests give a somewhat greater inclination than 45°, so that the crushing stress is somewhat greater than twice the actual shearing stress.

We may then conclude, both from theory and from tests, that the shearing strength of concrete, in the sense here used, is nearly one-half the crushing strength. It is in fact so large that it will need to be considered only in exceptional cases.

Some writers used the term "shearing stress" to mean quite a different thing from that discussed above, namely, the complex action which occurs in the web of a beam. In this case there exist direct tensile and compressive stresses which at the neutral axis are equal in intensity to the vertical and horizontal shearing stresses. The limit of distortion in the concrete will be reached, and failure will occur, when the tensile strength of the material is exceeded. Such a failure may perhaps be called a shearing failure, but is more strictly a failure in tension in a diagonal direction, and is so considered in this work. Treated as a shearing failure the strength should be very nearly the same as the tensile strength of the material determined in the usual way. In practice the diagonal tensile

* Bulletin No. 8, Univ. of Ill., 1906.

stresses in a beam must often be considered, but shearing stresses, as such, will be dangerous only in exceptional circumstances, such as exist where a heavy load is applied close to a support.

23. Elastic Properties of Concrete.—*Stress-strain Curve in Compression.*—In the design of combination structures, such as those of steel and concrete, it is necessary to know the relative stresses under like distortions. These will depend upon the moduli of elasticity of the two materials. For purposes of safe design we need to know also the elastic-limit strength.

Fig. 2 represents typical stress-strain curves for concrete in compression. Curves C, D, E, and F were obtained at the University of Wisconsin from tests on cylinders 6 inches in diameter by 18 inches high. The concrete was 1:2:4 limestone concrete 30 days old. The ultimate strengths ranged from 1500 to 2300 lbs/in^2. Curves A and B are typical curves selected from the Watertown Arsenal tests already quoted, and represent 1:2:4 and 1:3:6 concrete respectively.

Unlike the elastic line for steel, the line for concrete is slightly curved almost from the beginning, the curvature gradually increasing towards the end. There is, however, no point of sharp curvature as for ductile materials. A release of load at a moderate stress, such as 500 to 600 lbs/in^2, will usually show a small set indicating imperfect elasticity. A second application of the load will, however, give a straighter line than the first and there will be much less permanent set following the release of load. After a few repetitions of load there will be no further set and the stress-strain line will become a straight line up to the load applied. There is a limit of stress, however, beyond which repeated applications of load will continue to add to the permanent deformation and the specimen will ultimately fail. The general behavior under repeated stress is indicated in Fig. 3, from tests on concrete similar to those represented in Fig. 2. For a very exhaustive study of this subject the reader is referred to the work of Bach.*

---

* Zeit. V. dt. Ing., 1895, etc.

24. *Modulus of Elasticity in Compression.*—The stress-strain line being curved almost from the beginning, the proper

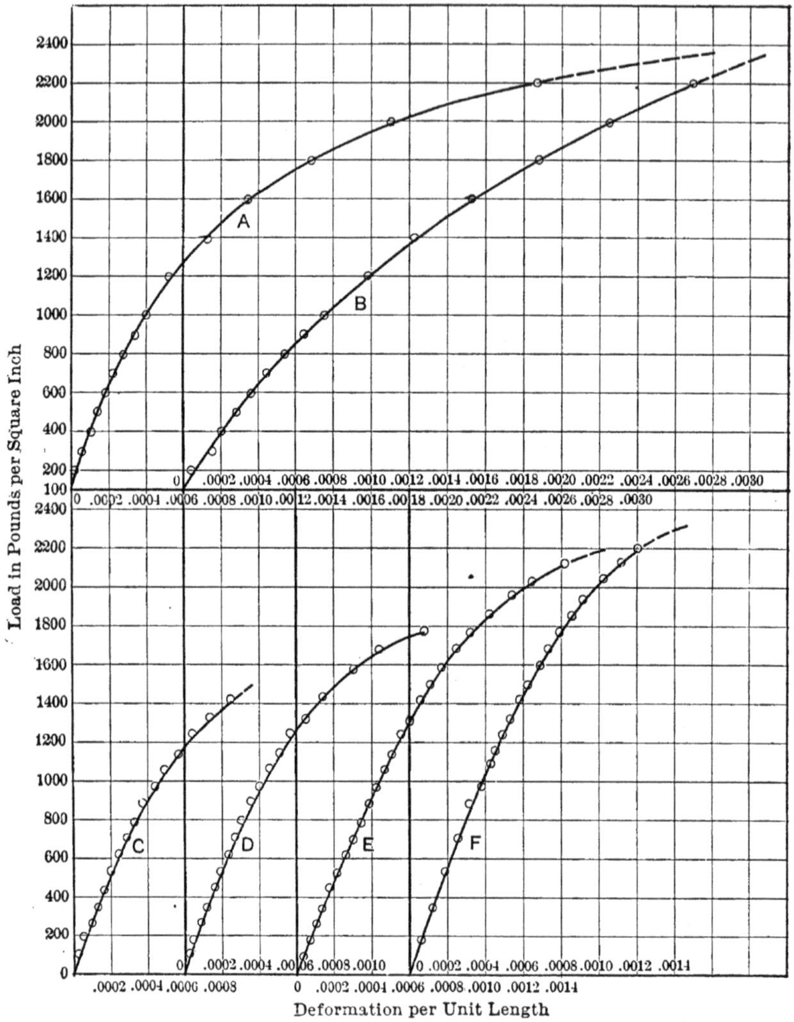

Fig. 2.—Compressive Stress-strain Diagrams of Concrete.

method of calculating the modulus of elasticity needs to be considered. Fig. 4 is a typical stress-strain diagram for compression (somewhat simplified), B and C being points where the loads have been removed and reapplied. For very low stresses, up to perhaps 300 to 400 lbs/in² (a low working stress), the variation of the curve from a straight line is so small

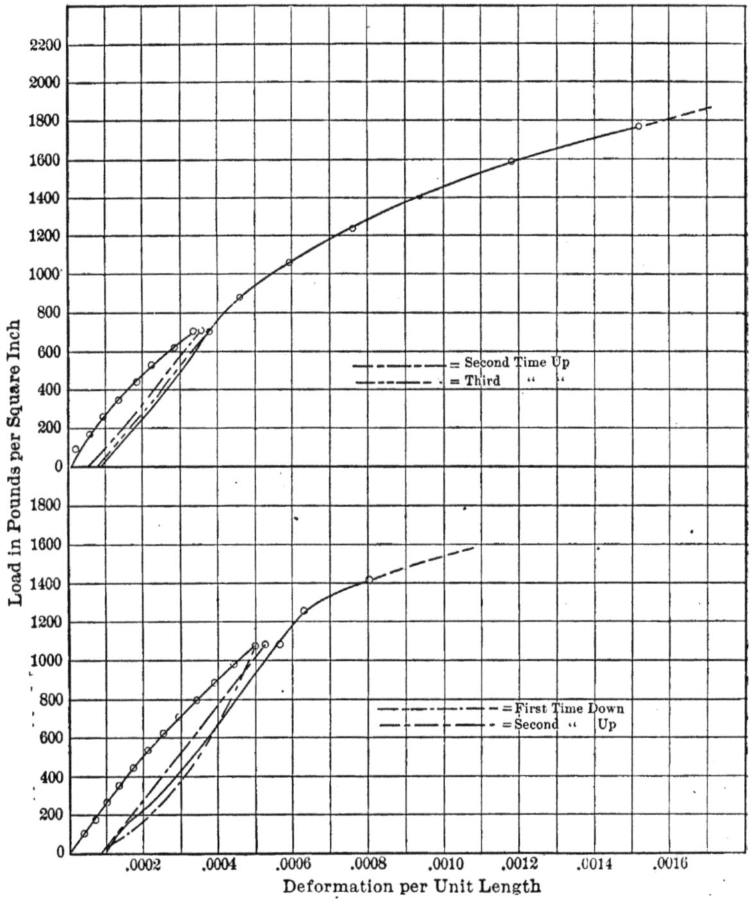

FIG. 3.—Stress-strain Diagram under Repeated Loads.

that it may be considered as straight, and an average straight line may be drawn, as OT, and its slope taken as the modulus of elasticity. This line may be considered the same as the tangent at the origin. For higher stresses, reaching to a point along the curved portion such as point B, it is usual to deduct the permanent set Oa from the deformation Ob and divide the stress by the remaining elastic deformation ab. This gives the slope of the line aB, and may be considered to represent the law of elastic deformation for stresses within the limit of the stress bB after the first few applications of load. A modified "elastic" curve, $OB'C'$, can thus be drawn by deducting from the deformation for each load the subsequent set, giving a steeper curve and one more nearly approaching a straight line. On the basis

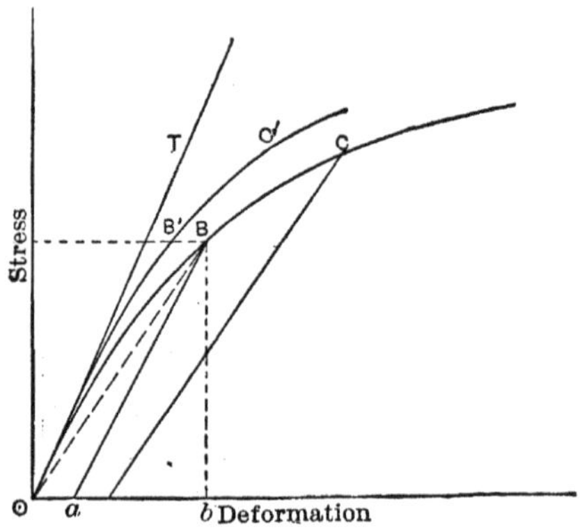

FIG. 4.

of this "elastic" curve the modulus of elasticity for stresses up to any given maximum would then be equal to that maximum stress divided by the elastic deformation at that stress.

There being no general agreement as to the exact definition of the word "modulus" for such materials as concrete, the method which should be employed in calculating its value should depend upon the purpose for which it is to be used. The principal use of the modulus of elasticity in reinforced-concrete design is to determine the relative stresses carried by the concrete and the steel in compression members, and to find the neutral axis in beams. After the neutral axis is once found the modulus does not enter into the calculations.

Consider the action in the case of a column. Assuming no initial stress in the steel or concrete, suppose that the column is loaded so as to cause a shortening equal to *Ob*, Fig. 4. The stress in the concrete will be *bB*, and that in the steel will be equal to the deformation *Ob* multiplied by its modulus of elasticity. Upon removal of the load there may be a permanent set *Oa*, which means that there is some residual compression in the steel (with an equal amount of tension in the concrete). A second application of the load will cause a deformation *ab*, but, measuring from the original position, the deformation is *Ob*, and this again fixes the stress in the steel. Hence, for the determination of the relative stresses in steel and concrete, the modulus for the concrete should be the ratio of *Bb* to *Ob*, or the slope of the chord *OB*.

In the case of a beam the stresses in the concrete at any section will vary from zero at the neutral axis to the value *Bb*, for example, at the extreme fibre. At intermediate points the stresses follow approximately the law of the curve *OB*. In this case a chord *OB* does not exactly represent the facts, but the error is small, and it is the best line to use if the rectilinear variation of stress be assumed. If a curvilinear law is used, then the modulus is supposed to be the slope of the tangent at the origin. In neither case is it correct to use the slope of the line *aB*.

In referring to these various methods of calculating the modulus, the slope of the tangent *OT* is generally called the "initial modulus." The slope of the line *OB* may be called the "secant modulus."

The value of the modulus for concrete varies greatly as determined by different experimenters and for different kinds of concrete. As a rule the denser and older the concrete the higher the modulus.

Among the most careful experiments are those by Bach,* in which he repeated the loads at each increment until there was practically no increase of set.

---

* Zeit. V. dt. Ing., 1895.

The following are some average results:

Kind of Concrete.	Modulus of Elasticity, lbs/in².		
	Based on Elastic Deformation.		Based on Total Deformation.
	At 114 lbs/in².	At 570 lbs/in².	At 570 lbs/in².
1:2½:5 (broken stone).......	4,660,000	3,590,000	3,440,000
1:2½:5 (gravel).............	3,170,000	2,520,000	2,200,000
1:3:6 (broken stone)........	3,870,000	2,990,000	2,570,000
1:3:6 (gravel)..............	3,000,000	2,240,000	2,110,000

The specimens were 25 cm. in diameter and 100 cm. high and were from three to four months old.

The average values of the modulus obtained in the Watertown Arsenal tests mentioned in Art. 19 were as follows:

TABLE NO. 2.

MODULUS OF ELASTICITY OF CONCRETE.

WATERTOWN ARSENAL TESTS, 1899.

Mixture.	Brand of Cement.	Modulus of Elasticity between Loads of 100 and 600 lbs/in², Based on Elastic Deformation.		
		7–10 Days.	1 Month.	3 Months.
1:2:4	Saylor...............	1,667,000	2,500,000	3,571,000
	Atlas................	2,778,000	3,125,000	4,167,000
	Alpha................	1,000,000	2,083,000	4,167,000
	Germania.............	2,500,000		3,571,000
	Alsen................	2,500,000	2,778,000	2,778,000
	Average...........	2,089,000	2,621,000	3,651,000
1:3:6	Saylor...............	2,273,000	2,778,000	4,167,000
	Atlas................	1,667,000	3,125,000	2,778,000
	Alpha................		2,083,000	3,571,000
	Germania.............	2,273,000	2,273,000	2,778,000
	Alsen................	1,667,000	2,273,000	2,778,000
	Average...........	1,970,000	2,506,000	3,214,000

These results were calculated by using the total deformation minus the set. If the total deformation be used the values would be reduced in most cases 10 to 20 per cent.

Tests made at the University of Wisconsin show a large variation of the modulus with variation in the quality of the concrete. Tests in 1906 on 30 prisms of 1:2:4 concrete, 30 days old, gave an average value of 2,560,000 lbs/in^2 at a stress of 600 lbs/in^2, using total deformation. Similar tests in 1907 gave an average value of 3,500,000 lbs/in^2, varying from 2,800,000 to 3,800,000. The respective average compressive strengths were 1780 and 1940 lbs/in^2, the latter concrete being considerably denser than the former.* Values still higher have been found by some experimenters, but generally the calculations have been made with reference to the tangent at the origin or to the elastic deformation.

Considering the various results obtained and the significance of total deformation it would appear that for working loads the modulus for ordinary concrete ranges from 2,500,000 to 3,500,000 lbs/in^2, depending upon the mixture and the age of the concrete. As will be shown subsequently, however, the value selected should also depend upon the purpose for which it is to be used, and that for most calculations relating to strength a value of 2,000,000 is more satisfactory than a higher value.

25. *Elastic Limit.*—As stated in the preceding article, concrete shows a permanent set under small loads so that, in the usual sense, the material can hardly be said to have an elastic limit. There appears to be, however, a limit to the stress which can be repeated indefinitely without continuing to add to the deformation, and this limit may be taken as the elastic limit for practical purposes. From experiments by Bach and others, this limit seems to be from one-half to two-thirds the ultimate strength. In repeated-load experiments on neat cement and on concrete made by Professor J. L. Van Ornum † it has been shown that the maximum load which may be repeated an indefinite number of times without rupture does not much exceed 50% of the ultimate strength (Art. 117). These results show a close relation to those obtained by Bach,

* Bulletins No. 1 and 2, Vol. 4, Univ. of Wis.

† Trans. Am. Soc. C. E., Vol. LI, p. 443. Proc. Am. Soc. C. E., Dec. 1906.

and it may therefore be concluded that the limit of permanent elasticity for repeated loads is from 50 to 60% of the ultimate strength.

26. *Comparison of Stress-strain Curve with the Parabola.*—As the parabola is often used in theoretical analyses to represent the stress-strain curve it will be useful to compare some typical curves with the parabola. The form of parabola used has its axis vertical and its vertex at the point of the curve representing the ultimate strength. In Fig. 5 the curves shown in Fig. 2 are compared with parabolas (shown in dotted lines). In the case of curves *C, D, E,* and *F* the agreement is very close.

27. *Stress-strain Curve for Tension.*—Comparatively few tests have been made on the elasticity of concrete in tension. Bach found for 1.4 concrete an average value of the modulus of 3,800,000 lbs/in^2 at a stress of 80 lbs/in^2, and 3,100,000 at a stress of 135 lbs/in^2. The ultimate tensile strength was 185 lbs/in^2. The modulus in compression for the same concrete was 3,850,000 at 80 lbs/in^2.* Professor Hatt has determined values ranging from 2,000,000 to 5,000,000 lbs/in^2, which were generally about equal to the values in compression.† These and other tests indicate that the initial moduli in tension and in compression are about the same, and as the working limit in tension is very low they may be assumed as equal. The relative strength and deformation of concrete in compression and tension is illustrated by a typical curve in Fig. 6.

28. **Coefficient of Expansion.**—Experiments by Professor W. D. Pence ‡ on 1·2.4 concrete gave an average value of the coefficient of expansion of .0000055 per degree Fahrenheit, there being little variation among the several tests. Tests made at Columbia University on 1.3:6 concrete gave values of about .0000065. Other experiments have shown somewhat higher results. A value of .000006 may be assumed.

* Mitt. Ueber Forsch. a. d. Gebiet des. Ing., 1907, Heft 45–47.

† Proc. Am. Soc. Test. Mat., 1902.

‡ Jour. West Soc. Eng., Vol. VI, 1901, p 549.

29. Contraction and Expansion in Hardening.—Many experiments have been made relative to shrinkage and swelling

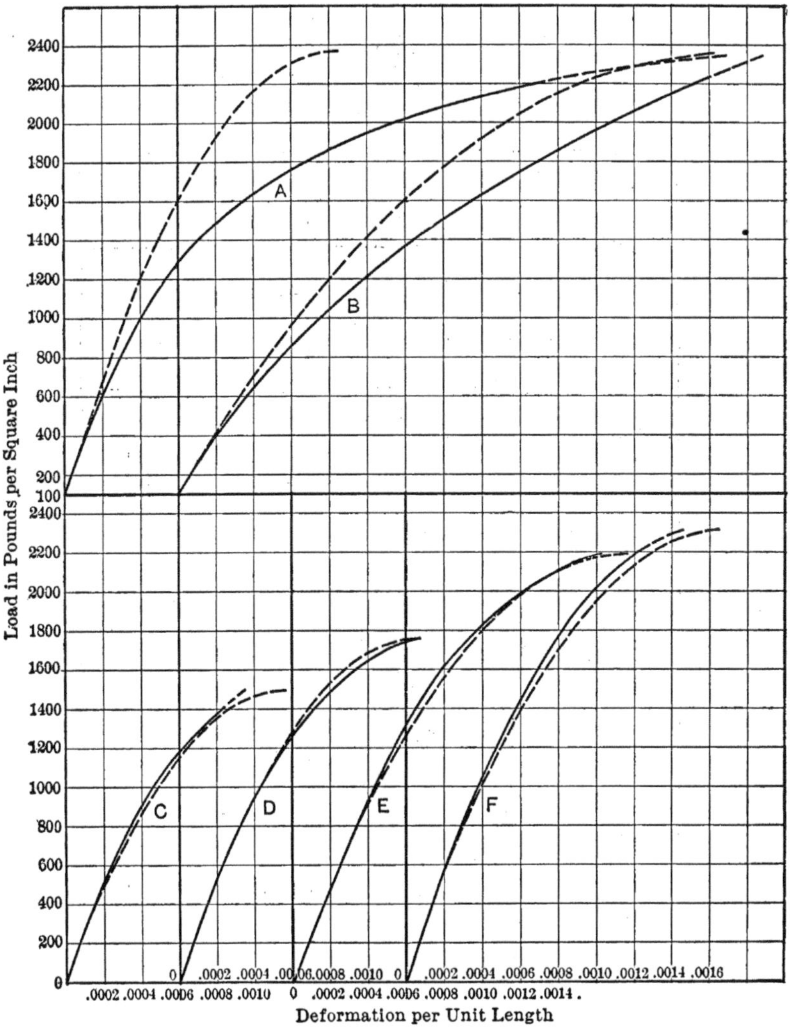

Fig. 5.—Stress-strain Curves Compared with Parabolas.

of cement-mortar in hardening. In general the results show that when hardened in air there will be more or less shrinkage, but when hardened in water there is likely to be some swelling, although results on this point are not entirely consistent. The richer the mortar (or concrete) the greater the change in dimensions. Experiments by Considère and others indicate that 1:3 plain mortar will shrink .05 to .15% when hardened

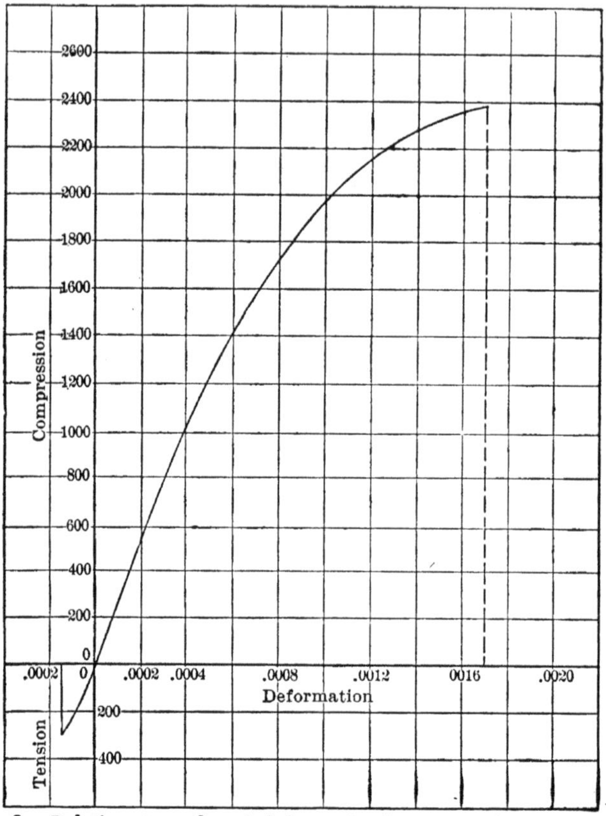

FIG. 6.—Relative strength and deformation in compression and tension.

in air for 2 to 4 months, and neat cement from two to three times as much. Considère found the shrinkage of a 1:3 mortar reinforced with $5\frac{1}{2}\%$ of steel to be only .01%, or one-fifth the amount his tests showed on plain mortar. The few tests available show that the shrinkage of concrete is less than that of mortar, and it would appear that the shrinkage should be nearly proportional to the amount of cement per unit volume, as the sand and stone are unaffected.

30. Weight of Concrete.—The weight of concrete of the usual proportions will vary from 140 to 150 lbs/ft^3, depending upon the degree of compactness and the specific gravity of the materials. Variation of proportions will affect the weight but little if the proper ratio of sand and stone be maintained, but a wet concrete when dried out will weigh less than a well-compacted concrete containing originally less water. For practical purposes an average value of 145 lbs/ft^3 may be taken. The addition of reinforcing steel in the usual proportions will add from 3 to 5 pounds, so that the weight of reinforced concrete may be taken at 150 lbs/ft^3.

31. Properties of Cinder Concrete.—The following table of results indicates fairly well the strength and modulus of elasticity of cinder concrete. The age of the specimens varied from 30 to 100 days. Cinder concrete will weigh from 110 to 115 lbs/ft^3.

TABLE NO. 3.

CRUSHING STRENGTH AND MODULUS OF ELASTICITY OF CINDER CONCRETE.

WATERTOWN ARSENAL TESTS, 1898.

Mixture.			Average Crushing Strength, lbs /in ^{2}.		Average Modulus of Elasticity between Loads of 100 and 600 lbs /in ^{2}.
Cement.	Sand.	Cinders.	One Month.	Three Months.	
1	1	3	1540	2050	2,540,000
1	2	3	1098	1634	
1	2	4	904	1325	
1	2	5	724	1094	1,040,000
1	3	6	529	788	

REINFORCING STEEL.

32. General Requirements.—In general, reinforcing steel must be of such form and size as to be readily incorporated into the concrete so as to make a monolithic structure. To provide the necessary bond strength and to distribute the steel where needed without concentrating the stresses on the concrete too greatly, requires the use of the steel in comparatively small sections. This requirement, as well as that of economy and convenience, leads to the use of the steel in the form of rods or bars. These will vary in size from about $\frac{1}{4}$ to $\frac{3}{8}$ inch for light floors up to $1\frac{1}{2}$ to 2 inches as maximum sizes for heavy beams or columns. Under certain conditions a riveted skeleton work is preferred for the steel reinforcement, but this is usually where for some reason it is desired to have the steelwork self-supporting or where it is to carry an unusually large proportion of the load.

33. Forms of Bars.—Plain round rods have been used generally in Europe for many years, and also very largely in this country, adhesion being depended upon for the transmission of stress. Square rods show about the same adhesive strength as round rods, but are not so convenient to use or so readily obtained. Flat bars are undesirable, as their adhesion to the concrete is much below that of round or square bars.

Many special forms of bars have been devised, the principal object of which is to furnish a bond with the concrete independent of adhesion,—a "mechanical bond" as it is usually called. Some of the most common types of such bars are illustrated in Fig. 7. Fig. (*a*) is the twisted square bar invented by Mr. Ransome and called by his name. It is usually twisted cold. Figs. (*b*), (*c*), and (*d*) illustrate various well-known types of deformed bars which are shaped in the rolling. Fig. (*e*) illustrates the Kahn bar, formed by turning up strips sheared from the thin part of the bar. Many other devices are employed to a greater or less extent to provide a mechanical bond, and a great variety of combinations of forms

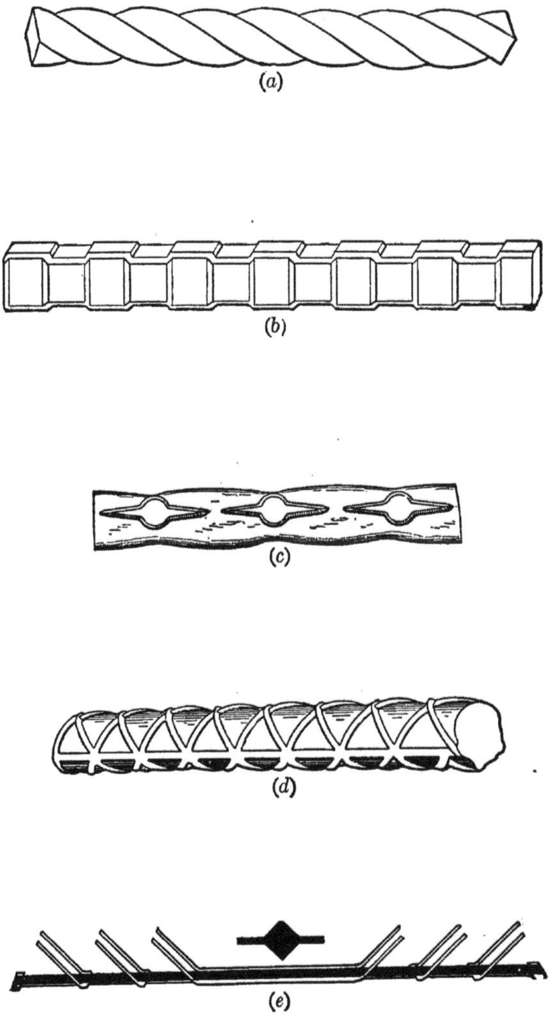

Fig. 7.—Deformed Bars.

are used in the construction of beams, floors, and columns as patented "systems". It is the purpose here to mention only the most common types of bar element.

34. General Quality of the Steel.—Steel used in reinforced work is not usually subjected to as severe treatment as that used in ordinary structural work. Bars must be capable of being bent to the desired form, but this is the only treatment to which the ordinary bars are subjected. In many concrete structures the impact effect is also likely to be less than in all-steel structures; consequently it is considered that a somewhat less ductile material may safely be used, but to what extent these considerations should permit the use of steel of cheaper grade or of higher elastic limit is an open question on which there is much difference of opinion. The question of elastic limit as related to working stresses and stresses in the concrete is discussed in Chapter V.

35. Tensile Strength.—As regards tensile strength the steel used is generally the ordinary mild or medium steel, or is a special high elastic limit material. The latter is seldom employed where plain bars are used. The ultimate strength and elastic limit of these two general grades of material are about as follows:

	Elastic Limit.	Ultimate Strength.	Elongation in 8 ins.
Medium steel..........	35–40,000 lbs/in²	60–70,000 lbs/in²	22–25%
High elastic limit steel..	50–60,000 "	80–100,000 "	12–18%

In some forms of rods used the elastic limit is artificially raised by cold working.

36. Modulus of Elasticity.—The modulus of elasticity of all grades of steel is very nearly the same and will be taken at 30,000,000 lbs/in².

37. Elastic Elongation.—As bearing upon deformations the elongation of the steel at its elastic limit will be here noted. Using the above value of the modulus of elasticity the elongation per unit length of the two grades of steel at their elastic limit will be as follows:

Medium.....................	0.0010–0.0013
High......................	.0015– .0020

38. Coefficient of Expansion.—The coefficient of expansion of steel may be taken at .0000065 per 1° F.

PROPERTIES OF CONCRETE AND STEEL IN COMBINATION.

39. Adhesion of Concrete and Reinforcing-bars. — The high value of the tangential adhesion, or bond, of concrete to steel rods embedded therein has long been known and has been utilized in the placing of anchor-rods, etc. It is somewhat remarkable, however, that only recently has this property been made use of in the design of combination structural forms. Experience has shown this adhesion to be sufficiently reliable and permanent to be utilized in such combination structures, and plain smooth bars have been entirely successful. Bars of irregular section in which adhesion is not entirely depended upon for the bond are also used to a large extent. Some form of mechanical bond is necessary where the adhesion area is deficient, and some engineers consider such a bond desirable in all cases.

Numerous tests have been made by various experimenters to determine the adhesion between concrete and plain rods of different forms, with results varying from about 200 to about 750 lbs/in^2. The adhesive strength is largely frictional resistance and varies greatly with the roughness of the bars. It also varies with the quality of the concrete and the method of conducting the test. Usually the test is made by embedding the rod in a block of concrete and pulling it therefrom, the rod being stressed in tension and the concrete in compression. This causes a maximum of elongation in the steel at the point where it enters the concrete, while the concrete is subjected to a maximum compression at this same point. This brings very unequal stresses upon the adhering surfaces, tending to a progressive separation until the entire rod has started to slip, after which friction alone holds the rod. This unequal action is greater the deeper the embedment. If the rod is *pushed* out, both rod and concrete are compressed, although not the same amount at the same point. Tests made in this way should therefore give higher results than where the rods are pulled out. Experimental results accord in general with these principles.

Neither of these methods of testing is entirely satisfactory. What is desired is the bond strength when the concrete and the rods are stressed as in a beam. In this case both the concrete and the rod are under tensile stress throughout and therefore subjected to deformations similar in character and approximately equal in amount. They are a minimum at the end of the beam and a maximum at the centre. These conditions tend to make the bond strength in a beam greater than that in the usual test specimen, but other differences tend in the opposite direction. It has been found, in a series of tests by Mr. Withey described further on, that the local compression to which the concrete in the ordinary block specimen is subjected tends materially to increase the bond strength, apparently by pressing the concrete more firmly against the rod. This effect is so marked in some cases that the net results from the usual tests are considerably higher than those made directly on beams.

Results of Tests by Direct Tension.—Table No. 4 contains in condensed form the results of some of the most important tests made by direct tension.

Individual results show little or no effect due to differences in size of rod, but the adhesion for flat bars is much less than for round or square bars.

In general the stronger the concrete the greater the bond strength. Feret found an increase of strength at age of two years of about 50% over that at three months, and a maximum value for quite wet concrete; further, that a small amount of corrosion increased the value. Bach found smaller values the greater the depth, the average value for a 4-inch minimum depth in 1:4 gravel concrete, three months old, being 470 lbs/in^2. He also found greater values when the rods were pushed out than when pulled out

TABLE No. 4.

BOND TESTS BY DIRECT TENSION.

Authority.	Kind of Concrete.	Steel Rods. Kind.	Steel Rods. Size, Inches.	Depth Embedded, Inches.	Adhesive Strength, lbs/in².
Feret; *Ciment Armé*, p. 755.	1:2:4	Plain round	0.8	$2\frac{3}{4}$	237
	1:2:5	" "	0.8	$2\frac{3}{4}$	190
	1:3:4½	" "	0.8	$2\frac{3}{4}$	237
	1:3:6	" "	0.8	$2\frac{3}{4}$	195
Emerson; *Eng. News*, Vol. LI, 1904, p. 222.	1:3	Plain round	$\frac{1}{2}$	6	512
	1:3	Plain flat	$\frac{1}{4}\times 1$	6	293
	1:2:4	Plain square	1×1	10	587
	1:3:6	" "	1×1	10	478
Talbot; *Bull. No. 8, Univ. of Ill.*, 1906.	1:2:4	Plain round	$\frac{1}{2}$ and $\frac{5}{8}$	6	438
	1:2:4	" "	$\frac{1}{2}$ and $\frac{5}{8}$	12	409
	1:3:5½	" "	$\frac{1}{2}$ and $\frac{5}{8}$	6	364
	1:3:5½	" "	$\frac{1}{2}$ and $\frac{5}{8}$	12	388
	1:3:5½	Cold rolled shafting	1 and $\frac{1}{2}$	6	146
	1:3:5½	Mild steel flat	$\frac{3}{16}\times 1\frac{1}{2}$	6	125
	1:3:6	Tool-steel round	$\frac{3}{4}$	6	147
Withey; *Bull. Univ. of Wis.*, 1907.	1:2:4	Plain round	$\frac{3}{16}$ to $\frac{3}{4}$	6	400
	1:2:4	" "	$\frac{3}{16}$ to $\frac{3}{4}$	8	310
Van Ornum; *Eng. News*, Vol. LIX, 1908, p. 142.	1:2:4	Plain round	$\frac{1}{2}$ to $1\frac{1}{4}$	25 diam.	410
	1:2:4	" "	$\frac{1}{2}$ to $1\frac{1}{4}$	40 diam.	390

Results of Tests on Beams.—In an important series of tests by Mr. Withey * at the University of Wisconsin, test beams

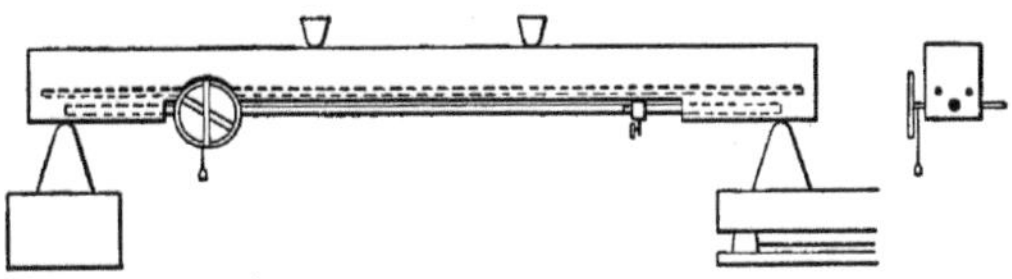

FIG. 7a.

were arranged as shown in Fig. 7a. The stresses in the exposed

* *Engineering Record*, 1908, LVII, p. 798.

rods were determined by means of extensometers. The conditions were very similar to those obtaining in an ordinary beam, but the beam was prevented from failing by the upper auxiliary rods. Table No. 4A gives the principal results of these tests. The table also contains results of comparative tests made at the same time by the usual direct tension method. The last column gives the ratio of the two results. The rods were of ordinary mild steel and were free from rust.

TABLE No. 4A.

BOND TESTS ON BEAMS.

UNIVERSITY OF WISCONSIN, 1907.

Concrete 1:2:4; age 60 days (Nos. 54–57, 28 days).

Test No.	Diameter of Rod, Inches.	Bond Strength, lbs/in².	Average of Group, lbs/in². (A).	Average Bond Strength by Direct Tension. lbs/in². (B)	Ratio, B : A.
38 39 40	$\frac{3}{8}$	345 298 190	278	394	1.42
41 42 43	$\frac{1}{2}$	361 312 186	286	455	1.54
7 8 9	$\frac{5}{8}$	362 264 201	276		
44 45 46	$\frac{3}{4}$	207 289 295	264	502	1.90
47 48 49	1	136 174 180	163	487	2.99
54 55 56 57	$\frac{5}{8}$	228 248 222 247	236		
36 37	$\frac{5}{8}$	254 278	266	467	1.76

These tests indicate that the bond is not affected by size of rod except in the case of the 1-inch size. This difference is undoubtedly due to other factors not explained. Excepting the tests on this size the results are quite uniform, the average of all the 60-day tests being 275 lbs/in^2, with maximum variations of 32% below and 32% above the average. The results obtained by direct tension are very much higher, averaging about 475 lbs/in^2, or 75% greater. Other tests to determine the effect of age of concrete gave values as follows, each result being the average from 3 beams; $\frac{5}{8}$-inch rods were used:

Age.	Bond Strength, lbs/in^2.
7 days	216
28 days	253
60 days	276
6 months	316

It was also found that practically the same results were obtained for three different consistencies, wet, medium, and dry, the averages being respectively 250, 235, and 275 lbs/in^2. In actual practice conditions would be less favorable to the dry mixture owing to the greater difficulty of securing a concrete free from voids.

Table 4B contains results of important tests on beams by Bach,* in which the primary cause of failure was the slipping of the bars. Calculating the corresponding bond stress by eq. (1) of Art. 92, there results the figures given in the table. The average result for the rectangular beams with straight rods only was 291 lbs/in^2, and with stirrups, 330 lbs/in^2. For the T-beams the results for straight bars range from 158 to 195 lbs/in^2. The last four beams contained bent bars, only one bar being continued straight to the end. Calculating the bond stress at the end of the beam with reference to this single rod gives an average value of 493 lbs/in.^{2}

* Mit. Ueber Forsch. a. d. Gebiet des Ing., 1907, 45-47.

Table No. 4b.

BOND TESTS ON BEAMS.

(Bach.)

Concrete 1:4; age 6 months; beams loaded at quarter points.

No.	Kind of Beam.	Reinforcement.	Calculated Bond Stress at beginning of Slip, lbs/in².	Average for Group, lbs/in².
2 3 4 5	Rectangular Beams	Straight rods only	312 300 271 281	291
161		Straight rods and stirrups	330*	330
162	T-Beams	Straight rods only	158	158
163 164		Straight rods and stirrups	182 208	195
165		1 straight, 4 bent	408	493
166		1 straight, 4 bent, with stirrups	498	
167		1 straight, 4 bent, with stirrups	545	
168		1 straight, 4 bent	522	

* Average of three.

Excepting where bent rods were used these results are about the same as those previously quoted. Stirrups appear to increase the bond resistance somewhat, an effect that has been reported by others. The high results obtained on beams with bent rods is important in connection with the design of web reinforcement, as explained in Chapter V. Inasmuch as the actual bond strength must have been practically the same as in the other tests these results show that where some of the bars are bent up so as to reinforce the web of the beam the bond stress on the remaning straight bars near the end of the beam is much less than the theoretical values obtanied by the usual formulas. In this case the real bond stress, as shown by the other tests, appears to be only about half the calculated values.

Frictional Resistance.—In bond tests it is found that after the adhesion has failed the rod still offers much resistance to movement due to friction alone. This frictional resistance varies from 50 to about 80% of the initial bond strength. Hatt found a frictional resistance, after starting, of 50 to 70% of the initial strength, and Mörsch reports such resistance as about two-thirds the initial. Talbot determined the frictional resistance in a large number of tests, finding it to range quite uniformly from about 55 to 65% of the initial bond strength for both plain round and flat rods. In the case of cold-rolled steel the friction was only 40% of the bond strength. Withey found an average result of 80% for 6-inch embedment and 83% for 8-inch embedment. Van Ornum obtained average frictional resistances of 83% for mild steel and 62% for high steel.

Conclusions as to Bond Strength.—A study of the various results leads to the conclusion that for ordinary round or square bars, not too smooth, the bond strength may be taken at from 200 to 300 lbs/in^2, depending upon richness of mixture, age of cement, and roughness of bar, with a frictional resistance of about two-thirds this amount; a much smaller value must be taken for very smooth bars and also for flat bars.

40. Mechanical Bond.—The bond strength of bars with indented surfaces depends upon the adhesive resistance and the compressive (or shearing) strength of the concrete. Under heavy stresses there is also a tendency for the concrete to split, owing to the tensile stresses developed by the wedging action of the bars. The initial movement in the case of indented bars is probably due to a slight crushing of the concrete under the projections. The bars cannot be pulled *through* the concrete. without shearing off an area equal to the total area of the indented portion and, in addition, overcoming considerable friction or adhesion. If one-half the area is indented, the ultimate bond strength can then be placed equal to one-half the shearing strength (see Art. 22), or about one-fourth the compressive strength of the material. For a 1:2:4 concrete this would equal 500 to 600 lbs/in^2. In tests of such bars, failures have usually occurred by the splitting of the specimen or the breaking of the bar, but the results indicate that the actual bond strength is fully equal to these figures.

It appears from tests reported by Mr. T. L. Condron * that the maximum bond strength is not developed until a slight movement has taken place. This is particularly true with deformed bars. In the case of plain round and square bars embedded from 12 to 38 diameters the maximum strength was very nearly reached under a movement of $^1/_{100}$ inch, as measured at the free end of the bar. For twisted bars the resistance continued to increase slightly under movements up to $^1/_{16}$ inch or more, while indented bars (the Thacher and corrugated bars) showed steadily increasing resistance under increased slip up to rupture. The actual bond stress for $^1/_{100}$ inch of movement was 400–600 lbs/in^2 for the Thacher and the corrugated bars, 250–400 lbs/in^2 for the twisted bars, and 175–300 lbs/in^2 for the smooth bars.

Efficiency of Hooked Ends.—It is a common practice among designers to bend the ends of reinforcing rods into short hooks generally consisting of right-angle bends. The efficiency of such hooks in increasing the bond strength has been investigated by Bach. Using rods from $\frac{3}{4}$ to 1 in. in diameter and a length of bend of about 3 diameters and an embedded length of 20 ins., he found that the initial slip was only slightly retarded,—about as much as would be caused by the same length of straight bar,—but the ultimate bond strength was much increased, this increase averaging about 50%. When hooked ends are used they should consist of bends through 180° with a short length of straight rod beyond the bend. Such hooks are found to be very effective.

41. Ratio of Moduli of Elasticity, E_s:E_c.—So long as the adhesion between steel and concrete is unimpaired the distortion of the two materials will be equal. Their stresses will then be proportional to their moduli of elasticity for the load in question, or as the ratio of $E_s : E_c$. Taking E_s at 30,000,000 and E_c at from 2,000,000 to 3,000,000, the ratio will vary from 15 to 10. In practice various values of this ratio are used,

* Jour. West. Soc. Eng., 1907, Vol. XII, p. 100.

depending upon the kind of concrete and the judgment of the designer. It should also depend upon the relative load for which the calculations are made and, to a certain extent, as explained in subsequent chapters, upon the methods of calculation employed. As will be seen in Chapter IV, the value 15 corresponds closely to actual determinations of neutral axes in beams. It is the value commonly used in German regulations and is also specified in the building laws of many American cities.

Equal ratios of moduli may be assumed for both tension and compression.

42. Tensile Strength and Elongation of Concrete when Reinforced.—We have seen that plain concrete has an ultimate tensile strength of about 200 lbs/in^2 and a total elongation of perhaps $^1/_{10000}$ part, corresponding to a value of 2,000,000 for E_c. Steel stretches this amount under a stress of 30,000,000/1000=3000 lbs/in^2. Again, the safe working tensile stress of concrete is about 50 lbs/in^2, and if we use a value of $E_s/E_c=15$, the corresponding stress in the steel will be but 750 lbs/in^2. From these relations it is evident that in reinforced tension members we must either use very low and uneconomical working stresses for steel, or else expect the concrete to be of no assistance in carrying stress.

In studying the behavior of reinforced concrete under tension, and especially when constituting the tensile side of a beam, results of some experiments indicate that the concrete in this condition elongates more before final rupture occurs than when not reinforced, and that the resistance of the concrete is nearly constant and at its maximum value for some time previous to rupture. The first to announce this principle was Considère, whose tests indicated that the ultimate stretch of reinforced concrete was as much as ten times that of plain concrete. Kleinlogel,* however, was unable to check these results, he finding an elongation of practically

* Beton u. Eisen, No.2, 1904.

the same amount as for plain concrete. In experiments of this sort it is extremely difficult to determine just when the concrete begins to crack. The steel forces it to elongate practically uniformly, even after rupture begins, so that a crack will open up very slowly and will therefore remain almost invisible for some time.

In some experiments made at the University of Wisconsin in 1901–2 a very delicate method of detecting incipient cracks was accidentally discovered. It was found that beams cured in water which were only partially dried before testing would, when tested, show very fine hair-cracks at an early stage, and moreover, by watching closely, it was observed that preceding the appearance of a crack there would appear a dark wet line across the beam. Such a line would soon be followed by a very fine crack. A larger series of tests were undertaken in the following year by a different set of experimenters, who observed the same phenomenon. Careful measurements of extension showed that these streaks or "water-marks", as they were named, occurred at practically the same deformation at which the concrete ruptured when not reinforced. Some of the results are given in Table No. 5.* The beams were of 1:2:4 mixture by weight and were 6″×6″ in cross-section by 60 inches span.

That these water-marks were incipient cracks was determined by sawing out a strip of concrete along the outer part of the beam. Fig. 8. is a photograph showing the results of this experiment. Very close observation also in many cases showed hair-like cracks appearing very soon after the appearance of the water-marks.

Comparing the observed and calculated elongations of the reinforced concrete with those of the plain concrete at rupture it will be seen that the initial cracking in the former occurs at an elongation practically the same as reached by the latter at rupture.

* Bulletin No. 4, Engineering Series, Univ. of Wis., 1906.

Table No. 5.

TESTS OF BEAMS SHOWING EXTENSIBILITY OF CONCRETE.

No.	Age.	Method of Loading.	Proportionate Extension. At First Water-mark.	Proportionate Extension. At First Visible Crack.	Compressive Strength of Cubes, lbs/in².
8	3 months	At third points	.00011	.00064	4250
10	"	"	.00024	.00046	2500
22	"	"	.00025	.00065	2775
26	"	"	.00016	.00056	3000
30	"	"	.00012	.00064	2600
7	1 month	At center	.00015	.00036	3500
5	"	"	.00020	.00031	3500
13	"	"	.00009	.00011	2350
23	"	"	.00020	.00060	2500
35	"	"	.00013	.00053	3150
				At rupture	
2*	"	"		.00013	3000
1*	"	"		.00010	2500

* Nos. 2 and 1 were plain concrete beams. The extensions of the beams loaded at the third points were measured by extensometers; those of the center loaded beams were calculated from deflections.

Fig. 8.

Since these experiments were made a very careful series of observations have been made by Bach.* He noted the same "water-marks," and his series of tests on 85 beams of rectangular and T-section, of 1:4 concrete from 6 to 8 months old, gave very uniform results. Water-marks appeared at elongations of from .00006 to .00010 part. He was also able to detect these on plain concrete beams at the same elongation and at loads about 80% of the breaking load; and on several tension specimens he found the elongation, just previous to rupture, to be .00006 to .00010 part. He was able to observe the first well-defined crack at elongations of .00012 to .00014. He concludes that reinforced concrete will begin to crack at the same elongation as plain concrete.

It should be said that in many cases the first "water-marks" do not extend entirely across the tension face of the beam, so that the concrete as a whole possesses some tensile strength.

The presence of these minute cracks of course seriously affects the tensile strength of the concrete, and as they appear at an elongation corresponding to a stress in the steel of 5000 lbs/in^2 or less, it would seem that no allowance should be made for the tensile resistance of the concrete where the usual working stresses are used for steel. In some cases, however, the stresses in the steel are necessarily very low, in which case it may be proper to consider the tensile resistance of the concrete. This limit may be placed at about 2000 lbs/in^2, corresponding to an elongation of .00006 part and a stress of 150 to 175 lbs/in^2 in the concrete.

In practical design the most important question which arises is how far a concrete may be cracked without exposing the steel to corrosive influences. In this respect experience indicates that the minute cracks which appear in the early stages of the tests are of no practical consequences.

43. Relative Contraction and Expansion.—Temperature changes affect both the steel and the concrete. But as the coefficient of expansion of steel is .0000065 and of concrete

* Zeit. Ver. Dt. Ing., 1907.

.000006, the two materials will be but slightly stressed because of any difference in their rates of expansion.

The effect of shrinkage in hardening is more serious. As shown in Art. 29, the hardening of concrete is accompanied by more or less contraction if in air, or expansion (to a less degree) if in water. Concrete which is unrestrained either by steel reinforcement or by exterior attachment will shrink or swell proportionally and no stresses will thereby be developed. If restrained by reinforcing material only, a shrinkage will develop tensile stresses in the concrete and compressive stresses in the steel.

If it be assumed that concrete when reinforced tends to shrink the same amount as plain concrete, and that such shrinkage is prevented only so far as the stresses developed in the steel react upon the concrete and cause an opposite movement, then it will be found, using the ordinary values of the modulus of elasticity, that the stresses developed in both the concrete and the steel will be large. These stresses would be determined as follows:

Let c = coefficient of contraction of the concrete;
f_c = unit stress in concrete (tensile);
f_s = unit stress in steel (compressive);
p = steel ratio;
$n = E_s/E_c$.

Then the net contraction per unit length as measured by the concrete will be $c - f_c/E_c$, and as measured by the steel will be f_s/E_s. These values are equal. Also, for equilibrium, $f_c = pf_s$. From these equations we get

$$f_c = cE_c \frac{np}{1+np} \quad \ldots\ldots\ldots \quad (1)$$

and

$$f_s = \frac{f_c}{p}. \quad \ldots\ldots\ldots\ldots \quad (2)$$

If, for example, $c = .0003$, $E_c = 2{,}000{,}000$, $n = 15$, $p = 1\%$, then $f_c = 80$ lbs/in² tension and $f_s = 8000$ lbs/in² compression. If $p = 2\%$, $f_c = 140$ and $f_s = 7000$ lbs/in².

It is doubtful if such large initial stresses actually occur in reinforced concrete due to shrinkage in hardening.

The experiments of Considère on the actual contraction of reinforced concrete, already quoted in Art. 29, indicate that the deformation is less than the above theory would call for. For example, the observed contraction of .01% in reinforced mortar would call for a stress in the steel of only about 3000 lbs/in^2, and in the concrete of only 30 to 60 lbs/in^2. In slowly hardening, with the steel in place, there is probably a gradual adjustment in the concrete which results in less internal stress than the experiments on plain concrete would indicate. Where the structure is restrained by outside supports which are relatively more rigid than the reinforcing steel, the stresses in the concrete become greater and may easily reach the limit of the tensile strength, thus causing cracks. (For further discussion of reinforcement under such conditions, see Chapter V, Art. 142.)

CHAPTER III.

GENERAL THEORY.

44. Kinds of Members.—Structural members are, for convenience, usually divided into *tension members, compression members,* and *beams,* according as the forces to be resisted produce in the member simple tension, simple compression, or simple bending. Bending moment is often accompanied by tension or compression, producing what are called *combined stresses* of *bending and tension,* or *bending and compression.* Since reinforced concrete is not used for plain tension members the analysis will be confined to the beam, both under plain bending and under combined stresses, and to the compression member or column. The flat slab supported on four sides will be considered as a special case of beam. In reinforced-concrete construction the beam is the most important element, being used under a great variety of conditions.

45. Relation of Stress Intensities in Concrete and Steel. In the following discussion it will be assumed that the concrete and steel adhere perfectly and therefore deform equally. Nearly all reinforced-concrete construction is dependent upon this equal action of the two materials, although simple adhesion is not always entirely depended upon. Many types of deformed, or roughened, bars are used so as to give the steel a grip independent of the adhesion, and in other cases bars are bent or anchored at the ends, but in all cases it is assumed that the materials adhere perfectly and therefore deform equally. Many tests show that under proper design this is for all practical purposes true.

Since the modulus of elasticity of a material is the ratio of stress to deformation, it follows that for *equal* deformations the stresses in different materials will be as their moduli of elasticity. If

f_s = unit stress in steel,
f_c = unit stress in concrete,
E_s = modulus of elasticity of steel, and
E_c = modulus of elasticity of concrete,

we have the fixed relation

$$f_s/f_c = E_s/E_c. \quad . \quad . \quad . \quad . \quad . \quad . \quad . \quad . \quad . \quad (1)$$

46. Distribution of Stress in a Homogeneous Beam.—To assist in forming correct notions of the action of steel reinforcement in a concrete beam, it will be desirable to consider, at the outset, the nature of the stresses due to bending moment in a plain concrete or homogeneous beam of any material. Considering a vertical section at any point there will exist in general certain normal stresses (tensile and compressive) and certain tangential or shearing stresses. A knowledge of these stresses on a vertical section, together with the well-known principle that the shearing stress at any point is of equal intensity vertically and horizontally, is sufficient for the designing of ordinary beams.

In accordance with the common theory of flexure, the normal stress on a vertical section varies in intensity as the distance from the neutral axis, and therefore the variation is represented by the ordinates to a straight line as in Fig. 9.

The shearing-stress intensity is a maximum at the neutral axis and is zero at the outer fibres. At any given point in the section it is given by the equation

$$v = VS/Ib, \quad . \quad . \quad . \quad . \quad . \quad . \quad . \quad . \quad . \quad (1)$$

in which V denotes the entire shear at the section containing the point under consideration, I the moment of inertia of the section with respect to the neutral axis, b the breadth of the section at the point, and S the statical moment of the part of the section above (or below) the point with respect to the neutral axis. For a rectangular beam the intensity of shear varies as the ordinates to a parabola, as shown in Fig. 10, the maximum value being 3/2 times the average, or equal to $\frac{3}{2}\cdot\frac{V}{bd}$.

If the stresses on *inclined* planes are analyzed, it is found that the normal and shearing stresses will not be the same as on vertical planes; and, furthermore, that wherever shearing stress exists on a vertical plane the *maximum* normal stress

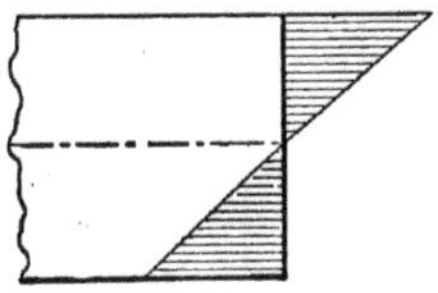

FIG. 9.

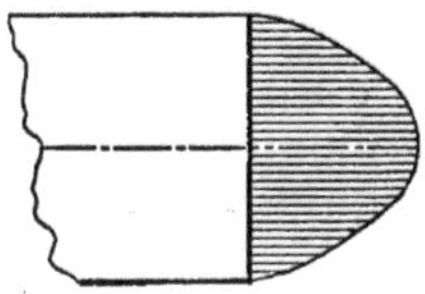

FIG. 10.

will not be on a vertical section, but on an inclined one. It is proved in treatises on mechanics that if f represents the horizontal unit tensile stress and v the vertical or horizontal unit shearing stress at any point in a beam, the maximum tensile stress will be given by the formula

$$t=\tfrac{1}{2}f+\sqrt{\tfrac{1}{4}f^2+v^2}, \quad . \quad . \quad . \quad . \quad . \quad . \quad (2)$$

and the direction of this maximum tension is given by the formula $\tan 2\theta=2v/f$, where θ is the angle of the maximum tension with the horizontal.

A study of these formulas shows that at all points in a beam where the shear is zero, the direction of the maximum tension is horizontal, as at points of maximum bending moment and along the outer fibres of the beam. Wherever the horizontal fibre stress is zero (at the neutral surface and at all sections of zero bending moment), the direction of the maximum tension is inclined 45° to the horizontal, and its intensity is equal to the unit shearing stress at the same place. Above the neutral axis of a section where the bending moment is not zero, the inclination of the maximum tension is greater than 45°, becoming 90° at the upper or compressive fibre. Fig. 11 illustrates the variation in normal stress, shearing stress, and maximum tensile stress throughout the entire depth of a rectangular beam. The outer normal or fibre stress is assumed at 200 lbs/in^2, and the shearing stress at the neutral axis at 150 lbs/in^2. The

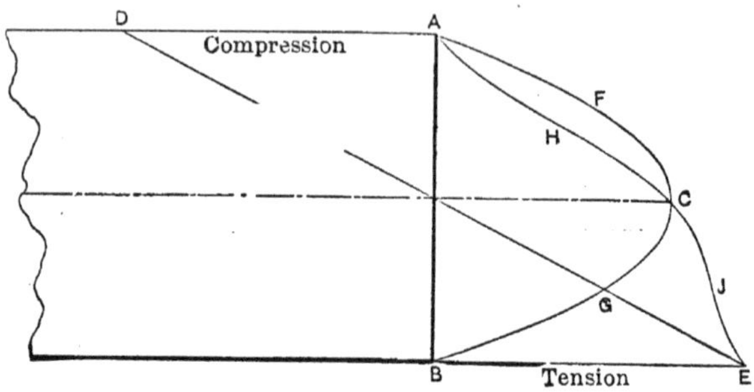

FIG. 11.—Showing Variation of Intensities of Normal Stress, Shear, and Maximum Tension.

variation in the fibre stress is shown by the straight line DE, and that in the shearing stress by the parabolic curve ACB. By means of eq. (2) the maximum tensile stresses have been computed; these are represented by the line $AHCJE$.

Fig. 12 illustrates the direction of the maximum tensile

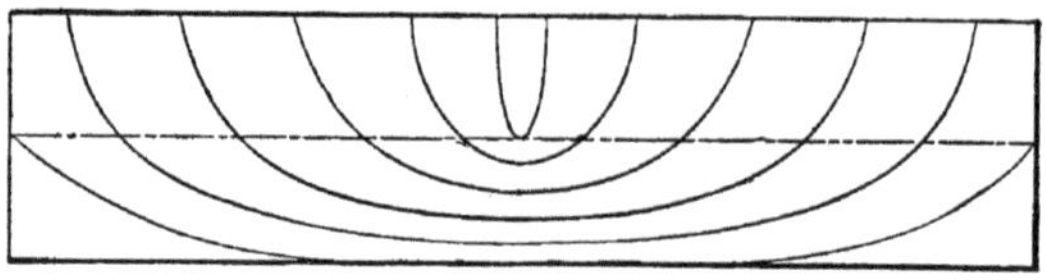

FIG. 12.—Lines of Maximum Tension.

stresses in a rectangular beam. The exact direction at any point depends upon the relation between shear and bending moment. Lines of maximum compression would run at right angles to the lines shown and lines of maximum shear at angles of 45° therewith.

47. Purpose and Arrangement of Steel Reinforcement.— The purpose of steel reinforcement is to carry the principal tensile stresses, the concrete being depended upon for the compressive and shearing stresses, its resistance to such stresses being large. If no steel were present the concrete would tend to rupture on lines perpendicular to the direction of maximum tension, as shown in Fig. 12, and hence we may conclude that the ideal tension reinforcement would require the steel to be distributed in the beam along the lines of maximum tension. At the centre of the beam, or place of maximum moment, this direction is horizontal for the entire depth of the beam, and horizontal rods placed near the lower edge of the beam constitute proper and sufficient reinforcement. As we approach the ends of the beam, where the shear is large, the intensity of the inclined tensile stresses becomes of importance, and in many cases these stresses require special attention. Horizontal rods at the bottom are still necessary, but do not entirely reinforce the concrete against tension, so that special consideration must be given to reinforcement in the body of the beam. The arrangement of this reinforcement demands careful consideration.

For purposes of discussion, the subject of beams will first be treated with reference only to the horizontal reinforcement. The inclined tensile stresses will be considered separately.

48. The Common Theory of Flexure and its Modification for Concrete.—The common theory of flexure is based on two main assumptions, namely, (1) a plane cross-section of an unloaded beam will still be plane after bending (Navier's hypothesis); (2) the material of the beam obeys Hooke's law, which is, briefly stated, "stress is proportional to strain". From the first assumption it follows that—*The unit deformations of the fibres at any section of a beam are proportional to their distances from the neutral surface.* In the case of simple bending (all forces at right angles to the beam) the neutral axis lies at the centre of gravity of the section; in the case of bending combined with direct tension or compression, the neutral axis may lie in the section or be merely an imaginary line without the section. From the second assumption it follows that—*The unit stresses in the fibres at any section of a beam also are* *proportional to the distances of the fibres from the neutral surface.* This may be called the linear law of the distribution of stress.

The linear law is the basis of all practical flexure formulas excepting some for reinforced-concrete beams. It is true that wrought iron and steel are the only important structural materials which closely obey Hooke's law, and they only within their elastic limits. But under working conditions these materials are not stressed beyond these limits, and so the formulas ordinarily hold. Timber, stone, and cast iron can hardly be said to obey Hooke's law, yet for working conditions the common flexure formulas for these materials are roughly correct and they are in general use.

In the case of those materials which do not obey Hooke's law, as concrete, and for all materials when stressed beyond their elastic limit, the common theory does not strictly apply. An exact analysis requires the use of the actual tension and compression stress-strain diagrams for the materials up to the limit of the actual stresses involved. It will be assumed still that plane sections remain plane during bending so that deformations will be proportional to the distances of the fibres from the neutral surface. The experiments by Talbot,* though not conclusive, bear out this assumption in the more important case of reinforced beams. Experiments by Schüle,† however, seem to show that original plane sections do not remain plane. Nevertheless Navier's hypothesis will probably remain a basis of flexure formulas for reinforced-concrete beams.

The variation of the normal stress on the cross-section can then be represented graphically in the following manner: Let Fig. 13*a* be the stress-strain diagram, compression above the x axis and tension below, for the material in question as determined by direct compression and tension tests. These curves are plotted with unit stresses as abscissas and unit strains as ordinates. Let Fig. 13*b* represent the beam, cut

* Univ. of Ill. Bull., Vol. II, No. 1, p. 28.

† Mitteilungen der Materialprüfungs-Anstalt am Polytechnikum in Zürich, Vol. X (1906), p. 40.

on section AB where the stresses are to be investigated. The neutral axis is at N. Since the deformations of the fibres are proportional to the distances of the fibres from the neutral axis, these distances themselves, $N1$, $N2$, $N3$, etc., will represent to some scale the deformations. If the unit deformation at point 1 is then represented by $N1$ the corresponding stress can be determined from the diagram of Fig. 13a, using the proper scale in both cases. Lay off the distance $1a$ to represent that stress. Proceeding similarly for all points and connecting, we have the stress curve $A'NB'$, which is nothing more than

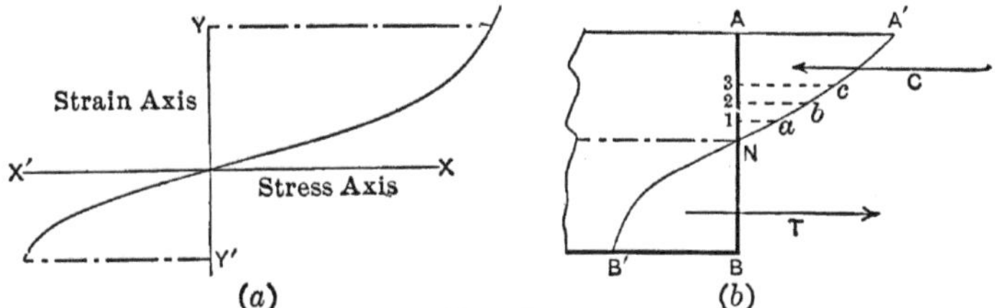

FIG. 13.

a portion of the diagram of Fig. 13a plotted to a different scale.

49. Resisting Moment and Inefficiency of Concrete Beams. — For use in the following and other discussions on flexure three important principles from the mechanics of beams are now recalled:

(1) For beams rectangular in section, the average unit tensile and compressive fibre stresses on any cross-section are represented by the average abscissas in the tensile and compressive parts of the stress diagram, NBB' and NAA', respectively (Fig. 13b). Also the whole tension T and whole compression C on the cross-section are proportional to the *areas* NBB' and NAA'; hence, according to some scale, the areas represent T and C respectively.

(2) The resultant tension T and resultant compression C act through the centroids of the tensile and compressive areas in the stress diagram.

(3) When all the forces (loads and reactions) applied to

the beam act at right angles to it, then the resultant tension T equals the resultant compression C; hence the two stresses constitute a couple—"the resisting couple".

Fig. 14 is a stress-strain diagram of a gravel concrete for both tension and compression. For any section of a beam made of this concrete, the stress diagram is a certain part of the stress-strain diagram, the exact part depending on the loading. Suppose that the loads produce in the lower fibre at the section in question a unit stress represented by bb' say, then T is represented by Nbb' and C by an area Naa' determined from the principle that it must equal the area Nbb'. Hence the stress diagram is $aa'Nb'b$, and the unit stress on the upper fibre is represented by aa'. Furthermore, ab represents the depth of the beam, and N the position of the neutral axis. Likewise, when the unit stress on the lower fibre is BB' (the ultimate tensile strength) and the beam is on the point of failing, T is represented by the area NBB', and C by the equal area NAA'; hence the stress diagram for the failure stage is $AA'NB'B$, and the unit stress on the upper fibre is AA'.

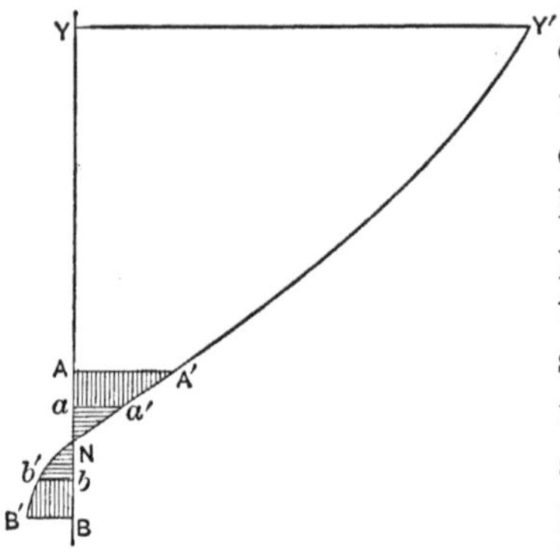

FIG. 14.

50. *Resisting Moment.*—The resisting moment of a section is the moment of the resisting couple which acts at that section. Its value is the product of the tension (or compression) and the distance between the centroids of these stresses. For example, at the failure stage of the beam above referred to the average unit tensile stress scales 128 lbs/in², and $\overline{NB}=0.6\overline{AB}=0.6d$, d denoting depth of beam. Hence if b denotes the breadth of the section,

$$C=T=128\times0.6d\times b=76.8bd.$$

The vertical distance between the centroids of the shaded parts (NAA' and NBB') of the diagram is $0.64\overline{AB}$; hence the arm of the resisting couple is $0.64d$, and the computed ultimate resisting moment of a beam made of the concrete under consideration is $76.8bd \times 0.64d = 49.2bd^2$ in-lbs., b and d to be expressed in inches.

Partly to test the correctness of the theory of flexure of concrete beams, Professor Mörsch* made three beams 15×20 cm. in section and several tension and compression specimens of the same mix of concrete. From tests on the specimens he obtained a stress-strain diagram from which he computed the probable resisting moment of the beams to be $3.45bd^2 = 3.45 \times 15 \times 20^2 = 20{,}700$ kg-cm. The average of the actual resisting moments of the beams (determined from tests to destruction) was 22,100 kg-cm., an agreement to be regarded as highly satisfactory.

The working resisting moment of a rectangular beam can be computed from the stress-strain diagram for the material in this same manner. Fortunately, engineers are not called upon to compute resisting moments by this method. It is here set forth principally as a means of introducing important ideas bearing on reinforced-concrete beams.

51. *Inefficiency of Concrete Beams.*—When a beam of the concrete above referred to is loaded to the breaking point, the greatest unit compressive stress in the beam is the stress AA', which is in this case about 375 lbs/in^2. This is very low compared to the ultimate compressive strength (2500 lbs/in^2), and the difference indicates a wasteful use of concrete.

The unshaded portion of the stress-strain diagram (Fig. 14) is also significant in this connection, for it indicates the unused compressive strength of the concrete above the neutral surface when the tensile strength of that below is fully developed and the beam is about to fail.

Another way to express the inefficiency of a concrete beam

* Der Eisenbetonbau.

is to compare its ultimate resisting moment with that which it would have if the tensile strength and elastic properties were the same as the compressive. On this supposition the tensile stress-strain diagram would be like the compressive; and for the concrete of Fig. 14, the ultimate C and T are represented by the area NYY', and the arm of the resisting couple by twice the vertical distance of the centroid of the area NYY' above N. Actual measurement of the area and distance gives $C=775bd$ and $arm=0.64d$; hence the ideal ultimate resisting moment is $775bd \times 0.64d = 496bd^2$ as against $49.2bd^2$, the actual value.

To supply the deficiency in tensile strength of concrete is the main purpose of steel reinforcement. A comparatively small amount of steel (rods or bars whose combined sectional area is from 1 to 2 per cent of the total sectional area of the beam) properly embedded will so strengthen the tensile side of the beam that the great strength of the compressive side can be utilized. The exact amount of steel required in any case depends on the elastic properties of the concrete and steel.

52. Varieties of Flexure Formulas.—Many formulas have been proposed for the strength of reinforced-concrete beams. The differences among them arise principally from three sources, namely: (1) The method of applying the factor of safety, (2) the law of distribution of the compressive fibre stress in the concrete, and (3) the value of the tensile fibre stress in the concrete. In regard to:

(1) Two views are held as to the proper method of applying the factor of safety. For example, to ascertain the safe load for a given beam, some engineers assume working strengths for the concrete and steel, with which, by means of a suitable flexure formula, they compute the safe load directly; other engineers compute the breaking load of the beam by a suitable formula and then, with reference to this load, they decide upon the safe load. (The pros and cons of these two methods are discussed in Art. 118.) Formulas for working conditions (for use in the first method) are explained in Arts. 54–9; those

for ultimate conditions (for use in the second method) in Arts. 60–4; and those for both conditions in Arts. 65–70.

(2) As already explained in Art. 48, the distribution of the compressive fibre stress can be represented by a portion of the stress-strain diagram for the concrete. As shown in Art. 23, the stress-strain curve for concrete up to and even beyond working stresses is nearly straight, and the most widely used flexure formulas for working conditions are based on the assumption that the stress-strain curve is practically straight up to working stresses. Formulas of Arts. 54–9 and all other flexure formulas of this book (except those of Arts. 60–70) are based on this assumption. When the curvature of the stress-strain curve has been taken into account, it has generally been assumed to be an arc of a parabola, the vertex

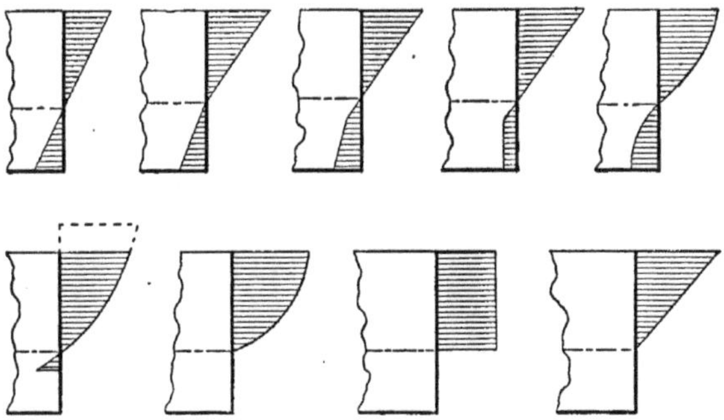

FIG. 15.—Distribution of Fibre Stress in Concrete According to Various Assumptions.

being taken, by some, at the end of curve (the ultimate strength end) and, by others, beyond that point. The formulas of Arts. 60–70 are based on a parabolic stress-strain curve, the vertex being at the end.

(3) As explained in Art. 42, when a reinforced-concrete beam is being loaded, the concrete adjoining the steel fails (cracks) probably always before the stress in the steel reaches 5000 lbs/in^2, and when the stress reaches working values the cracks will have extended well-nigh to the neutral surface. The amount of tension remaining in the concrete at the section

of the crack is comparatively small, and this tension being near the neutral surface, the resisting moment due to it is also small compared to that due to the tension in the steel. In a certain formula for ultimate resisting moment in which this residual tension in the concrete is allowed for, the value of the term expressing the contribution of this tension is less than ½ per cent of the total moment. It is the almost universal practice to neglect this tension entirely in flexure formulas; this practice is followed in this book.

An idea of the variety of flexure formulas proposed can be gained from Fig. 15, which shows nine distributions of fibre stress in the concrete according to as many different formulas.

53. Notation.—Fuller explanations of some of these symbols are given in subsequent articles where the formulas are derived; see also Fig. 16.

f_s denotes unit fibre stress in steel;
f_c " " " " " concrete at its compressive face;
e_s " " elongation of the steel due to f_s;
e_c " " shortening of the concrete due to f_c;
E_s " modulus of elasticity of the steel;
E_c " " the concrete in compression;
n " ratio E_s/E_c;
T " total tension in steel at a section of the beam;
C " total compression in concrete at a section of the beam;
M_s " resisting moment as determined by steel;
M_c " resisting moment as determined by concrete;
M " bending moment or resisting moment in general;
b " breadth of a rectangular beam;
d " distance from the compressive face to the plane of the steel;
k " ratio of the depth of the neutral axis of a section below the top to d;
j " ratio of the arm of the resisting couple to d;
A " area of cross-section of steel;
p " steel ratio, A/bd.

54. Flexure Formulas for Working Loads Based on Linear Variation of the Compression and Neglecting Tension in the Concrete.—The loads being working loads, the unit stress in the steel is within the elastic limit, and the unit stresses in the concrete vary as the ordinates to the compressive stress-strain curve for concrete up to working stresses. This curve is nearly straight; it will be assumed straight to simplify the formulas. The resulting errors are small, as is explained in Art. 70.

55. *Neutral Axis and Arm of Resisting Couple.*—It follows from the assumption of plane sections that the unit deformations of the fibres vary as their distances from the neutral axis; hence, $e_s/e_c = (d-kd)/kd$ (see Fig. 16). Also $e_s = f_s/E_s$ and $e_c = f_c/E_c$; hence, introducing the abbreviation n,

$$\frac{f_s}{nf_c} = \frac{d-kd}{kd} = \frac{1-k}{k}. \quad . \quad (a)$$

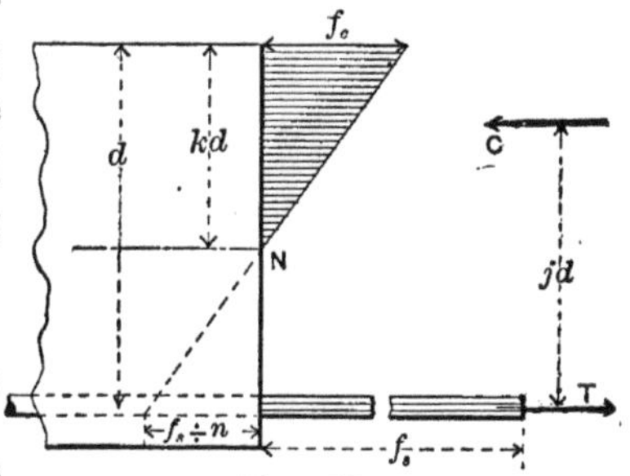

FIG. 16.

When the loads and reactions are vertical—beam horizontal—the total tension and compression on the section are equal, i.e.,

$$f_s A = \tfrac{1}{2} f_c bkd. \quad . \quad . \quad . \quad . \quad . \quad . \quad . \quad . \quad (b)$$

Eliminating f_s/f_c between equations (a) and (b) and introducing the abbreviation p gives $2pn(1-k) = k^2$; this if solved for k gives

$$k = \sqrt{2pn + (pn)^2} - pn. \quad . \quad . \quad . \quad . \quad . \quad (1)$$

This formula shows that the neutral axes of all beams of a given concrete and of a given percentage of reinforcement are at the same proportionate depth, k, for all working loads. The lower group of curves in Fig. 17 gives k for different values of p and n; thus for $p=0.015$ (percentage of steel $=1.5$) and $n=15$, $k=0.48$. The curves show that k increases as p or n increases.

The distance of the centroid of the compressive stress from the compressive face of the beam is $\frac{1}{3}kd$; therefore the arm of the resisting couple, TC, is given by

$$jd=d-\tfrac{1}{3}kd, \quad \text{or} \quad j=1-\tfrac{1}{3}k. \quad . \quad . \quad . \quad . \quad (2)$$

As k increases, j decreases, but not in the same ratio. Fig. 17 shows how j changes with p for four different values of n. It should be noticed that j does not vary much with p, and that for $n=15$ and p between 0.75 and 1.0%—common values—the average value of j is about $\frac{7}{8}$.

56. *Resisting Moment for Given Working Stresses f_s and f_c.*—If the beam is under-reinforced, its resisting moment depends on the steel and its value then is

$$M_s=T\cdot jd=f_sA\cdot jd=f_spjbd^2. \quad . \quad . \quad . \quad . \quad (3)$$

If over-reinforced, the resisting moment depends on the concrete and its value then is

$$M_c=C\cdot jd=\tfrac{1}{2}f_cbkd\cdot jd=\tfrac{1}{2}f_ckjbd^2. \quad . \quad . \quad . \quad . \quad (4)$$

To find the resisting moment in a given case, these values of M must be compared, and the lesser one taken; but it may be noticed that a comparison of the quantities f_sp and $\frac{1}{2}f_ck$ is sufficient to determine which of the values is the lesser.

For approximate computations one may use the average values $j=\frac{7}{8}$ and $k=\frac{3}{8}$; then formulas (3) and (4) become respectively

$$M_s=f_sA_s\cdot\tfrac{7}{8}d, \quad . \quad . \quad . \quad . \quad . \quad . \quad . \quad (3)'$$

$$M_c=f_c\cdot\tfrac{1}{6}bd^2. \quad . \quad . \quad . \quad . \quad . \quad . \quad . \quad (4)'$$

57. *Unit Fibre Stresses for a Given Bending Moment.*—Formulas for these may be obtained from equations (3) and (4) by solving them for f_s and f_c respectively; M will denote bending moment. Or, one may reason as follows: Since the resisting moment is Tjd,

$$T=\frac{M}{jd} \quad \text{and} \quad f_s=\frac{T}{A}; \quad . \quad . \quad . \quad . \quad . \quad . \quad (5)$$

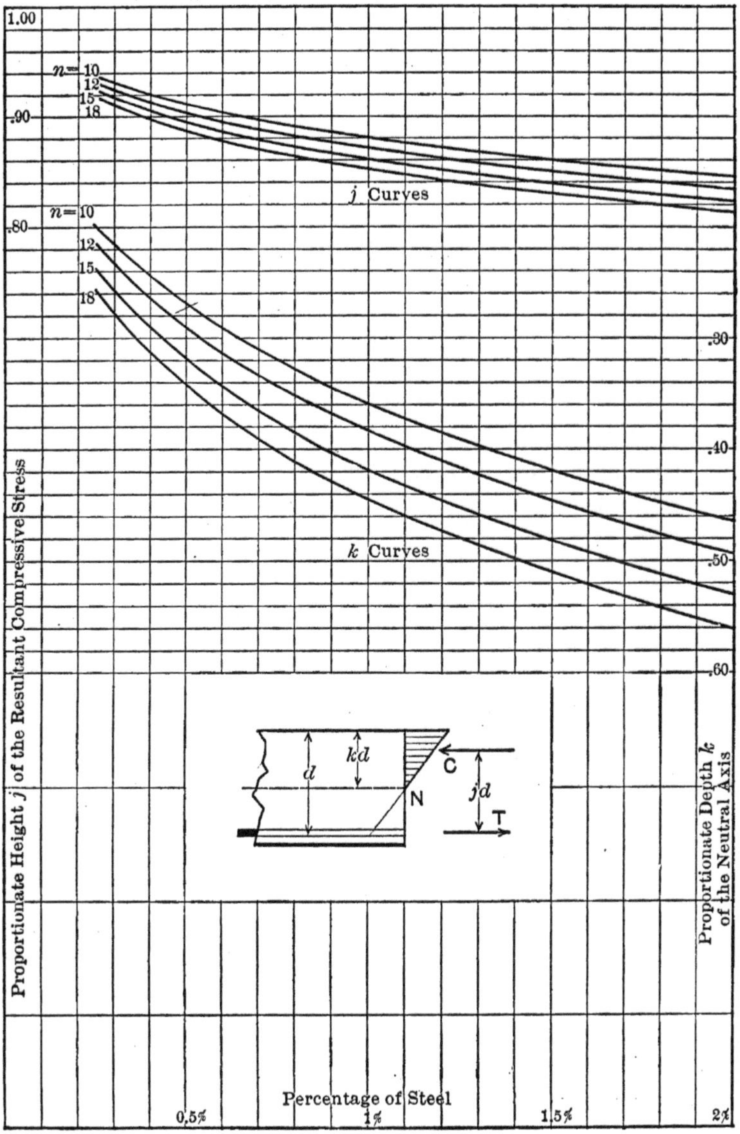
1.00
.90
.80
n= 10
12
15
18
j Curves
k Curves
.30
.40
.50
.60
kd
d
C
N
jd
T
Proportionate Height j of the Resultant Compressive Stress
Proportionate Depth k of the Neutral Axis
Percentage of Steel
0.5%
1%
1.5%
2%

FIG. 17.

also, since f_c equals twice the average unit compressive stress on the section, and $C=T$,

$$f_c=\frac{2T}{kbd}=\frac{2f_sp}{k}. \quad . \quad . \quad . \quad . \quad . \quad . \quad . \quad . \quad (6)$$

Approximating as before, i.e., using average values $j=\frac{7}{8}$ and $k=\frac{3}{8}$, formulas (5) and (6) become respectively

$$T=\frac{M}{\frac{7}{8}d} \text{ and } f_s=\frac{T}{A}, \quad . \quad . \quad . \quad . \quad . \quad . \quad (5)'$$

and

$$f_c=\frac{2T}{\frac{3}{8}bd}=\frac{16}{3}f_sp. \quad . \quad . \quad . \quad . \quad . \quad . \quad . \quad . \quad (6)'$$

58. *Determination of Amount of Steel and Cross-section of Beam for a Given Bending Moment.*—If k be eliminated between equations (a) and (6), the following formula for steel ratio results:

$$p=\frac{1/2}{\frac{f_s}{f_c}\left(\frac{f_s}{nf_c}+1\right)}. \quad . \quad . \quad . \quad . \quad . \quad . \quad . \quad (7)$$

It shows that for given concrete and ratio of working stresses, p has the same value for all sizes of beams. Fig. 18 gives graphically the proper values of p for different ratios f_s/f_c and four different values of n.

If a value of p less than that given by (7) is adopted then the cross-section, or bd^2 rather, should be determined from the first of equations (8); if greater, from the second. (These are (3) and (4) solved for bd^2 respectively.)

$$bd^2=\frac{M}{f_spj}, \quad bd^2=\frac{M}{\frac{1}{2}f_ckj} \quad . \quad . \quad . \quad . \quad . \quad (8)$$

Values of k and j can be obtained from (1) and (2) or Fig. 17; then inserting an assumed value of b, d can be obtained by direct solution of the formula.

For Approximate Design.—To determine the percentage of steel, use (6)′ in this form, $p=\frac{3}{10}f_c/f_s$. If a smaller percentage than this is decided upon, use the first of equations (8)′ to determine b and d; and if a larger then the second one.

$$bd^2=\frac{M}{\frac{7}{8}f_sp},\quad bd^2=\frac{M}{\frac{1}{6}f_c}. \quad . \quad . \quad . \quad . \quad . \quad (8)'$$

59. *Diagrams and Examples.*—Some numerical examples illustrating the preceding principles will now be given, and

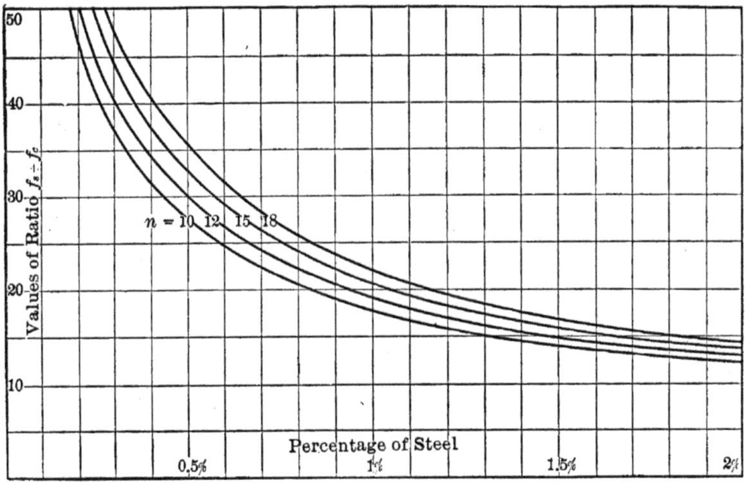

FIG. 18.

then some diagrams will be explained by means of which computations in such examples can be wholly avoided or nearly so.

(1) A concrete beam is 10×16 inches in cross-section and the tension reinforcement consists of four $\frac{3}{4}$ inch steel rods, their centres being two inches above the lower face of the beam. The working stress of the concrete being 600 lbs/in² and that of the steel 15,000, what is the safe resisting moment of the beam?

Solutions. The cross-section of one steel rod is 0.442 in², hence $A=1.768$; and as $b=10$ and $d=14$, $p=1.768/140=0.0126$. Therefore, n being taken as 15, from (1) $k=0.453$; also from (2) $j=0.849$. As determined by the steel, the resisting moment is (see eq. 3)

$$M_s=15,000\times1.768\times0.849\times14=315,000 \text{ in-lbs.}$$

As determined by the concrete, the resisting moment is (see eq. 4)

$$M_c = 300 \times 10 \times 0.453 \times 14 \times 0.849 \times 14 = 227{,}000 \text{ in-lbs.}$$

The safe resisting moment is the latter value.

The approximate formulas, (3)′ and (4)′, give respectively

$$M_s = 15{,}000 \times 1.768 \times \tfrac{7}{8} \times 14 = 325{,}000$$

and $$M_c = 600 \times \tfrac{1}{6} \times 10 \times 14^2 = 196{,}000 \text{ in-lbs.}$$

The approximate formula relating to the steel always gives a closer result than the other.

(2) Suppose that the beam of the preceding example is 19 in. deep and is subjected to a bending moment of 350,000 in-lbs. Compute the greatest unit stresses in the steel and concrete.

Solutions. The steel ratio is $1.768/170 = 0.0104$; and with $n = 15$, eq. (1) gives $k = 0.424$, and eq. (3) gives $j = 0.859$. Therefore $T = 350{,}000/0.859 \times 17 = 24{,}000$ lbs., and $f_s = 24{,}000/1.768 = 13{,}600$ lbs/in². Also see eq. (6), $f_c = 48{,}000/0.424 \times 10 \times 17 = 665$ lbs/in².

The approximate formulas (5)′ and (6)′ give respectively

$$f_s = 13{,}500 \quad \text{and} \quad f_c = 750 \text{ lbs/in}^2.$$

Again, of the approximate formulas, the one relating to the steel gives the closer result.

(3) A beam is to be figured to withstand a bending moment of 135,000 in-lbs., the working strength of the concrete and steel being taken at 700 and 12,000 lbs/in² respectively.

Solutions. For $n = 15$, eq. (7) gives $p = 0.0136$. With this value of p, eq. (1) gives $k = 0.462$, and hence $j = 0.846$. Eq. (8) now gives

$$bd^2 = \frac{135{,}000}{12{,}000 \times 0.0136 \times 0.846} = 978.$$

Many different values of b and d will satisfy the last equation. If b is taken as 7 in., then

$$d^2 = 978/7 = 140, \quad \text{or} \quad d = 12 \text{ in.}$$

Finally $$A = 0.0136(7 \times 12) = 1.14 \text{ in}^2.$$

The approximate formula 6′ gives for a suitable steel ratio $p = \tfrac{3}{16}\,700/12{,}000 = 0.0109$. Adopting 0.011, then 8′ gives $bd^2 = 135{,}000/\tfrac{1}{6}\,700 = 1157$. Taking $b = 7$ in. as before, $d^2 = 1157/7 = 165.3$, or $d = 12.8$, 13 in. say. Finally $A = 0.011 \times 7 \times 13 = 1.00$ in².

The construction of the diagrams (Plates I–IV, pages 275 to 278) referred to will now be explained and then their use. It will be convenient to have names for the quantities $f_s pj$ and $\tfrac{1}{2} f_c kj$

(see eqs. 3 and 4) and single symbols for them. We shall call them *coefficients of resistance* relative to the steel and the concrete and will denote them by R_s and R_c respectively; that is,

$$(a)\quad R_s = f_s pj \quad \text{and} \quad (b)\quad R_c = \tfrac{1}{2} f_c kj.$$

Then the formulas for resisting moments of a given beam with particular working strengths f_s and f_c may be written thus:

$$M_s = R_s bd^2 \quad \text{and} \quad M_c = R_c bd^2. \quad . \quad . \quad . \quad . \quad (1)$$

Similarly for any particular beam subjected to a bending moment M,

$$R_s = R_c = M/bd^2. \quad . \quad . \quad . \quad . \quad . \quad . \quad . \quad (2)$$

Likewise for any particular bending moment and working strengths f_s and f_c, the necessary section is given by

$$bd^2 = M/R, \quad . \quad . \quad . \quad . \quad . \quad . \quad . \quad . \quad (3)$$

R being the smaller of the two coefficients of resistance.

In the four diagrams values of p are given at the upper and lower margins and values of R_s and R_c at the sides. The diagrams are drawn for four different values of n, viz., 10, 12, 15, and 18, as noted on the plates.

The f_s curves of the diagrams are merely the plots, or graphs, of equation (a) for certain values of f_s as marked on the curves. The f_c curves are the graphs of equation (b) for various values of f_c as marked. For example, when $n=15$, $f_s=14{,}000$, $f_c=600$, and $p=1\%$ (see page 277), $R_s=120$ and $R_c=108$.*

The foregoing three examples will now be solved by means of the diagram, page 277 ($n=15$).

(1) The percentage of steel being 1.26, we first find that value on the lower margin; then trace vertically, stopping at the first of the two curves $f_c=600$ and $f_s=15{,}000$; then trace horizontally to either side

* These diagrams are modeled after those contributed by Prof. French in Trans. Am. Soc. C. E., Vol. LVI, 1906, pp. 362–4.

margin and read off the value $R=115$. Finally $M=115\times10\times14^2=$ 225,400 in-lbs.

(2) $R=M/bd^2=350{,}000/10\cdot17^2=121$, and the percentage of steel is 1.04. We enter the diagram with these values of R and p, find the intersection of the horizontal and vertical lines through these values respectively, and from the steel and concrete curves adjacent to this intersection estimate f_s to be 13,750 and f_c 675 lbs/in².

(3) We first find the intersection of the curves $f_c=700$ and $f_s=12{,}000$; from that point tracing down we find $p=1.35\%$, and tracing horizontally we find $R=137$. Then $bd^2=M/R=135{,}000/137=986$, from which b and d may be decided upon, and then finally the amount of steel.

60. Flexure Formulas for Ultimate Loads, Based on Parabolic Variation of Compression and Neglecting Tension in Concrete.—It is assumed that the amount of reinforcement is sufficient to develop the full compressive strength of the concrete without straining the steel beyond its yield point; or otherwise expressed, failure occurs by crushing of the concrete, the stress in the steel being still within the yield point. Then the parabola representing the variation of compression is a full parabola (see Art. 26), the upper end (see Fig. 19) being the vertex.

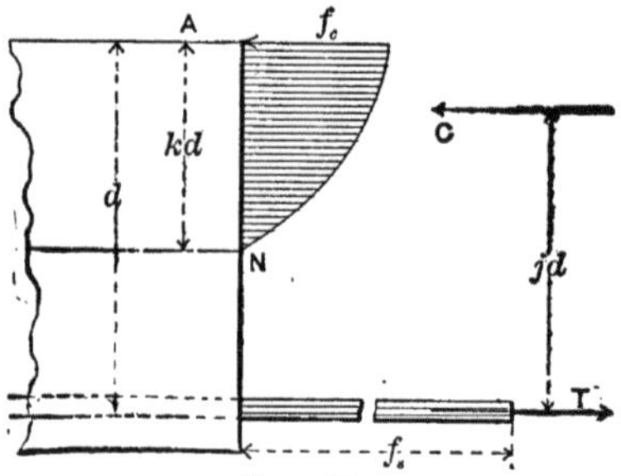

FIG. 19.

If the amount of steel in a beam is such that the ultimate strength of the concrete and the elastic limit of the steel would be reached simultaneously if the beam were subjected to a gradually increasing load, then this will be called the ideal amount—no better term seems available—but this amount may not be the best in a given case.

In the present connection, the two following properties of a parabola like that of Fig. 19 are useful: (1) The average abscissa of the parabolic arc equals two-thirds the greatest, f_c; (2) the distance from the centroid of the parabolic area to its top equals three-eighths the total height, kd.

61. *Neutral Axis and Arm of Resisting Couple.*—The "initial modulus of elasticity" of the concrete (Art. 24) is denoted by E_c in the present article. It is represented by the tangent of the angle between the vertical through N and the tangent to the stress-strain curve at N. And since NA represents e_c, it follows from a well-known property of the parabola that $f_c = \frac{1}{2}E_c e_c$. Also $f_s = E_s e_s$, and from the assumption of plane sections it follows that $e_s/e_c = (d-kd)/kd$. Eliminating e_s/e_c from the above equations, and introducing the abbreviation n, gives

$$\frac{f_s}{2nf_c} = \frac{1-k}{k}. \quad \text{(a)}$$

When the loads and reactions are vertical—beam horizontal—the total tension and the total compression on the section are equal, i.e.,

$$f_s A = \tfrac{2}{3} f_c bkd. \quad \text{(b)}$$

Eliminating f_s/f_c between equations (*a*) and (*b*) and introducing the abbreviation p, gives $3pn = k^2/(1-k)$; this if solved for k gives

$$k = \sqrt{3pn + (\tfrac{3}{2}pn)^2} - \tfrac{3}{2}pn. \quad \text{(1)}$$

This formula shows that the neutral axes of all beams of a given concrete and of a given percentage of reinforcement are at the same proportionate depth, k, for their respective ultimate loads. The lower group of curves (Fig. 20) gives k for different values of p and n; thus for $p = 2\%$ and $n = 15$, $k = 0.60$. The curves show that k increases as p or n increases.

The distance of the centroid of the compressive stress from the compressive face of the beam is $\frac{3}{8}kd$; therefore the arm of the resisting couple TC is given by

$$jd = d - \tfrac{3}{8}kd, \quad \text{or} \quad j = 1 - \tfrac{3}{8}k. \quad \text{(2)}$$

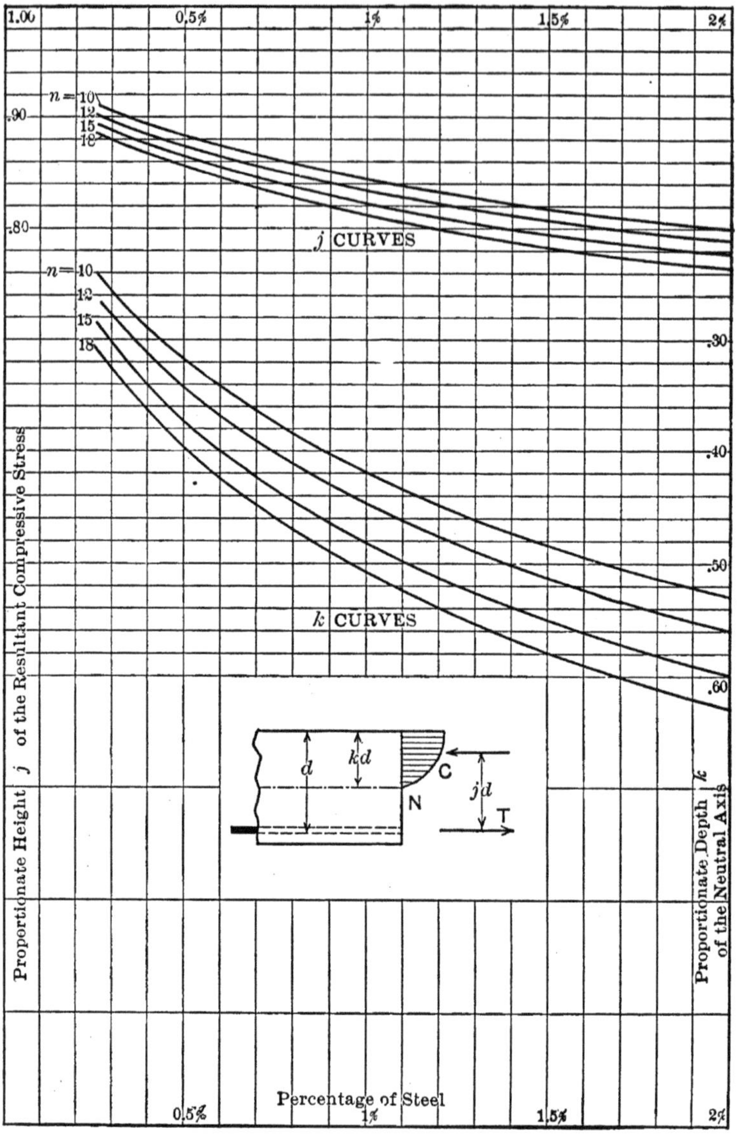

1.00
0.5%
1%
1.5%
2%
.90
.80
$n=10$
12
15
18
j CURVES
$n=10$
12
15
18
.30
.40
.50
k CURVES
.60
d
kd
C
N
jd
T
Proportionate Height j of the Resultant Compressive Stress
Proportionate Depth k of the Neutral Axis
Percentage of Steel
0.5%
1%
1.5%
2%

FIG. 20.

Plainly, as k increases j decreases, but not at the same rate. The upper group of curves in Fig. 20 gives j for different values of p and n; thus for $p=2\%$ and $n=15$, $j=0.775$. It should be noticed that j does not vary much with p, and that for $n=15$ and p greater than 1% the average value of j is about 0.80.

62. *Ultimate Resisting Moment for a Given Ultimate Strength f_c.*—Remembering the assumption made at the outset in regard to the amount of steel (Art. 60), it will be understood that the ultimate resisting moment always depends on the concrete; the value is

$$M_c=C\cdot jd=\tfrac{2}{3}f_cbkd\cdot jd=\tfrac{2}{3}jkf_cbd^2. \quad . \quad . \quad . \quad . \quad (3)$$

It should be remembered that this equation gives the ultimate resisting moment only if when the unit stress in the concrete is at the ultimate that in the steel is not beyond the elastic limit.

If the beam has the "ideal amount" of reinforcement before referred to, then the ultimate resisting moment can be computed from the steel by means of

$$M_s=T\cdot jd=f_sA\cdot jd=f_spjbd^2, \quad . \quad . \quad . \quad . \quad (4)$$

in which f_s denotes elastic limit of steel.

For approximate computations one may use the average values $j=0.80$ and $k=0.52$; with these, formulas (3) and (4) become respectively

$$M_c=0.278f_cbd^2, \quad . \quad . \quad . \quad . \quad . \quad . \quad . \quad (3)'$$
$$M_s=T0.8d=0.8f_spbd^2. \quad . \quad . \quad . \quad . \quad . \quad (4)'$$

63. *Determination of Amount of Steel and Cross-section of Beam for a Given Ultimate Bending Moment.*—When a beam contains the "ideal amount" of steel, the values of M given by (3) and (4) are equal; hence, $f_s/f_c=2k/3p$. If the value of k as given by equation (a) be inserted in this equation, then the following formula for the "ideal steel ratio" results:

$$p=\frac{2/3}{\frac{f_s}{f_c}\left(\frac{f_s}{2nf_c}+1\right)}. \quad \quad (5)$$

This shows that p depends only on the ultimate strength of concrete and elastic limit of steel, and not at all on the size of beam. Fig. 21 gives graphically the "ideal ratio" p for

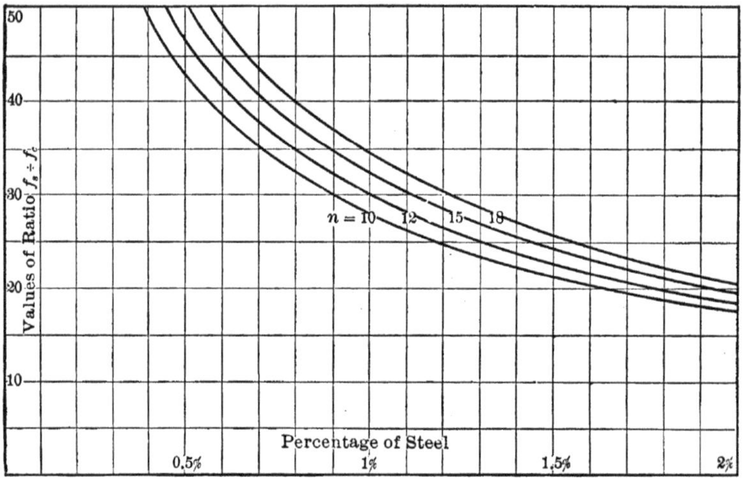

FIG. 21.

different values of the ratio f_s/f_c and four values of n; thus for $f_s=34{,}000$, $f_c=1700$, and $n=15$, $p=1.93\%$.

If, in any given case, the steel ratio as given by (5), or a higher value, is adopted, then the concrete would crush without straining the steel beyond the elastic limit, and the ultimate resisting moment of the beam is given by (3), which value equated to the ultimate bending moment, M, to be provided for, gives $\frac{2}{3}f_cjkbd^2=M$, or

$$bd^2=\frac{M}{\frac{2}{3}f_cjk}. \quad \quad (6)$$

From this d may be computed for any assumed value of b. If a lower value than that given by equation (5) is adopted for p, then under a gradually increasing load the stress in the steel would reach the elastic limit before the concrete would crush, and the formulas of this article could not be used to compute the ultimate resisting moment of the beam. See Art. 67 for solution of this case.

Approximating as before, $j=0.80$ and $k=0.52$, and eq. (6) becomes

$$bd^2=\frac{3.6M}{f_c}. \quad \ldots \ldots \ldots \quad (6)'$$

64. *Diagrams and Examples.*—Two numerical examples will now be given to illustrate the foregoing principles, and then a diagram will be explained by means of which computations in such examples can be wholly or partially avoided.

(1) A concrete beam is 10×16 inches in cross-section and the tension reinforcement consists of four $\frac{3}{4}$-in. steel rods, their centers being two inches above the lower face of the beam. The ultimate compressive strength of the concrete being 2000 and the elastic limit of the steel 40,000 lbs/in² compute the ultimate resisting moment of the beam.

Solutions. Here $p=0.0126$, and for $n=15$, eq. (1) gives $k=0.52$ and (2) gives $j=0.805$. Hence

$$M_c=\tfrac{2}{3}\,0.805\times0.52\times2000\times10\times14^2=1{,}096{,}000 \text{ in-lbs.}$$

It remains to test whether the stress in the steel would be within the elastic limit, the beam being subjected to a bending moment of 1,096,000 in-lbs. This is done by dividing the bending moment by the arm of the resisting couple, which gives the whole tension in the steel, and then this tension by the area of the steel; thus

$$\frac{1{,}096{,}000}{0.805\times14}=97{,}300 \text{ lbs.}=T$$

and

$$\frac{97{,}300}{1.768}=55{,}000 \text{ lbs./in}^2=f_s$$

This result being beyond the stated elastic limit, eq. (3) does not apply to the problem in hand. (The ultimate resisting moment can be computed by other methods. See ex. 2, page 80.)

(2) A beam is to be figured to safely withstand a bending moment of 135,000 in-lbs., the ultimate compressive strength of the concrete being taken at 2000 and the elastic limit of the steel at 40,000 lbs/in².

Solution. With $n=15$, eq. (5) gives as the "ideal steel ratio," since $f_s/f_c=20$,

$$p=\frac{2/3}{20(\frac{20}{15}+1)}=0.02.$$

For this value of p, eq. (1) gives $k=0.598$, and (2) gives $j=0.775$. With a factor of safety of 3, the ultimate bending moment is 405,000 in-lbs., and eq. (6) gives

$$bd^2=\frac{405,000}{\frac{2}{3}\times2000\times0.775\times0.598}=656 \text{ in}^3.$$

Trying 6 inches for b, then $d^2=109.3$ or $d=10.5$ in.; also $A=0.02\times6\times10.5=1.26$ in².

The "coefficients of resistance" on the parabolic theory are f_spj and $\frac{2}{3}f_cjk$ (see equations 4 and 3), and using the symbols R_s and R_c, as in Art. 59,

$$R_s=f_spj \quad \text{and} \quad R_c=\tfrac{2}{3}f_cjk.$$

The f_s curves of the diagram (Plate V, page 279) are graphs of the first equation for certain values of f_s as marked on the curves and $n=15$. (The curves for $n=12$ differ very little from these.) The f_c curves are graphs of the second equation for various values of f_c as marked; the full curves are for $n=15$ and the dotted for $n=12$.

In using the diagram to determine (1) the ultimate resisting moment of a given beam for a specified ultimate compressive strength of the concrete, or (2) a steel ratio and size of beam to withstand a given ultimate bending moment with specified compressive strength of concrete, these formulas respectively should be borne in mind:

$$M=Rbd^2 \quad \text{and} \quad bd^2=M/R.$$

The foregoing two examples will now be solved by means of the diagram.

(1) The percentage of steel being 1.26, we first find that value on the lower margin of the diagram, and then trace vertically to the line marked $f_c=2000$. We note that the point thus found is above the line

$f_s=40{,}000$, the elastic limit of the steel of the beam, and hence conclude that the amount of steel in this beam is insufficient to develop the full compressive strength of the concrete without straining the steel beyond the elastic limit. If the elastic limit of the steel were as high as 55,000 lbs/in², we would trace horizontally from the point as found above to either side of the diagram and read $R=552$. Then $M=Rbd^2=552\times10\times14^2=1{,}087{,}000$ in-lbs., which is the ultimate resisting moment of this beam with the high elastic limit steel.

(2) We first find the intersection of the curves $f_c=2000$ and $f_s=40{,}000$; from that point tracing down we find $p=2\%$, and horizontally we find $R=620$. Then $bd^2=M/R=405{,}000/620=654$, from which b and d can be decided upon, and finally the amount of steel.

65. Flexure Formulas for any Load up to Ultimate, Based on Parabolic Variation of Compression and Neglecting Tension in Concrete (After Talbot).—It is assumed that the stress in the steel is not above the yield point. The parabola representing the variation of compressive stress is not a "full one", that is, its top is not the vertex, see Fig. 22, unless

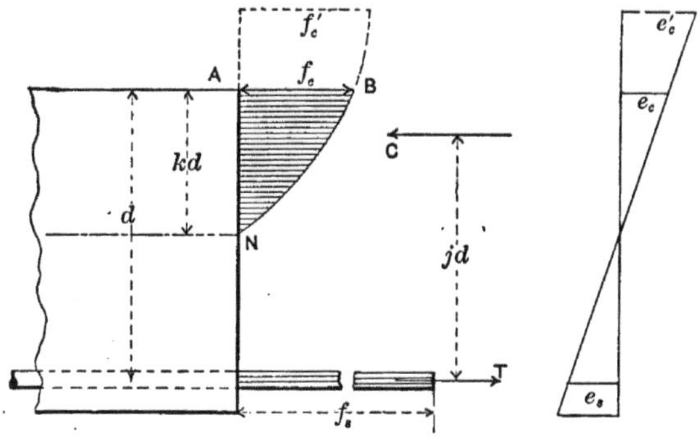

FIG. 22.

the maximum concrete stress is at the ultimate value. As heretofore f_c and e_c will denote the unit stress and strain respectively at the compressive face of the concrete, and as in Art. 61, E_c will denote the initial modulus of elasticity of the concrete. In this article f_c' and e_c' will denote these same quantities at the ultimate stage of the concrete, and q will be used as an abbreviation for e_c/e_c'. It can be shown from the properties of a parabola that: (1) The average abscissa to the parabola NB is $(3-q)/3(2-q)$ times the greatest abscissa

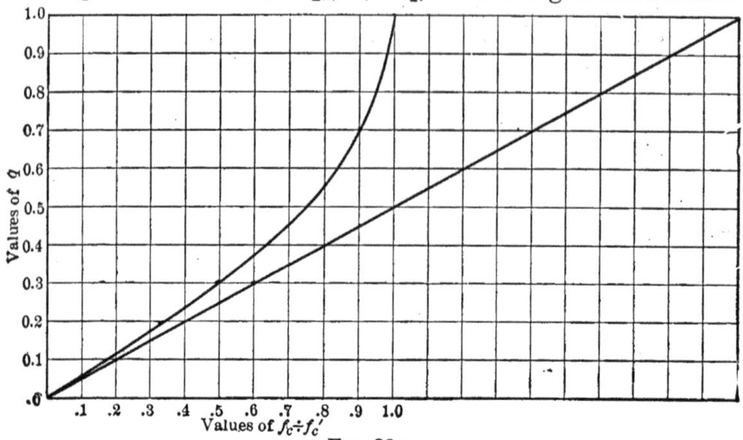

FIG. 23a.

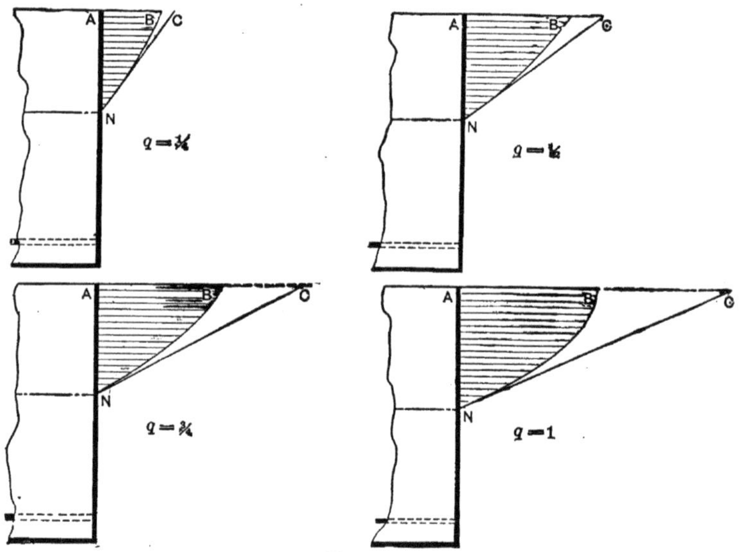

FIG. 23b.

f_c; (2) the distance from the centroid of the parabolic area to the top AB is $(4-q)/4(3-q)$ times its height, kd; and (3)

$$f_c = f_c'(2-q)q = \tfrac{1}{2}(2-q)E_c e_c. \quad . \quad . \quad . \quad . \quad (a)$$

Fig. 23*a* shows graphically the relation between q and the ratio f_c/f_c'; thus when $q=\frac{1}{4}$ (the concrete is strained to one-fourth its limit of compression) the unit stress in the concrete is about 0.45 of the ultimate strength.

The lines NB in Fig. 23*b* show the distributions of compressive stress at a section of a beam when q is $\frac{1}{4}$, $\frac{1}{2}$, $\frac{3}{4}$ and 1 respectively as marked. In each case N is the neutral axis and AB represents the unit stress on the remotest fiber. When q is $\frac{1}{4}$, the distribution is almost linear.

66. *Neutral Axis and Arm of Resisting Couple.*—As in Arts. 55 and 61, $e_s/e_c=(d-kd)/kd$, and $f_s=E_se_s$. Eliminating e_s/e_c from these two equations and (a), and introducing the abbreviation n, gives

$$\frac{f_s}{nf_c}=\frac{2(1-k)}{k(2-q)}. \quad . \quad . \quad . \quad . \quad . \quad . \quad . \quad . \quad (b)$$

When the loads and reactions are vertical—beam horizontal—the total tension and total compression on the section are equal, i.e.,

$$Af_s=bkdf_c(3-q)/3(2-q). \quad . \quad . \quad . \quad . \quad . \quad . \quad (c)$$

Eliminating the ratio f_s/f_c between equations (b) and (c), and introducing the abbreviation p, gives $6pn(1-k)=k^2(3-q)$, which solved for k furnishes the following formula:

$$k=\sqrt{2\frac{3pn}{3-q}+\left(\frac{3pn}{3-q}\right)^2}-\frac{3pn}{3-q}. \quad . \quad . \quad . \quad . \quad (1)$$

It shows that the neutral axes of all beams of a given concrete and a given percentage of reinforcement are at the same proportionate depth, k, for any particular stage of loading as given by q. The lower group of curves in Fig. 24 shows how k depends on p and n for $q=\frac{1}{4}$, the value taken by Talbot as closely corresponding to the working stage. The lower

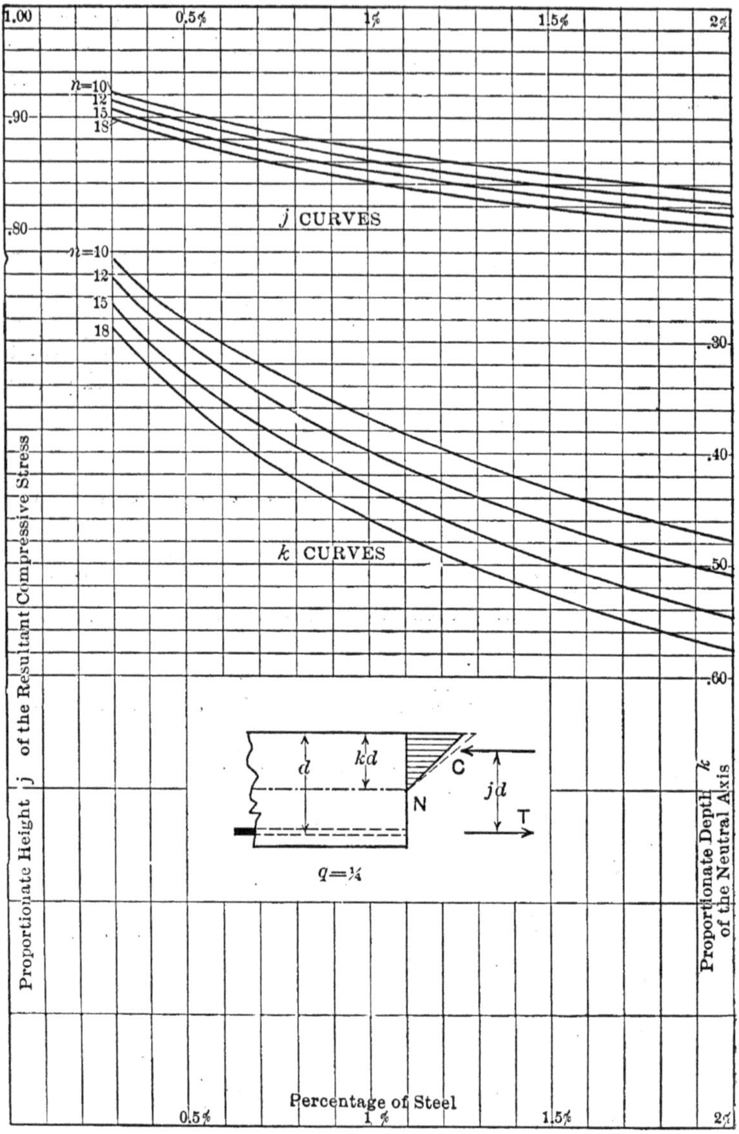
1.00
0.5%
1%
1.5%
2%
n=10
12
15
18
.90
.80
j CURVES
n=10
12
15
18
.30
.40
k CURVES
.50
.60
d
kd
C
N
jd
T
q=¼
Proportionate Height j of the Resultant Compressive Stress
Proportionate Depth k of the Neutral Axis
Percentage of Steel
0.5%
1 %
1.5%
2%

FIG. 24.

group of curves in Fig. 25 shows how k depends on q (that is, on the stage of loading) for several values of p, n being taken as 15. Thus when $p=0.01$ and $q=0$ nearly (load very small), $k=0.42$; and when $q=1$ nearly (ultimate load), $k=0.48$.

The distance of the centroid of the compressive stress from the top of the beam is $kd(4-q)/4(3-q)$; hence the arm of the resisting couple is given by $jd=d-kd(4-q)/4(3-q)$ or

$$j=1-\frac{k(4-q)}{4(3-q)}. \quad . \quad . \quad . \quad . \quad . \quad . \quad . \qquad (2)$$

The upper group of curves in Fig. 24 shows how j depends on p and n, for the stage $q=\frac{1}{4}$. The upper group of curves in Fig. 25 shows how j depends on q for several values of p, n being taken as 15. It should be noticed that j does not change much for considerable changes in q.

67. *Resisting Moment for Given Values of f_c and f_s.*—Whether the resisting moment is determined by the concrete or steel depends on the percentage of reinforcement; in a general way the higher percentages make the moment depend on the concrete and the lower on the steel. As depending on the concrete, the resisting moment is given by

$$M_c=Cjd=\frac{3-q}{3(2-q)}jkf_cbd^2. \quad . \quad . \quad . \quad . \quad . \quad . \qquad (3)$$

The value of q to be used here must correspond with the f_c used, the relation between q and f_c being given by (a) of Art. 65; or by Fig. 23a. As depending on the steel, the resisting moment is

$$M_s=Tjd=f_sAjd=f_spjbd^2. \quad . \quad . \quad . \quad . \quad . \quad . \qquad (4)$$

68. *Determination of Fibre Stresses f_s and f_c for a Given Bending-Moment.*—Formulas for these can be obtained by solving (3) and (4) for f_c and f_s respectively; thus

$$\left.\begin{aligned} f_c&=\frac{3(2-q)}{(3-q)}\frac{M}{jkbd^2} \\ f_s&=\frac{M}{pjbd^2} \end{aligned}\right\} \quad . \quad . \quad . \quad . \quad . \quad . \quad . \quad . \qquad (5)$$

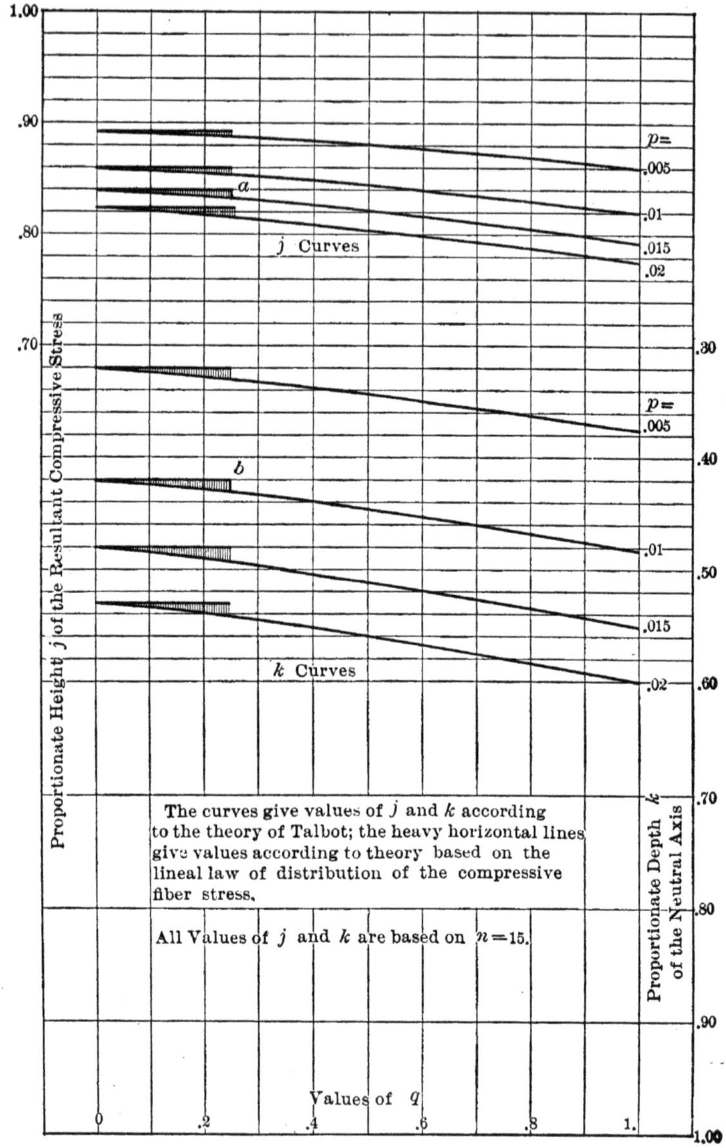
1.00
.90
.80
.70
p=
.005
.01
.015
.02
a
j Curves
.30
p=
.005
.40
b
.01
.50
.015
k Curves
.02
.60
.70
.80
.90
1.00
Proportionate Height j of the Resultant Compressive Stress
Proportionate Depth k of the Neutral Axis
The curves give values of j and k according to the theory of Talbot; the heavy horizontal lines give values according to theory based on the lineal law of distribution of the compressive fiber stress.
All Values of j and k are based on n=15.
Values of q
0
.2
.4
.6
.8
1.

FIG. 25.

Neither f_c nor f_s can be determined directly from these, for each formula contains q (j and k depend on q), which is an unknown in the problem in hand. An estimated value of q must be used for a trial solution of (5), and then with the value of f_c thus found a better value of q may be obtained from (a) or from Fig. 23a, which value may be used in a second trial solution.

69. *Determination of Amount of Steel and Cross-section of Beam for a Given Bending Moment.*—In order that the maximum unit compression in the concrete, f_c, and the unit stress in the steel, f_s, may have certain definite values when the beam is subjected to a given bending moment, a certain definite percentage of steel must be used. This percentage is such as makes the values of the resisting moment as determined by steel and concrete equal. Thus equating values of M from equations (3) and (4) and simplifying,

$$p=(3-q)kf_c/3(2-q)f_s.$$

Inserting in this the value of k furnished by (b) gives

$$p=\frac{3-q}{3(2-q)}\,\frac{1}{\frac{f_s}{f_c}\left(\frac{2-q}{2n}\frac{f_s}{f_c}+1\right)} \quad . \quad . \quad . \quad . \quad . \quad . \quad (6)$$

In this also the value of q used should correspond to the value of f_c adopted as working stress. The curves of Fig. 26 give values of p for different values of f_s/f_c up to 50, q being taken at $\frac{1}{2}$.

If in any given case a value for p less than that given by (6) is adopted, then the resisting moment is given by equation (4), which equated to the bending moment to be provided for gives $f_s pjbd^2=M$, or

$$bd^2=\frac{M}{f_s pj} \quad . \quad . \quad . \quad . \quad . \quad . \quad . \quad . \quad . \quad . \quad (7)$$

If a greater value of p is adopted, then the resisting moment is given by (3), which if equated to the bending moment gives $jkf_c bd^2(3-q)/3(2-q)=M$, or

$$bd^2=\frac{3(2-q)}{3-q}\,\frac{M}{jkf_c}. \quad . \quad . \quad . \quad . \quad . \quad . \quad (8)$$

From the proper one of these, d can be computed for any assumed value of b.

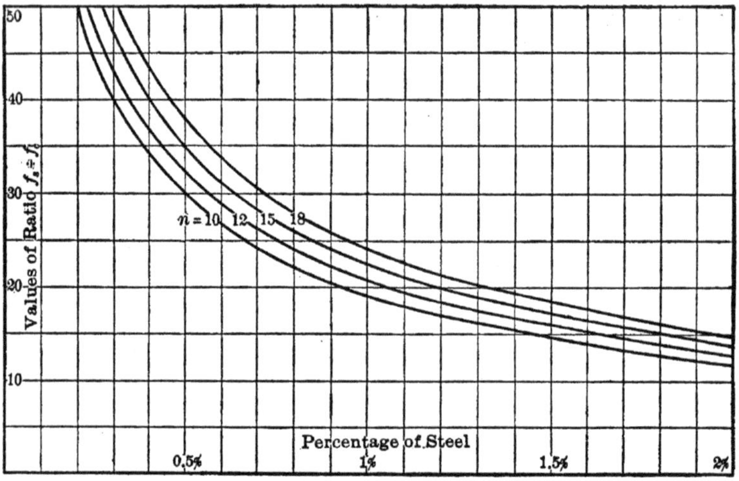

FIG. 26.

Examples.—(1) It is required to solve example 1, Art. 59, by the methods of this article, it being supposed that for the working stress $f_c=600$ lbs/in², $q=\frac{1}{4}$.

Solution. As shown in the solution of the example referred to, $A=1.768$ in², and $p=0.0126$; therefore from eq. (1) or Fig. 24, n being taken as 15, $k=0.466$, and from eq. (2) or Fig. 24, $j=0.842$. Then from eqs. (3) and (4)

$$M_c=\tfrac{11}{21}\times0.842\times0.466\times600\times10\times14^2=242{,}000 \text{ in-lbs.}$$

and

$$M_s=15{,}000\times1.768\times0.842\times14=313{,}000 \text{ in-lbs.}$$

(2) It is required to solve example 1 of Art. 64 by the methods of this article.

Solution. As disclosed by the solution in Art. 64, the stress in the steel will reach the elastic limit before that in the concrete would reach the ultimate strength; hence the ultimate resisting moment depends on the steel. The stress existing in the concrete when the steel is stressed to the elastic limit is unknown; so is q. Supposing that this stress in concrete is $\frac{3}{4}$ the ultimate strength, $q=0.5$ (see Fig. 23a); then, since $p=0.0126$, and n is taken as 15, $k=0.48$ and $j=0.83$ (see Fig. 25), and eq. (4) gives $M_s=820{,}000$ in-lbs. For a bending moment of this value, the stress in the concrete would be (with the above values of q, j, and k) 1260 lbs/in² (see eq. 5). Now for the ratio 1260/2000, q is about 0.4, k, 0.75, and j, 0.825. Since this value of j is practically like the one used in the trial computation, the ultimate resisting moment may be taken as 820,000 in-lbs.

(3) It is required to solve example 2 of Art. 59 by the methods of this article, supposing the ultimate compressive strength of concrete to be 2500 lbs/in².

Solution. This problem can only be solved by trial because it is necessary to know q at the outset, and q depends on a quantity sought, f_c. Supposing that the load is about a safe one, then q equals about $\frac{1}{4}$. With this value, n equal to 15, and p equal to 0.0104 (already found on page 60), $k=0.43$, and $j=0.85$ (see Fig. 24). Then eq. (5) gives $f_c=630$ lbs/in². Now q depends on the ratio of the working stress in the concrete to its ultimate strength; for the approximate value, 630, the ratio is 0.25, and eq. (a), or Fig. 23, gives $q=0.15$. With this value eq. (1) gives $k=0.432$, eq. (2), $j=0.854$ (see also Fig. 25), and eq. (5), $f_c=635$ lbs/in². This value is so near the first that $q=0.15$ must be practically correct, and $j=0.854$ may be used to determine the stress in the steel. For this, eq. (5) gives $f_s=13{,}700$ lbs/in².

(4) It is required to solve example 3 of Art. 59 by the methods of this article, the ultimate compressive strength of the concrete being taken at 2000 lbs/in².

Solution. For the ratio 700/2000, q is about 0.2 (see Fig. 23a.) With $n=15$ eq. (6) gives $p=0.018$. For this value of p, we may use either (7) or (8) to compute the dimensions of the section. Choosing (7) we need first a value of j, which may be obtained from (2) and (1), or closely enough from Figs. 24 or 25; the figures give $j=0.82$, and eq. (7) gives $bd^2=763$. With $b=7$ (as in Art. 59) d is 10.5 in.

70. Comparison of Flexure Formulas after Talbot with (1) those for working conditions as given in Art. 54–59, and (2) those for ultimate conditions as given in Art. 60–64:

(1) The heavy horizontal lines of Fig. 25 give values of j and k, according to the linear law (Art. 54), and the curved lines

those after Talbot. For $q=0.25$ and $p=0.015$, the difference between the two values of j is represented by ad and the difference between the two values of k by b. For all values of q up to 0.25 or 0.30 the first difference is small, and so the values given by the two formulas for f_s must be nearly the same. The second difference is larger, and the two formulas for f_c will not agree so closely. An exact comparison will now be made.

Art. 56 gives (see eqs. 3 and 4).

$$f_s' = \frac{M}{pj'bd^2} \quad \text{and} \quad f_c' = \frac{2M}{j'k'bd^2}.$$

(The primes are used to distinguish the symbols from the corresponding ones in the other formulas.) Comparing these with eqs. (3) and (4), Art. 62, one gets

$$\frac{f_s'}{f_s} = \frac{j}{j'} \quad \text{and} \quad \frac{f_c'}{f_c} = \frac{2(3-q)}{3(2-q)} \frac{jk}{j'k'}.$$

As already explained, q rarely exceeds $\frac{1}{4}$ for working conditions; with this value and $n=15$, the following table gives the ratios f_s'/f_s and f_c'/f_c for five percentages of steel. For values of q less than $\frac{1}{4}$, the ratios are nearer unity; for $q=0$, they are all unity and the two sets of formulas are identical.

$p=$	$\frac{1}{4}$%.	$\frac{1}{2}$%.	1%.	1.5%.	2%.
f_s'/f_s	0.995	0.993	0.991	0.990	0.989
f_c'/f_c	1.092	1.091	1.090	1.088	1.086

The unit stresses in the steel as given by the two formulas are practically identical. Any error involved in the formulas for f_c', based on the linear law, is on the side of safety.

(2) For loads which stress the concrete to the ultimate limit, the stress parabola of Fig. 22 is full like that of Fig. 19, and $q=1$. The formulas of Arts. 65–70 for this stage and those of Arts. 60–64 are identical.

71. Flexure Formulas for T-beams.—The following discussion is based on the linear law of compression, and it neglects the tension in the concrete. The following additional notation is employed (see also Fig. 27):

b denotes width of flange;
d " effective depth of beam;
b' " width of web;
t " thickness of flange;
z " depth of resultant compression below top of flange;
p " steel ratio, A/bd.

It is necessary to distinguish two cases, namely, (1) the neutral axis is in the flange, (2) the neutral axis is in the web.

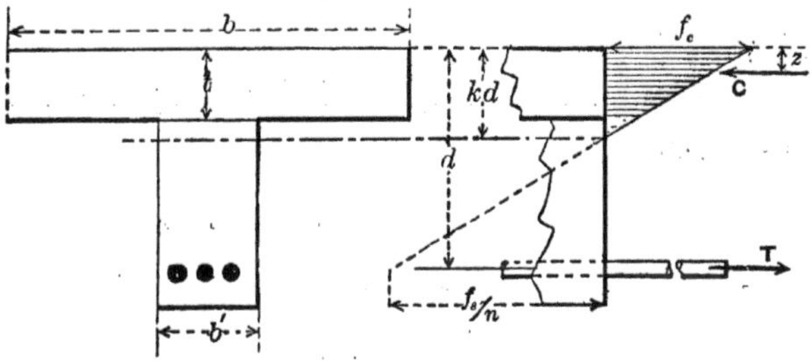

FIG. 27.

Which of the two cases is at hand in any particular computation may not be apparent at the outset. This may readily be determined, however, by means of certain formulas to be explained.

72. *Case I. The Neutral Axis in the Flange.*—All formulas of Arts. 54–58 (except approximate ones) apply to this case. It should be remembered that b of the formulas denotes flange—not web—width, and p (the steel ratio) is $A \div bd$, not $A \div b'd$ (see Fig. 27).

Approximate Formulas.—Evidently the arm of the resisting couple, CT, is always greater than $d-\frac{1}{2}t$; hence the following approximate formulas err on the side of safety:

$$M_s=f_sA(d-\tfrac{1}{2}t) \quad \text{and} \quad f_s=M/A(d-\tfrac{1}{2}t).$$

These give good results. There are no satisfactory corresponding formulas based on concrete; for determining the concrete stress use the formulas for rectangular beams. Usually the concrete stress will be relatively small, as the flange of a T-beam which comes under Case I is generally much stronger than the steel.

73. *Case II. The Neutral Axis is in the Web.*—The amount of compression in the web is commonly small compared with that in the flange and will be neglected in the analysis of this article. The formulas are thereby greatly simplified and the resulting error is generally very small. To provide for designs in which the web is very large as compared to the flange, formulas which take account of web compression are given in Art. 73*a*.

Neutral Axis and Arm of Resisting Couple.—Just as in Art. 55, eq. (*a*),

$$\frac{f_s}{nf_c}=\frac{1-k}{k}, \quad . \quad . \quad . \quad . \quad . \quad . \quad . \quad . \quad (a)$$

hence we have, in terms of f_s and f_c,

$$k=\frac{1}{1+f_s/nf_c}. \quad . \quad . \quad . \quad . \quad . \quad . \quad . \quad (1)$$

The average unit compressive stress in the flange is $\frac{1}{2}\left[f_c+f_c\left(1-\frac{t}{kd}\right)\right]=f_c\left(1-\frac{t}{2kd}\right)$, and the whole compression is $f_c\left(1-\frac{t}{2kd}\right)bt$. And since the whole tension and whole compression on the section are equal,

$$f_sA=f_c\left(1-\frac{t}{2kd}\right)bt. \quad . \quad . \quad . \quad . \quad . \quad . \quad . \quad (b)$$

Eliminating f_s/f_c between equations (*a*) and (*b*) we get an equation which when solved for k gives

$$k=\frac{nA+\frac{1}{2}bt\cdot\frac{t}{d}}{nA+bt}. \quad . \quad . \quad . \quad . \quad . \quad . \quad . \quad . \quad . \quad (2)$$

Substituting pbd for A we also derive the form

$$k=\frac{pn+\frac{1}{2}\left(\frac{t}{d}\right)^2}{pn+\frac{t}{d}}. \quad . \quad . \quad . \quad . \quad . \quad . \quad . \quad . \quad (3)$$

The arm of the resisting couple is $d-z$ (see Fig. 27). The distance z is equal to the distance of the centroid of the shaded trapezoid from the top of the beam, that is,

$$z=\frac{3k-2\frac{t}{d}}{2k-\frac{t}{d}}\cdot\frac{t}{3}. \quad . \quad . \quad . \quad . \quad . \quad . \quad . \quad . \quad (4)$$

We also have

$$jd=d-z, \quad . \quad . \quad . \quad . \quad . \quad . \quad . \quad . \quad (5)$$

and, by substitution from (3) and (4) we have, in terms of t/d and p,

$$j=\frac{6-6\frac{t}{d}+2\left(\frac{t}{d}\right)^2-\left(\frac{t}{d}\right)^3\Big/2pn}{6-3\frac{t}{d}}. \quad . \quad . \quad . \quad (6)$$

The neutral axis will be at the junction of web and flange when $t/d=k$.

On Plate VI, p. 280, are plotted curves for values of k and j for various values of p and of the ratio t/d. The value of n is taken at 15. This diagram, as well as eq. (6), shows that j is affected very little by changes in the amount of steel. The diagram also gives on the right hand margin the values of f_s/f_c, corresponding to the various values of k as determined from eq. (1). The curves for k and j end at points where

$k=t/d$. They become horizontal at these points and the values of k are equal to those for rectangular beams. (See Fig. 17, Art. 55).

Resisting Moment and Working Stresses.—If the beam is under-reinforced, its resisting moment depends on the steel; if over-reinforced, on the concrete. These two values of the moment are respectively

$$\left.\begin{aligned} M_s&=f_sA\cdot jd,\\ M_c&=f_c\left(1-\frac{t}{2kd}\right)bt\cdot jd \end{aligned}\right\} \quad . \quad . \quad . \quad . \quad . \quad . \quad (7)$$

If one is in doubt which of these to use when about to compute the resisting moment of a given beam with specified working stresses, then both values should be computed and the smaller taken as the resisting moment.

The unit stresses, f_s and f_c, produced by a certain bending moment M in a given beam can be computed by solving (7), for f_s and f_c or from

$$\left.\begin{aligned} &C=T=\frac{M}{jd}; \quad f_s=\frac{T}{A};\\ &f_c=\frac{f_s}{n}\cdot\frac{k}{1-k}=\frac{f_sp}{\left(1-\frac{t}{2kd}\right)\frac{t}{d}} \end{aligned}\right\} \quad . \quad . \quad . \quad . \quad . \quad (8)$$

Approximate formulas corresponding to (7) and (8) can be established as follows: From the stress diagram in Fig. 27, it is plain that the arm of the resisting couple is never as small as $d-\frac{1}{2}t$, and that the average unit compressive stress is never as small as $\frac{1}{2}f_c$, except when the neutral axis is at the top of the web. Using these limiting values as approximations for the true ones, we have as substitutes for (7) and (8)

$$\left.\begin{aligned} M_s&=Af_s(d-\tfrac{1}{2}t)\\ M_c&=\tfrac{1}{2}f_cbt(d-\tfrac{1}{2}t)\end{aligned}\right\} \quad . \quad . \quad . \quad . \quad . \quad . \quad . \quad (7)'$$

$$C=T=\frac{M}{d-\frac{1}{2}t},\quad f_s=\frac{T}{A},\quad f_c=\frac{2C}{bt}. \quad . \quad . \quad . \quad (8)'$$

The errors involved in these approximations are on the side of safety, for (7)′ gives values smaller than (7), and (8)′ larger ones than (8). Satisfactory approximate results may also be reached by assuming a fixed value of $\frac{7}{8}d$ for the arm of the resisting couple jd.

73a. *Formulas Taking into Account the Compression in the Web.*—When the web is very large compared to the flange it may be desirable to use more exact formulas than those already given. In this case the formulas for the position of neutral axis, arm of resisting couple, and moment of resistance become as follows:

$$kd=\sqrt{\frac{2ndA+(b-b')t^2}{b'}+\left(\frac{nA+(b-b')t}{b'}\right)^2}-\frac{nA+(b-b')t}{b'} \qquad (9)$$

$$z=\frac{b(kdt^2-\frac{2}{3}t^3)+b'\left[(kd-t)^2\left(t+\frac{kd-t}{3}\right)\right]}{bt(2kd-t)+b'(kd-t)^2}; \quad . \quad . \quad . \quad . \quad . \quad . \quad (10)$$

$$jd=d-z; \quad . \quad . \quad . \quad . \quad . \quad . \quad . \quad . \quad (11)$$

$$\left.\begin{aligned} M_s&=f_sA\cdot jd\\ M_c&=\frac{f_c}{2kd}\left[(2kd-t)bt+(kd-t)^2b'\right]jd\end{aligned}\right\} \quad . \quad . \quad . \quad (12)$$

Equations (12) also give f_s and f_c for given values of M.

74. *Problems of Design—Either Case I or II.*—In practice, various forms of problems will arise: (*a*) The dimensions may be given, to find the safe resisting moment of the beam or the stresses in the steel and concrete under a given load; (*b*) the dimensions of the flange may be given, together with the loading and specified working stresses, to determine suitable web dimensions and steel area; (*c*) the loading and working stresses may be given, to determine suitable proportions for the entire beam.

(*a*) Where all the dimensions are given, the value of k and j are found from eqs. (3) and (6), or from Plate VI, and thence the values of the moment of resistance from eqs. (7), or the fibre stresses from (7) or (8). If the value of k is found to be less than t/d then the problem falls under Case I and the formulas for rectangular beams apply, or the approximate formulas of Art. 72 may be used.

(*b*) Generally the flange has been predetermined as it is usually formed by a portion of the floor slab which is already designed. A suitable web must then be determined, together with the necessary amount of steel; and finally the fibre stress in the concrete must be calculated to ascertain if it is within the specified working limit. The depth and width of web are selected with reference to shearing strength, space for the necessary rods and other considerations, as fully explained in subsequent articles. The depth having been selected, the value of j is estimated and the amount of steel, A, approximately determined by eq. (7). The amount of steel being known, the value of j can be determined by eq. (6) and then, if necessary, the value of A corrected by eq. (7). The value of k should also be found from eq. (2) in order to ascertain if the beam falls under Case I or II. The stress in the concrete is then found from eq. (8).

In estimating the value of j use Plate VI, or a value of $\frac{7}{8}$ may be assumed, as for rectangular beams.

(*c*) When all parts of the beam are to be selected on the

basis of given working stresses it is convenient to first select suitable proportions for the web, as in Case (*b*). A flange thickness is then assumed such as to give satisfactory proportions between t and d. The value of t/d is then known and k and j can be determined from (1), (4), and (5). The area of steel and the breadth of flange is then found from eq. (7). The smaller the value of t the smaller will be the flange area required, but too slender proportions are to be avoided, as explained in Chapter V

Examples.—(1) A T-beam has the following dimensions: $b=48$ in., $t=6$ in., $d=22$ in., and $b'=10$ in.; the steel consists of six $\frac{3}{4}$-in. rods. If the working strengths of steel and concrete are 15,000 and 600 lbs/in² respectively, and $n=15$, what is the safe resisting moment of the beam?

Solution. The area of the steel is 2.65 in², and $p=2.65/(48\times22)=0.0025$. Supposing this beam to fall under Case I, we find k from Fig. 17 (or eq. (1), Art. 55) to be about 0.24, hence $kd=5.3$ in., and the neutral axis is in the flange, that is, the case was correctly guessed. Now $j=1-\frac{1}{3}k=0.92$; hence (see eqs. (3) and (4), Art. 56)

$$M_s=(15{,}000\times2.65)(0.92\times22)=806{,}000 \text{ in-lbs.},$$

and

$$M_c=300(5.3\times48)(0.92\times22)=1{,}545{,}000 \text{ in-lbs.},$$

The safe resisting moment hence depends on the steel, as it usually does in T-beams. The approximate formula gives $M_s=795{,}600$ in-lbs.

(2) Change t of the preceding example to 4 in. and find the safe resisting moment.

Solution. Evidently this beam now falls under Case II. Equation (2) gives $k=0.247$ and (4) $z=1.61$ in. From (7)

$$M_s=15{,}000\times2.65(22-1.61)=950{,}000 \text{ in-lbs.},$$

and

$$M_c=(600/5.44)3.44(48\times4)(22-1.61)=1{,}485{,}000.$$

The values of k and j may also be found from Plate VI.

The approximate formulas (7)′ give $M_s=716{,}000$ and $M_c=1{,}152{,}000$ in-lbs.

(3) Suppose that the diameter of the rods in example (1) is 1 in.,

and that the beam is subjected to a bending moment of 1,250,000 in-lbs. Compute the working stresses in the steel and concrete.

Solution. Equation (2) gives $k=0.306$ and $kd=6.73$, hence the beam falls under Case II. Equation (4) gives $z=2.22$ in., and (8)

$$f_s=\frac{1,250,000}{(22-2.22)4.71}=13,400 \text{ lbs/in}^2,$$

and

$$f_c=\frac{13,400}{15}\frac{0.306}{1-0.306}=395 \text{ lbs/in}^2.$$

The approximate formulas (8)′ give $f_s=13,960$ and $f_c=457$ lbs/in².

(4) The flange of a T-beam is 24 in. wide and 4 in. thick. The beam is to sustain a bending moment of 480,000 in-lbs., the working strengths of steel and concrete being respectively 15,000 and 500 lbs/in². What depth of beam and amount of steel will answer?

Solution. We will try $d=18$ in. Assume $jd=16$ in. Then eq. (7) gives $A=2$ in², and hence $p=2/(24\times18)=0.00462$. Then (6) gives $j=0.91$ and the corrected value of A is 1.95 in². Equation (3) gives $k=0.325$, and shows that the beam falls under Case II as assumed. The stress in the concrete is found from (8) to be $(15,000/15)\times0.325/0.675=480$ lbs/in², which is permissible.

(5) Let it be required to design a T-beam to sustain a bending moment of 480,000 in-lbs., the working stresses to be 15,000 and 600 lbs/in², respectively.

Solution. The same depth of web and thickness of flange will be assumed as in (4), as these proportions are reasonable. The amount of steel and width of flange are to be determined. From (1) we find $k=1\div(15,000/9000+1)=0.375$, and from (4) $z=1.72$ and $jd=18.0-1.72=16.28$ in. Then from (7),

$$A=1.97 \text{ in}^2 \quad \text{and} \quad b=17.5 \text{ in.}$$

74*a*. *Diagrams of M/bd^2 for Use in Designing.*—By reason of the additional variable (the flange thickness) involved in formulas relating to T-beams as compared to rectangular beams, it is not possible to arrange so simple a graphical solution of the resisting moment as is done by means of Plates I-IV, Chapter VI. Assuming, however, a single value of n

the values of the resisting moment or coefficient of resistance may readily be represented graphically. It will be convenient to consider as variables the values of f_s, f_c, and the ratio t/d, or thickness of slab to the effective depth of the beam. From eq. (7) we may write

$$\frac{M}{bd^2}=f_c\left(1-\frac{t}{2kd}\right)\cdot\frac{t}{d}\cdot j. \quad . \quad . \quad . \quad . \quad . \quad . \qquad (13)$$

In this equation k and j are functions of f_s and f_c, as appears from eqs. (1), (4), and (5). Plates VII–XI are plotted from this equation, assuming $n=15$ in all cases. Each plate contains values of M/bd^2 for a certain value of f_s and for various values of f_c and t/d. On the same diagram are also given values of k and values of j. The former are given by the dotted curve in the right-hand part of the diagram, and the value corresponding to a given value of f_c is to be read off on the axis for t/d. This value of k is, in fact, the value of t/d which brings the neutral axis just to the lower surface of the flange. These diagrams are particularly useful in solving problems under cases (*b*) and (*c*), Art. 74; they are not adapted to case (*a*). Plate VI gives all the information needed for this case.

Examples.—Examples (4) and (5) given in the previous article will now be solved by means of these diagrams.

(4) Use Plate IX. The value of $M/bd^2=480{,}000/(24\times 18^2)=61.6$. For this value of M/bd^2 and for $t/d=4/18$, we find from the diagram f_c=about 470 lbs/in² and $j=.91$. Then, as before, $A=480{,}000/(15{,}000\times .91\times 18)=1.95$ in²

(5) Use Plate IX. For $f_c=600$ and $t/d=4/18$, we find $M/bd^2=85$, whence $b=480{,}000/(85\times 18^2)=17.4$ in.; also, $j=.90$ and $A=1.97$ in².

(6) Using the same depth of beam as in Ex. (5) what will be the effect on the amount of concrete and steel of changing the flange thickness to 3 inches and to 5 inches? From Plate IX we find, for $t/d=3/18$, $M/bd^2=74$, whence $b=20.0$ in., $j=.92$ and $A=1.93$ in². For $t/d=5/18$, $M/bd^2=93$, $b=16.0$ in.. $j=.91$ and $A=1.95$ in². The volumes of the concrete in the flanges are, for the flange thicknesses of 3, 4 and 5 in., respectively, 60, 69.6, and 80 sq. in. The steel areas vary but little.

75. T-beams Double-reinforced.—T-beams are often continuous over their supports; at such places the bending moment is negative, and the flange is under tension and the lower part of the web under compression. Not only is tensile steel provided, but some steel is always left in the web (see Chap. VII); that is, the beam is reinforced in compression, and is said to be double-reinforced. For a discussion of double-reinforcement see the following articles—particularly Art. 79—in which there is explained a simple method for determining the effect of the compressive steel on the stress in the tensile steel and the compression in the concrete.

76. Beams Reinforced for Compression.—The compression in the concrete is assumed to follow the linear law and the tension in it is neglected; the formulas then apply to working conditions only. In addition to the notation already adopted (see page 58), let

A' denote the cross-sectional area of the compressive reinforcement;

p' denote the steel ratio for the compressive reinforcement, that is A'/bd;

f_s' denote the unit stress in the compressive reinforcement;

C' denote the whole stress in the compressive reinforcement;

d' denote the distance from the compressive face of the beam to the plane of the compressive reinforcement;

x denote the distance from the compressive face to the resultant compression, $C+C'$, on the section of the beam.

77. *Neutral Axis and Arm of Resisting Couple.*—From the stress diagram (Fig. 28) it appears that $f_s/nf_c=(d-kd)/kd$, or

$$f_s=n\frac{1-k}{k}f_c. \quad . \quad . \quad . \quad . \quad . \quad . \quad . \quad . \quad (1)$$

Similarly, $f_s'/nf_c=(kd-d')/kd$, or

$$f_s'=n\frac{k-d'/d}{k}f_c. \quad . \quad . \quad . \quad . \quad . \quad . \quad . \quad (2)$$

For simple flexure, the whole tension T and whole compression $C+C'$ are equal, hence

$$f_sA=\tfrac{1}{2}f_cbkd+f_s'A'. \quad . \quad . \quad . \quad . \quad . \quad . \quad (a)$$

Inserting the values of f_s and f_s' from (1) and (2) in (a) gives an equation which may be written thus:

$$k^2+2n(p+p')k=2n(p+p'd'/d), \quad . \quad . \quad . \quad . \quad (3)$$

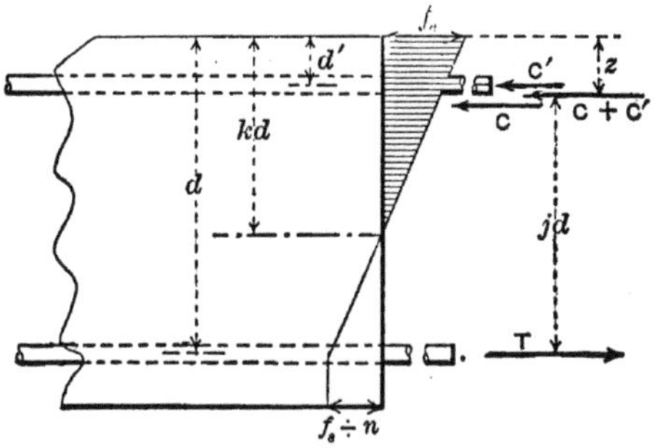

FIG. 28.

and from this the neutral axis of a given section can be located. The lower group of curves in Fig. 29 gives values of k for several values of p and all values of p' up to 2%; n is taken at 15 and d'/d as 1/10. Thus for $p=2\%$ and $p'=1.5\%$, $k=0.434$.

The arm of the resisting couple is the distance between T (see Fig. 28) and the resultant of the compressions C and C'. It follows from the principle of moments and the law of distribution of stress respectively that

$$z=\frac{\tfrac{1}{3}k+d'C'/dC}{1+C'/C}d, \quad \text{and} \quad \frac{C'}{C}=\frac{2p'n(k-d'/d)}{k^2},$$

from which z can be computed for any given section. Finally the arm $jd=d-z$ or

$$j=(1-z/d). \quad . \quad . \quad . \quad . \quad . \quad . \quad . \quad . \quad (4)$$

The upper group of curves in Fig. 29 gives values of j for several values of p and all values of p' up to 2%; n is taken at 15 and d'/d at 1/10. Thus for $p=2\%$ and $p'=1.5\%$, $j=0.875$.

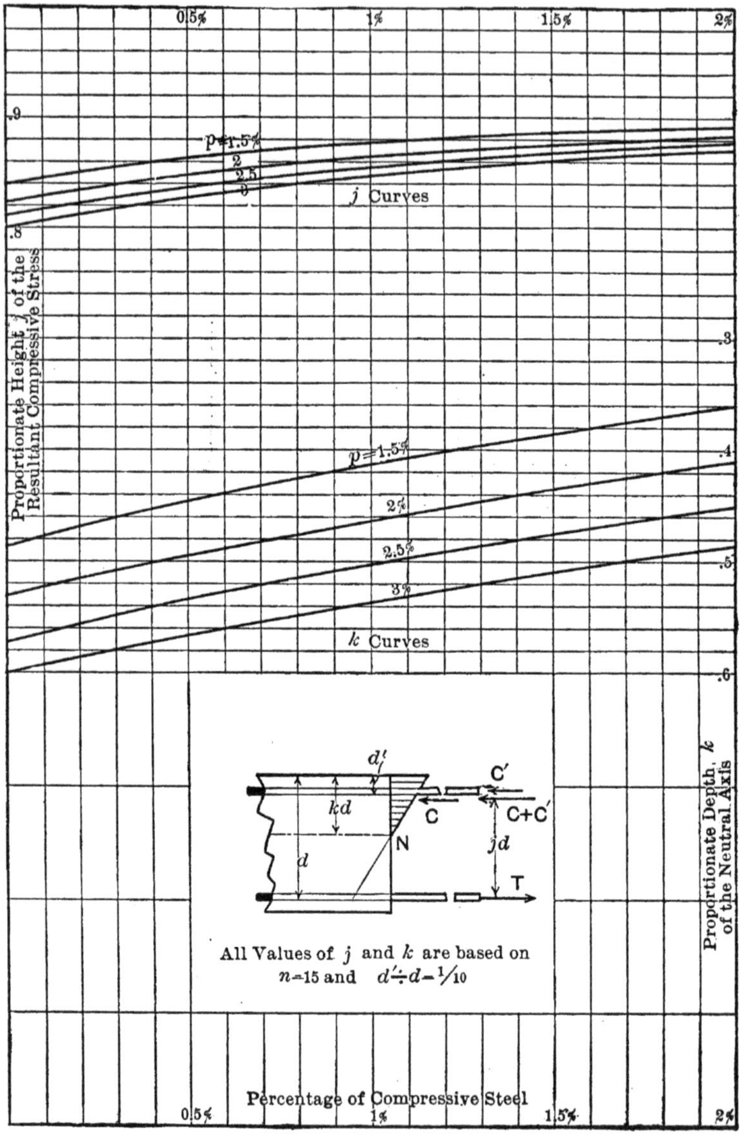
0.5%
1%
1.5%
2%
.9
.8
p=1.5%
2
2.5
3
j Curves
Proportionate Height j of the Resultant Compressive Stress
.3
.4
.5
.6
p=1.5%
2%
2.5%
3%
k Curves
d′
C′
kd
C
C+C′
N
d
jd
T
All Values of j and k are based on
n=15 and d′÷d=1/10
Proportionate Depth, k of the Neutral Axis
Percentage of Compressive Steel

FIG. 29.

78. *Resisting Moment and Working Stresses.*—If the tensile reinforcement is low, the resisting moment depends upon it, and is given by

$$M_s = f_s A j d = f_s p j b d^2. \quad . \quad . \quad . \quad . \quad . \quad . \quad . \quad (5)$$

If the compressive reinforcement is low, the resisting moment depends upon it and the concrete, and is given by

$$M_c = \tfrac{1}{2} f_c k(1 - \tfrac{1}{3}k) b d^2 + f_s' p' b d(d - d');$$

but f_s' bears a certain relation to f_c (see eq. 2), which inserted in the preceding equation gives finally

$$M_c = [k(\tfrac{1}{2} - \tfrac{1}{6}k) + np'(k - d'/d)(1 - d'/d)/k] f_c b d^2. \quad . \quad . \quad (6)$$

The unit fibre stress in the tensile steel produced by any bending moment M can be computed from

$$f_s = \frac{M/jd}{A} = \frac{M}{pjbd^2}, \quad . \quad . \quad . \quad . \quad . \quad . \quad . \quad (7)$$

and those in the concrete and compressive steel from f_s and equations (1) and (2) respectively.

Fig. 29 shows that the neutral axis is nearer the compressive steel ($k < 0.55$) unless the percentage of tensile reinforcement is quite high and the compressive low; thus for $p = 3\%$, the neutral axis is nearer the compressive steel unless p' is less than $3/4\%$, and when $p = 2\%$, it is nearer for all values of p'. Now since the unit stresses in the tensile and compressive steels are as the distances of the steels from the neutral axis, it follows that the unit stress in the compressive steel is generally less than that in the tensile, that is $f_s' < f_s$.

For approximate computations one might use the average values $j = 0.85$ and $k = 0.45$ in equations (5), (6), and (7); then they would become respectively ($n = 15$)

$$M_s = 0.85 p f_s b d^2, \quad . \quad . \quad . \quad . \quad . \quad . \quad . \quad . \quad . \quad (5)'$$

$$M_c = (0.19 + 10.5p') f_c b d^2, \quad . \quad . \quad . \quad . \quad . \quad . \quad (6)'$$

$$f_s = 1.18 M / p b d^2. \quad . \quad . \quad . \quad . \quad . \quad . \quad . \quad . \quad . \quad (7)'$$

79. *Determination of Amount of Compressive Reinforcement.*—This problem presents itself as follows: From the circumstances of the case, the beam needs so much tensile steel that the compressive concrete, if unreinforced, would be stressed too high, and it is necessary to employ compressive reinforcement to reduce the stress in the concrete; the percentage of reinforcement necessary to lower the stress a certain amount is desired.

An explicit formula for this percentage is too cumbersome for practical use, but a diagram (Plate VI, page 280) can be constructed from which the desired quantity can be easily determined. The construction of such a diagram will now be explained.

Let f_s and f_c denote the unit stress in the tensile steel and the concrete respectively, kd the depth of the neutral axis, and jd the arm of the resisting couple, CT, when there is no compressive reinforcement (see Fig. 16); also let f_s', f_c', $k'd$, and $j'd$ denote the same quantities when there is compressive reinforcement. Then

$$f_c=\frac{f_s kd}{n(d-kd)}=\frac{Mk}{jdAn(1-k)},$$

and

$$f_c'=\frac{f_s'k'd}{n(d-k'd)}=\frac{Mk'}{j'dAn(1-k')}.$$

From these the *relative* reduction in f_c due to the addition of compressive steel is found to be

$$\frac{f_c-f_c'}{f_c}=1-\frac{j}{j'}\frac{k'}{k}\frac{1-k}{1-k'}. \qquad (8)$$

Since j and k depend on p, and j' and k' on p', the equation furnishes the relation between relative reduction in concrete stress and the percentages of steel. The relative reduction $(f_c-f_c')/f_c$ depends largely on the percentage of compressive steel and for a given value of this percentage the reduction is practically the same for all ordinary percentages of tensile steel (from $\frac{1}{2}$ to 3%). Plate VI, page 280, gives values of this reduction for different values of compressive steel from 0 to 2%. As heretofore, values $n=15$ and $d'/d=1/10$ were used.

Addition of compressive steel reduces the stress in the tensile steel. The relative amount of this reduction is given by

$$\frac{f_s-f_s'}{f_s}=1-\frac{j}{j'}. \quad . \quad . \quad . \quad . \quad . \quad . \quad . \quad (9)$$

The group of curves (Plate XII) gives this reduction in per cent (right-hand margin) for different percentages of tensile and compressive steels as noted. (For illustration of the use of this diagram, see example (3) following.)

Examples.—(1) A beam of which $b=12$ in., $d=18$ in., and $d'/d={}^1/_{10}$ has $2\frac{1}{2}$% of tensile steel and 1% of compressive. If the working strengths of steel and concrete are 15,000 and 600 lbs/in² respectively, what is the safe resisting moment of the beam?

Solution. From Fig. 29, $k=0.5$ and $j=0.85$; therefore

$$M_s=15{,}000\times0.025\times0.85\times12\times18^2=1{,}238{,}000 \text{ in-lbs.,}$$

and

$$M_c=(0.5\times0.417+15\times0.01\times0.4\times0.9/0.5)600\times12\times18^2=736{,}000 \text{ in-lbs.,}$$

which is the safe resisting moment.

(2) Suppose that the beam of the preceding example were subjected to a bending moment of 1,000,000 in-lbs. What are the working stresses f_c, f_s, and f_s'?

Solution. As in example (1), $k=0.5$ and $j=0.85$; therefore (see eq. 7) $f_s=1{,}000{,}000/0.025\times0.85\times12\times18^2=12{,}100$ lbs/in². From equation (1), $f_c=(12{,}100\times0.5)\div(15\times0.5)=810$ lbs/in², and from equation (2), $f_s'=15(0.4/0.5)810=9720$ lbs/in².

(3) In a certain design of a beam it is necessary to use 2.5% of tensile steel and this would result in a stress of 1200 lbs/in² in the concrete; it is necessary to reduce this to 900 by adding compressive steel. How much additional steel is required?

Solution. (See Plate XII.) The desired reduction of the compressive stress is 25%. We find this value at the left side of the diagram, then trace horizontally to the concrete curve, and then down to the lower margin, reading there 0.9%, the required quantity. From the last point we trace up to the 2.5% steel curve and then to the right margin, where we note about 4.5% reduction in tensile steel stress due to 0.9% compressive steel.

80. Flexure and Direct Stress.—When the resultant, R, of the external forces acting on one side of a section of a beam is not parallel to the section, then, in general, there exist both direct and flexural stresses at the section. The exception obtains when the resultant passes through the centroid of the section (transformed, as explained below, if the section is reinforced unsymmetrically); in this exceptional case the fibre stress is wholly direct.

In concrete work, the direct stress is always compressive. Combination of direct compressive and flexural stress gives resultant fibre stress which is either (1) all compression or (2) part compression and part tension; these cases are discussed separately below. Whether a given R will produce fibre stress falling under case (1) or (2) depends on the eccentricity * of R, the relative amounts of steel and concrete at the section and on n. If the reinforcement is symmetrical, steel imbedded a depth equal to 1/10 the whole depth of beam, and n is 15, then for eccentricities *lower* than those given in the table, case (1) obtains, and for *higher*, case (2).

$p=$	0%	½%	1%	1½%	2%
$e/h=$	⅙	0.187	0.202	0.214	0.224

In addition to notations already adopted, the following will be used (see Fig. 30):

R denotes the resultant of all the external forces acting on a beam on either side of the section under consideration;

e denotes the eccentric distance of R; that is, the distance from the point where R cuts the section to the middle of the section;

N denotes the component of R normal to the section;

* By the eccentricity of R is meant the ratio of the distance between the centre of the section and the point where R pierces the section to the whole height of the section.

M denotes bending moment at the section; it equals Ne or the sum of the moments of all the external forces about the horizontal line through the middle of the section, but when the transformed section is used, the moment axis must be taken through its centroid;

A' denotes the area of the steel nearer the face of the concrete most highly stressed;

d' denotes the distance from that face to the plane of this steel;

A denotes the area of the steel at the other face;

d denotes the distance from the former face to the plane of this steel;

h denotes the whole height of the section;

p' denotes the steel ratio A'/bh;

p denotes the steel ratio A/bh;

u denotes the distance from the face most highly stressed to the centroid of the transformed section;

A_t denotes the area of the transformed section;

I_t denotes the moment of inertia of the transformed section with respect to its horizontal centroidal axis;

I_c denotes the moment of inertia of the section bh with respect to that axis; and

I_s the moment of inertia of the sections of the steel with respect to the same axis.

81. *Transformed Section.*—By the transformed section of a reinforced concrete beam is meant the actual section with the areas of the reinforcement replaced by concrete n-fold and in the planes of the reinforcement. Thus if Fig. 30a represents an actual section, 30b represents the section transformed, the areas of the upper and lower wings of the latter section being respectively n times the areas of the upper and lower reinforcements.

(A prism of steel of a given area and one of concrete of n times the area are equally rigid as regards simple tension or compression; hence a reinforced-concrete beam and a plain concrete beam whose section is that of the first transformed are equally stiff in so far as stiffness depends upon fibre stress, and in certain cases, as stated later, the fibre stress in the reinforced beam can be computed from those in the plain concrete beam. In those cases, the actual section and the transformed section are equivalent, ideally at least. Actually, the two beams are

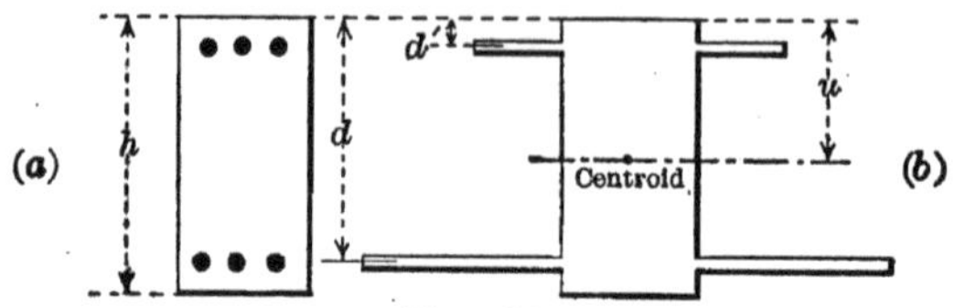

FIG. 30.

not equally strong because of dangerous stresses in the wings of the transformed section.)

Referring to Fig. 30 it will readily be seen that

$$A_t = bh + n(A + A'), \quad I_t = I_c + nI_s, \qquad (1)$$

$$u = \frac{h/2 + npd + np'd'}{1 + np + np'}, \qquad (2)$$

$$I_c = \tfrac{1}{3}b[u^3 + (h-u)^3] \quad \text{and} \quad I_s = A(d-u)^2 + A'(u-d')^2. \qquad (3)$$

If the reinforcement is symmetrical, then $u = h/2$ and

$$I_c = \tfrac{1}{12}bh^3 \quad \text{and} \quad I_s = 2A(\tfrac{1}{2}h - d')^2. \qquad (3)'$$

82. *Case I. The Fibre Stress is Wholly Compressive.*—(a) The unit fibre stress in the concrete can be computed just as though the beam were homogeneous, but the transformed section must be used in the computations if the beam is reinforced. The unit stresses in the steel will be n times those in the concrete in the planes of the reinforcements respectively. Thus the unit direct stress in the concrete is N/A_t; the unit flexural stress in the concrete highest stressed is Mu/I_t; that in the concrete adjoining the reinforcement highest stressed

is $M(u-d')/I_t$; and that in the concrete adjoining the other reinforcement is $M(d-u)/I_t$. The combined unit stresses are:

$$f_c = \frac{N}{A_t} + \frac{Mu}{I_t}, \quad \ldots \ldots \quad (4)$$

$$f_s' = n\frac{N}{A_t} + \frac{nM(u-d')}{I_t}, \quad \ldots \ldots \quad (5)$$

$$f_s = n\frac{N}{A_t} - \frac{nM(d-u)}{I_t}. \quad \ldots \ldots \quad (6)$$

These equations—and the stress diagram, Fig. 31—show that f_s is always less than f_s', and f_s' is always less than nf_c; hence the unit stresses in both steel reinforcements will always be safe if f_c is a safe value.

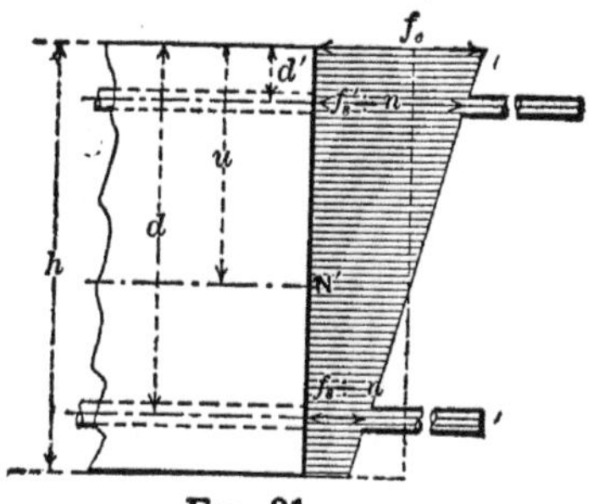

FIG. 31.

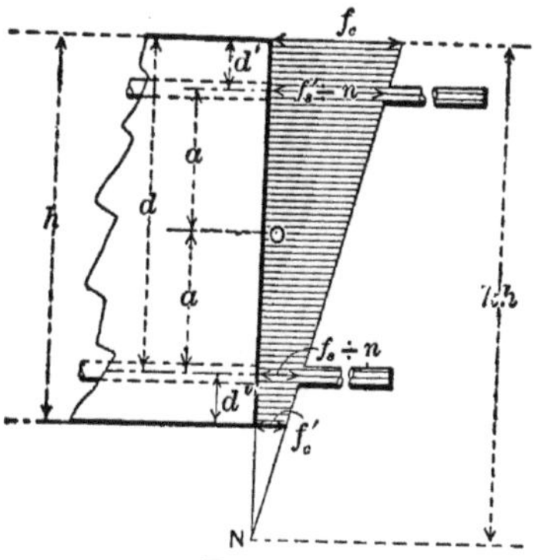

FIG. 32.

(b) The method employed for simple flexure, suitably modified, leads to formulas not involving the transformed section, as will now be explained.

From the stress diagram (Fig. 32) it will be seen that

$$f_s' = nf_c(1-d'/kh), \quad \ldots \ldots \quad (7)$$

$$f_s = nf_c(1-d/kh), \quad \ldots \ldots \quad (8)$$

and

$$f_c' = f_c(1-1/k). \quad \ldots \ldots \quad (9)$$

From the condition that the resultant fibre stress equals N,

$$\tfrac{1}{2}(f_c+f_c')bh+f_s'A'+f_sA=N;$$

and from the condition that the moment of the total fibre stress about the centroidal axis equals M,

$$\tfrac{1}{2}(f_c+f_c')bh\frac{h}{6(2k-1)}+f_s'A'\left(\frac{h}{2}-d'\right)-f_sA\left(\frac{h}{2}-d'\right)=M.$$

From these equations it is possible to compute the unit fibre stresses f_c, f_s, and f_s' in a given case.

When the reinforcement is symmetrical the equations simplify greatly, and they lead to the following formula:

$$12k(1+2np)e/h=1+24npa^2/h^2+6(1+2np)e/h; \quad . \quad (10)$$

they also give the following formula for f_c or M:

$$\frac{M}{bh^2f_c}=\frac{1}{12k}(1+24npa^2/h^2). \quad . \quad . \quad . \quad . \quad (11)$$

When $d'/h=1/10$, and $n=15$, Fig. 33 gives values of $1/k$ for different values of eccentricity and percentage of steel; thus for $e/h=0.1$ and $p=1.5\%$, $1/k=0.635$, hence $k=1.57$.

83. *Case II. There is Some Tension at the Section.*—(*a*) If the tension in the concrete is so small as to be permissible, and this tension is taken account of in the computations, then the unit fibre stresses in the concrete and steel, if reinforcement is present, may be computed by the method explained under Case I.*

The combined unit stress in the remote tensile fibre is given by

$$f_c'=\frac{M(h-u)}{I_t}-\frac{N}{A_t}, \quad . \quad . \quad . \quad . \quad . \quad (12)$$

* It is assumed that the linear law of variations of the unit flexural stresses holds for the tension as well as compression.

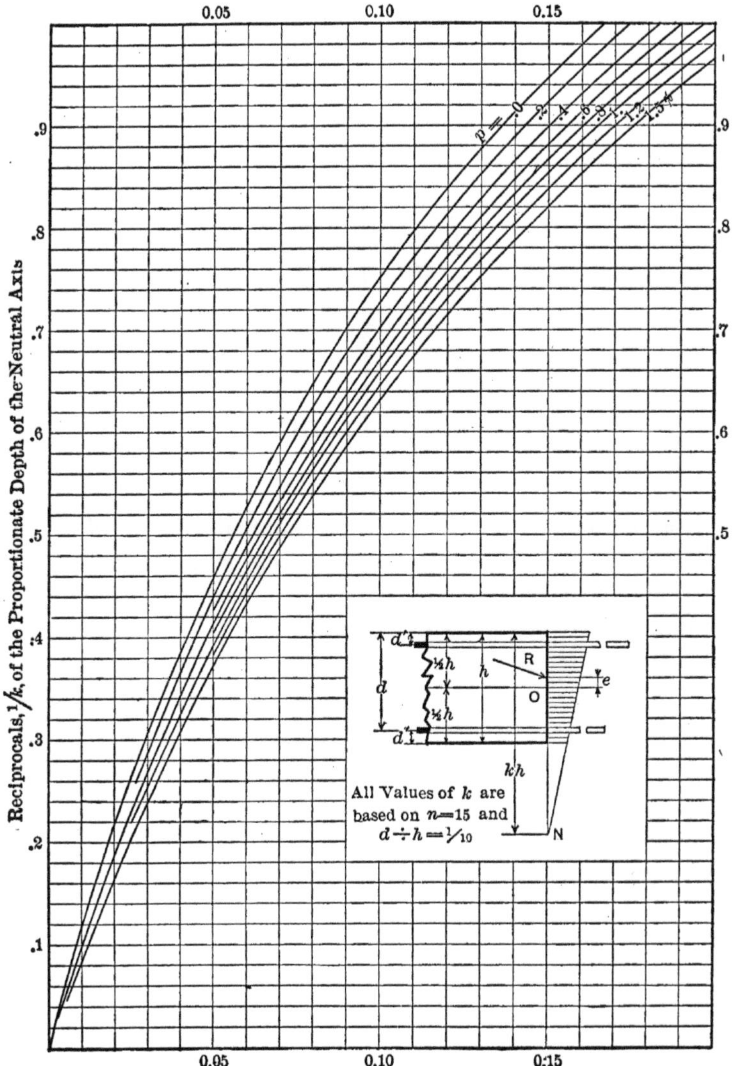
0.05
0.10
0.15
Reciprocals, 1/k, of the Proportionate Depth of the Neutral Axis
.1
.2
.3
.4
.5
.6
.7
.8
.9
p = 0
.2
.4
.6
.8
1.
1.2
1.5%
Eccentricity e/h
All Values of k are
based on n=15 and
d ÷ h = 1/10
d'
d
½h
h
R
O
e
kh
N

FIG. 33.

and f_s as given by (5) is compressive or tensile according as its value is positive or negative.

(*b*) If the tensile stresses are so high that it is advisable to neglect the tension in the concrete, then a method similar to that used heretofore in simple flexure is simplest. The transformed section is not used. O (Fig. 34) denotes a horizontal

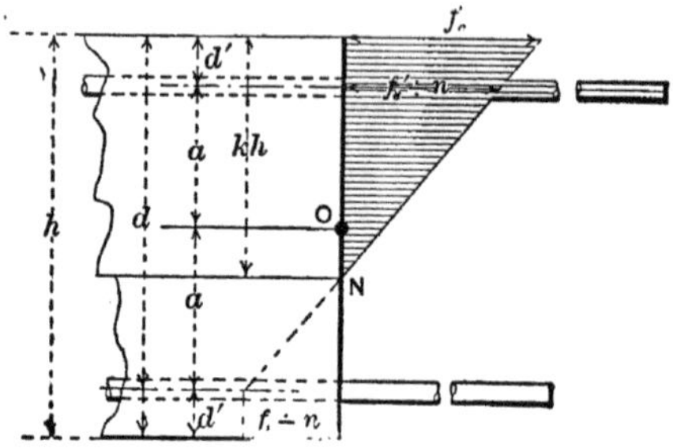

FIG. 34.

axis at mid-depth of the beam, M the moment sum of all the external forces on one side of the section with respect to that axis, and N, as before, the algebraic sum of the components of those forces perpendicular to the section. From the stress diagram, it follows that

$$f_s = nf_c\left(\frac{d}{kh} - 1\right) \quad . \quad . \quad . \quad . \quad . \quad . \quad . \quad (13)$$

and

$$f_s' = nf_c\left(1 - \frac{d'}{kh}\right). \quad . \quad . \quad . \quad . \quad . \quad . \quad . \quad (14)$$

Since the resultant fibre stress equals N,

$$\tfrac{1}{2}f_c bkh + f_s'A' - f_sA = N,$$

and since the moment of the fibre stress about the horizontal axis through O equals M,

$$\tfrac{1}{2}f_c bkh\left(\frac{h}{2} - \frac{kh}{3}\right) + f_s'A'\left(\frac{h}{2} - d'\right) + f_sA\left(d - \frac{h}{2}\right) = M.$$

From these four equations k, f_c, f_s, and f_s' can be determined for a given section, reinforcement, M, and N.

If the reinforcement is symmetrical, then the equations simplify. The value of k is given by

$$k^3-3\left(\frac{1}{2}-\frac{e}{h}\right)k^2+12np\frac{e}{h}k=6np\left(\frac{e}{h}+2\frac{a^2}{h^2}\right). \quad . \quad . \quad (15)$$

The greatest unit compressive fibre stress in the concrete is given by

$$\frac{M}{bh^2f_c}=\frac{1}{12}k(3-2k)+\frac{2pn}{k}\frac{a^2}{h^2}, \quad . \quad . \quad . \quad . \quad (16)$$

and the unit stresses in the steel are given by (7) and (8). From (7), or the stress diagram, it is plain that f_s' is less than nf_c even for unsymmetrical reinforcements.

When $d'/h=1/10$ and $n=15$, Fig. 35 gives values of k for different values of eccentricity and percentage of steel; thus for $e/h=1$, and $p=0.8\%$, $k=0.42$.

84. *Diagrams.*—To facilitate the application of equation (11) (Case I), and equation (16) (Case II), Plates XIII and XIV, pages 287 and 288, have been constructed.

In the first diagram, values of the eccentricity, e/h, are given at the upper and lower margins; the ordinates from the lower margin to any curve are values of $(1+24npa^2/h^2)/12k$ (see equation 11), and hence of M/bh^2f_c, for the value p marked on that curve. Thus when $e/h=0.1$ and $p=1\%$, $M/bh^2f_c=0.087$.

The dotted portions of the curves correspond to eccentricities which involve small tensile stress in the concrete and belong strictly to Case II. The values of the unit tensile stress f_c' can be calculated from equation (12) or from

$$\frac{f_c'}{f_c}=\frac{h-kh}{kh}=\frac{1}{k}-1, \quad . \quad . \quad . \quad . \quad . \quad . \quad (17)$$

$1/k$ being obtained from equation (10), or from an extension of the appropriate curve in Fig. 33.

In the second diagram, also, values of the eccentricity e/h are given at the upper and lower margins; the ordinates from the lower margin to any solid curve are values of

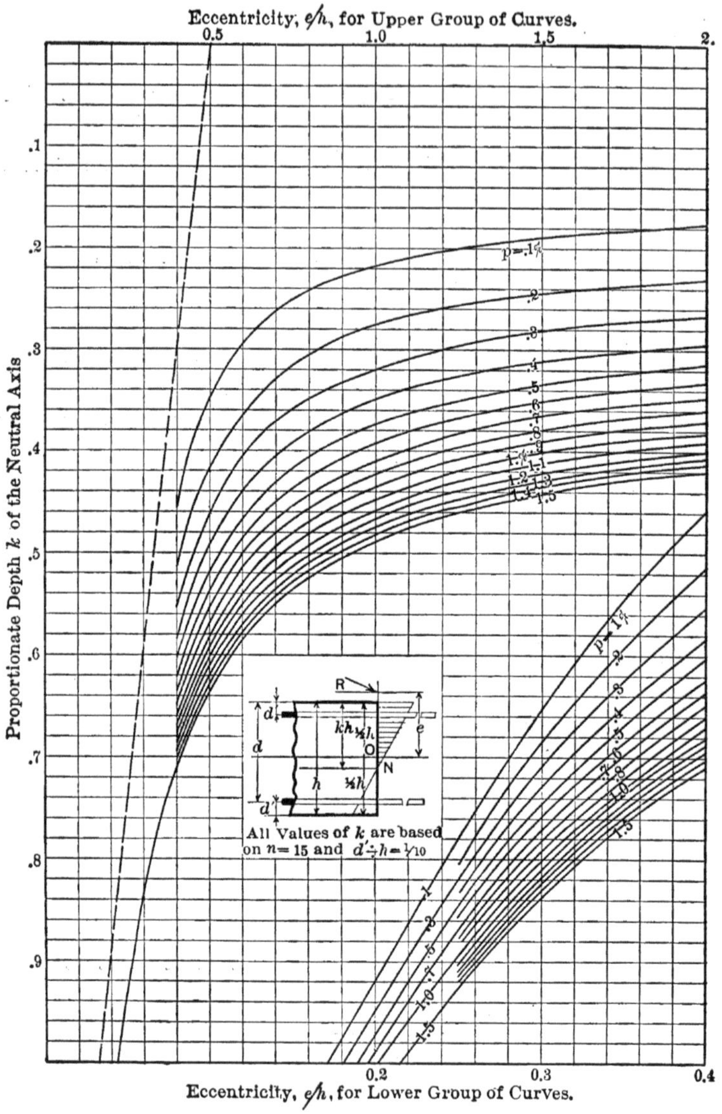
Eccentricity, e/h, for Upper Group of Curves.
0.5
1.0
1.5
2.
Proportionate Depth k of the Neutral Axis
.1
.2
.3
.4
.5
.6
.7
.8
.9
p=.1%
0.2
0.3
0.4
Eccentricity, e/h, for Lower Group of Curves.
All Values of k are based on n=15 and d'÷h=1/10

FIG. 35.

$\frac{1}{12}k(3-2k)+2pna^2/kh^2$ (see equation 16), and hence of M/bh^2f_c, for the value of p marked on that curve. Thus when $e/h=1$ and $p=1\%$, $M/bh^2f_c=0.187$.

The dotted curves in the second diagram enable one to estimate the ratio of the unit stress in the tensile steel to that in the concrete, f_s/f_c, for most eccentricities and percentages of steel. Thus when $e/h=1$ and $p=0.5\%$, we find $e/h=1$ at the top or bottom and then trace vertically to the 0.5% curve and note the point of intersection. This point falls between the curves $f_s/f_c=20$ and 25, and the ratio is about 21. For values of e/h and p, which bring the "point" to the left of the line $f_s/f_c=15$, f_s will be less than $15f_c$, and hence less than the working strength of steel for all ordinary allowable values of f_c. No similar curves for f_s/f_c appear on the first diagram because that ratio is always less than 15, and hence the unit stresses in the steel (both upper and lower) are within safe values for Case I, if f_c is safe.

85. *Examples.*—It is supposed in these that the steel is imbedded a depth of one-tenth the total height of the beam, and that $n=15$, so that the diagrams on pages 287 and 288 apply.

(1) A beam is 12 in. wide, 30 in. high, and contains 1% of steel above and an equal percentage below. At a particular section, the resultant R is 80,000 lbs., its inclination to the axis of the beam is 5°, and its eccentric distance is 4.5 in. Compute the unit fibre stresses in the concrete and steel (f_c, f_s, and f_s').

Solution. The eccentricity is $e/h=0.15$, and $M=80,000 \cos 5° \times 4.5=$ 358,650 in-lbs. The beam falls under Case I because this eccentricity gives a "point" on the 1% curve of page 287, but not on that of page 288. Tracing horizontally from the point we read $M/bh^2f_c=0.112$; hence

$$f_c=\frac{358,650}{24\times30^2\times0.112}=297 \text{ lbs/in}^2.$$

The unit stresses in the steel are less than $15f_c=4500$ lbs/in². Their exact values can be computed from equations (7) and (8); the value of k for use in them can be easiest obtained from the diagram on page 103.

(2) Change the eccentricity of the preceding example to 15 in. and solve.

Solution. The eccentricity is $e/h=0.5$, and $M=80,000 \cos 5° \times 15=$ 1,195,500 in-lbs. The beam falls under Case II (see page 288), and for the eccentricity 0.5 and 1% of steel the diagram gives $M/bh^2f_c=0.171$; hence

$$f_c=\frac{1,195,500}{12\times30^2\times0.171}=647 \text{ lbs/in}^2.$$

The intersection of the 1% curve and the 0.5 eccentricity line lies to the left of the curve $f_s/f_c=15$; hence the unit stress in the tensile steel is less than $15\times647=9470$ lbs/in². The exact value can be computed from equation 13; the value of k for use in it can be obtained easiest from the diagram on page 98

(3) The breadth of a beam is 12 in. and its height 24 in. At a certain section the bending moment is 450,000 in-lbs., and the eccentric distance is 4 in. The working strength of the concrete being 600 lbs/in², how much steel reinforcement, if any, is required?

Solution. The eccentricity is $e/h=\frac{1}{6}$, and hence the beam would be on the border between Case I and II even if no steel were used. With steel, the beam falls under Case I, and

$$\frac{M}{bh^2f_c}=\frac{450,000}{12\times24^2\times600}=0.1085.$$

Entering the diagram, page 287, with this value and tracing horizontally to the 0.167 eccentricity vertical, we find their intersection and note that it falls between the 0.6 and 0.8% curves; about 0.7% of steel therefore is required.

(4) In example (3) change the eccentric distance to 12 in. and solve.

Solution. The eccentricity is $e/h=\frac{1}{2}$, and the beam falls under Case II (see page 288). M/bh^2f_c has the same value as in example (3); hence entering the diagram with that value and tracing horizontally to the 0.5 eccentricity vertical, we find their intersection and note that it falls between the 0.2 and 0.3% curves; hence 0.3% is the required amount.

(At first thought it may seem that more steel is necessary in example (4) than in (3) because of the greater eccentricity in the former example. But it should be noted that the thrust N is much less in (4) than in (3), its values being $M/e=37,500$ and 112,500 lbs. respectively.)

86. Shearing Stresses in Reinforced Beams.—In Art. 46 the variation in shearing stress in a homogeneous beam was discussed and the general formula given for the intensity of shear at any point (see eq. (1)). In a reinforced beam the variation in shear differs from that in a homogeneous beam owing to the concentration of tensile stress in the steel. The general formula for shearing stress may, however, still be used if the transformed section be employed; that is, if the area of the steel be multiplied by n and considered equivalent to concrete at the same horizontal plane. The tension area of the concrete should be neglected. A simpler method for present purposes is the following: In Fig. 36 is represented a short portion of a beam where the total vertical shear is V. Let v=horizontal (or vertical) shearing stress per unit area at the neutral axis, and let b=width of beam. Other quantities are sufficiently indicated in the figure. $C=T$ and $C'=T'$.

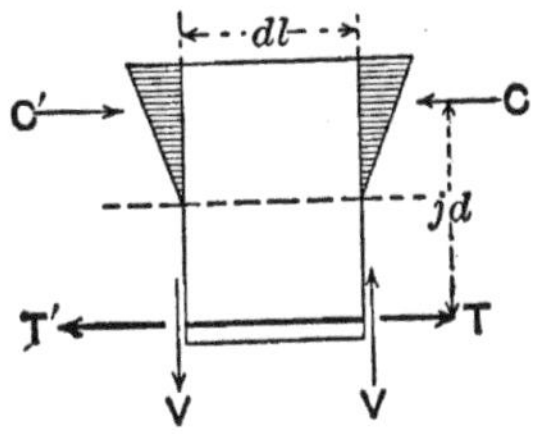

FIG. 36.

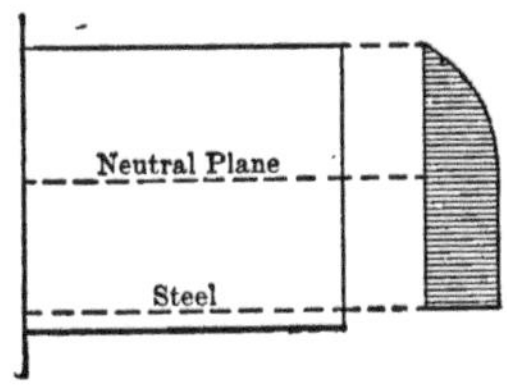

FIG. 37.

The total shearing stress on any horizontal plane between the steel and the neutral axis will be equal to $T'-T$ and the stress per unit area $=v=\frac{T'-T}{bdl}$. From equality of moments we have the relation $Vdl=(T'-T)jd$, whence is derived the expression

$$v=\frac{V}{bjd}. \quad \ldots \ldots \ldots \quad (1)$$

The shearing stress given by eq. (1) is the same at all points between the neutral axis and the steel; above the neutral axis the shear follows the parabolic law as in a homogeneous beam. Fig. 37 represents the law of variation for the case under discussion.

Using 7/8 for an approximate value of j (see Art. 55) we have approximately

$$v=\frac{8}{7}\frac{V}{bd}; \quad \ldots \ldots \ldots \quad (2)$$

that is, the shearing stress at the neutral axis (equal to the maximum) is one-seventh or about 14% more than the average value.

87. *Beams Reinforced for Compression.*—In beams reinforced for compression formula (1) will still apply, the value of jd being the distance between the tensile steel and the resultant of the compressive stresses as shown in Art. 77. In this case j is somewhat greater than in the previous case and v is more nearly equal to the average shearing stress $\frac{V}{bd}$.

88. *T-beams.*—Here again formula (1) still holds true, j retaining its general significance. As shown in Art. 73, j may be taken as closely equal to the distance from the steel to the centre of the flanges of the T; hence

$$v=\frac{V}{b'(d-\frac{1}{2}t)}. \quad . \quad . \quad . \quad . \quad . \quad . \quad . \quad . \quad (3)$$

It is to be noted that in the T-beam the shearing stresses are practically the same as in a rectangular beam of the same depth and having the same width as the stem of the T. The slab aids in reducing the shear only by its effect in increasing slightly the value of j.

89. *Working Formula.*—Since the value of j varies only within narrow limits it is quite as satisfactory for comparative purposes and for purposes of design to use the *average* value of the shearing stress,

$$v'=\frac{V}{bd}, \quad . \quad . \quad . \quad . \quad . \quad . \quad . \quad . \quad . \quad (4)$$

in which b is the breadth and d is the net depth of the beam. In T-beams b is the breadth of the stem and d is the total depth from top of beam to steel. The true maximum shear will generally be from 10 to 15 per cent higher than the average value thus determined.

90. *Effect of Shear on the Tensile Stresses in the Concrete.*—In Art. 46 it was shown that in a homogeneous beam the direction of the maximum tensile stresses is horizontal at the lower face and becomes more and more inclined as the neutral axis is approached, reaching an inclination of 45° at that place. In the reinforced beam we have assumed, for purposes of design, that there is no tension in the concrete. While such possible tension will add very little to the resisting moment of the beam it is desirable to consider it here in relation to the shearing stresses and the resultant effect on lines of probable rupture. The shearing stresses determined in the preceding article have been calculated on the assumption of no tensile stress in the concrete, but the effect of such tension on the distribution of the shear is very small and need not be considered.

To determine the amount and direction of the maximum inclined tensile stresses at any point, eq. (1), Art. 46, is still applicable. In this case large shearing stresses exist immediately above the steel, hence the maximum tensile stresses become considerably inclined as soon as we leave the line of the steel, the exact direction depending upon the relation between the shear and the horizontal tension. Exact calculations are impossible, since the actual horizontal tension in the concrete is unknown. While the steel is assumed to carry all tension the concrete will in fact be stressed in accordance with its deformation up to the point of ultimate deformation and rupture. Where the steel has a stress of its full working value of 12,000 to 15,000 lbs/in^2, the deformation will much exceed the ultimate deformation of the concrete and rupture must occur, but at points where the steel stress is low, as for example near the end of the beam, the concrete may be intact.

Suppose, for example, the stress in the steel is 3000 lbs/in^2. If the modulus of elasticity of the concrete in tension is 1,500,000 the stress in it will be $3000/20=150$ lbs/in^2, which is not far from its ultimate strength. Suppose further that the unit shearing stress in the lower part of the beam is 100 lbs/in^2. By eq. (2) of Art. 46 the resultant maximum tension will be $t=\frac{1}{2}(150)+\sqrt{\frac{1}{4}\cdot 150^2+100^2}=200$ lbs/in^2, and will have a direction inclined $26\frac{1}{2}°$ from the horizontal. This stress may exceed the ultimate strength of the concrete and the result will be an inclined crack. At points nearer the neutral axis the horizontal tensile stresses become less and the inclined tension approaches the value of the shearing stress and its inclination approaches 45°. The result of these inclined stresses is likely to be a progressive tension failure in an inclined direction which the horizontal rods are not very effective in preventing.

Excessive stresses of this kind are prevented in various ways. Obviously they will be reduced by keeping the horizontal tension small through the use of considerable horizontal steel at points of heavy shear, by keeping the shearing stresses low, and by various means of directly carrying the inclined stresses by special reinforcement.

91. *Ratio of Length to Depth for Equal Strength in Moment and Shear.*—For any given values of per cent of steel and of working stresses in shear and direct stress there is a definite ratio of length to depth of beam which will give equal strength in moment and shear. The strength of beams of greater relative length will be determined by their moment of resistance, while that of shorter beams by their shearing resistance. The ratio of length to depth for equal strength depends on the method of loading.

For Single Concentrated Loads.—In this case the shear V, due to a given load W, is $\frac{1}{2}W$, and the moment M is $\frac{1}{4}Wl$. Hence

$$W = 2V = 4M/l. \quad \ldots \ldots \quad (a)$$

From Art. 89 we have $V = v'bd$ and from Art. 56 $M_s = f_s jpbd^2$, in which v' = safe average shearing stress and f_s = working stress in steel. Substituting, we have

$$2v'bd = \frac{4f_s pjbd^2}{l},$$

from which

$$\frac{l}{d} = \frac{2f_s jp}{v'}. \quad \ldots \ldots \ldots \quad (1)$$

For a Uniformly Distributed Load a similar process gives the ratio

$$\frac{l}{d}=\frac{4f_s jp}{v'}. \qquad (2)$$

For Beams Loaded with Equal Loads at the Third Points,

$$\frac{l}{d}=\frac{3f_s jp}{v'}. \qquad (3)$$

In the case of continuous girders these formulas will apply if l be taken as the length between points of inflection.

Taking, for example, $p=0.01$, $v'=50$ lbs/in², and $f_s=15{,}000$ lbs/in², and using an average value of 7/8 for j, we have the following ratios for $\frac{l}{d}$:

For concentrated loads $\frac{l}{d}=5.25$.

For uniformly distributed loads $\frac{l}{d}=10.5$.

For double concentrated loads $\frac{l}{d}=7.87$.

92. **Bond Stress.**—The stress on the bond between steel and concrete (Fig. 36, Art. 86) will be equal to $T'-T$ on the length dl.

If U denote the bond stress per lineal inch, we then have

$$U=\frac{T'-T}{dl},$$

whence we derive

$$U=\frac{V}{jd}. \qquad (1)$$

The bond stress per unit area will be equal to U divided by the sum of the perimeters of the steel sections. Or, if $o=$ perimeter of one bar, $\Sigma o=$ sum of perimeters and $u=$ bond stress per unit area, we have

$$u=\frac{V}{\Sigma o\cdot jd}. \qquad (2)$$

92a. *Bond Stress for Compressive Reinforcement.*—The question of bond stress for compressive reinforcement will seldom come into consideration. In this case it may also be calculated by formulas based on the shear at the section, but it is necessary to take account of the compression carried by the concrete, and the formula is the general formula for horizontal shear, involving the statical moment of the concrete area and the equivalent steel area about the neutral axis.

By this general formula the horizontal shear per lineal inch at any horizontal plane is proportional to the statical moment of the effective area outside such plane about the neutral axis. At the neutral axis this shear per lineal inch has been shown to be $vb=V/jd$. The shear between the compression rods and the concrete, or the bond stress desired, will therefore be equal to $\frac{V}{jd}\times\frac{\text{Statical moment of equivalent steel area}}{\text{Total statical moment of compression area}}$. The total moment of the compression area equals the total moment of the tension area, which is $nA(1-k)d$, where A is the total area of tensile steel. (See Art. 76 for notation.) The moment of the compressive steel is equal to $nA'(kd-d')$. Hence, for bond stress of compressive steel,

$$U=\frac{V}{jd}\times\frac{A'(kd-d')}{A(1-k)d}. \quad . \quad . \quad . \quad . \quad . \quad . \quad (3)$$

That is, the bond stress per lineal inch for the two steel areas will be proportional to the areas times their distances from the neutral axis. Since the compressive steel will generally be nearer the neutral axis than the tensile steel it follows that if the compression bars are no larger in diameter than the tension bars, the bond stress will always be less than that in the tension bars.

92b. *Variation of Bond Stress in Beams.*—The theoretical results assume that there is perfect adhesion and that the stress in the steel is taken over by the concrete at all points, as called for by the usual theory. They show that the bond stress is a simple function of the shear and varies therewith. Thus, in a beam supporting a single concentrated load, the shear and bond stress varies, as shown in Fig. 37*a*. In this case the bond stress is uniform from load to end of beam. For a distributed load the variation is as shown in Fig. 37*b*. The value is a maximum at the ends and decreases towards the centre. Thus if a reinforcing rod carries a certain stress S at

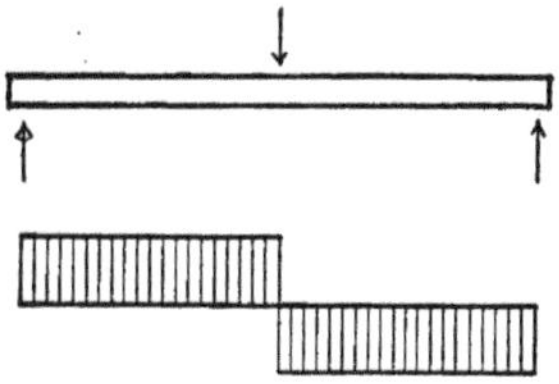

FIG. 37*a*.

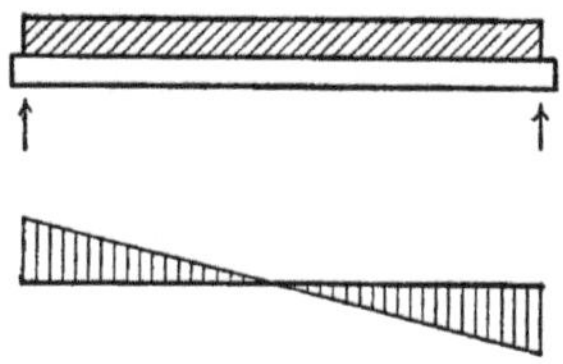

FIG. 37*b*.

the centre, in the first case the bond stress per lineal inch will be uniform and may be calculated by the formula $U=S\div\frac{l}{2}$. Or, by eq. (1), it is $U=\frac{V}{jd}=\frac{\frac{1}{2}w}{jd}$. But $S=\frac{M}{jd}=\frac{\frac{1}{4}wl}{jd}$. Hence $U=S\div\frac{l}{2}$, as above given. In the second case the shear and bond stress is not uniform and hence the bond stress is not to be calculated by dividing the stress S by $l/2$. It is a maximum at the end where it is equal to twice the average value. In practice, a beam will have some length beyond the theoretical centre of bearing, and the rods will extend entirely through the beam. This doubtless modifies the bond stress near the end, somewhat as shown by the dotted lines in Fig. 37*c*, some stress being carried by the rod beyond the theoretical centre of bearing, thus reducing the maximum bond stress below the theoretical value. In the case of continuous

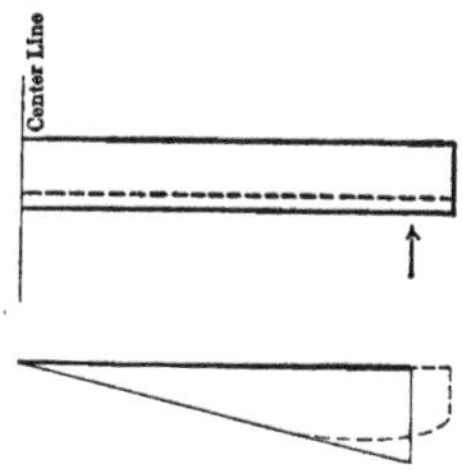

FIG 37*c*.

girders the bond stress is still a maximum at the support and is measured by the shear, although the stress may be compressive (the *increment* of stress is still of the same sign). Continuous rods would tend to modify the bond stress, as shown in Fig. 57*b*, Art. 123, thus reducing the maximum on both sides of the support, and the maximum compression in the steel. The stress in the concrete will be increased. For discussion of anchored rods see Art. 123.

92*c*. Deflection of Reinforced Concrete Beams.—Deflection formulas for homogeneous beams can be interpreted semi-rationally to make them applicable to reinforced concrete beams. So interpreted they yield results in fair agreement with actual measured deflections. (See Art. 112*a*.)

General Theory.—As is well known, a concrete-steel beam under full working load contains one or more cracks at or near the section of maximum bending moment or else the condition there is near the cracking stage, and to compute the maximum unit fibre stresses at such section, engineers rightly assume the presence of a tension crack, and, in effect, that it has extended to the neutral axis. Since the deflection depends on the stress at all sections, and the cracked sections are comparatively very few, a deflection formula should be based on the intact section. It may be thought that a cracked section influences the deflection more than an intact one, the idea is correct, but the effect of incipient cracking on the deflection is not as great as on the fibre stress at the section. These effects are entirely different in "order of magnitude," the first is not noticeable at all in careful measurements on deflections due to increasing loads, whereas the latter certainly would be if fair measurements of fibre stress *at* a section of a beam were possible. To simplify certain relations, it will be assumed that the depth of the intact section for use in the deflection formula extends from the top of the beam to the centre of the steel; this in effect assumes all sections cracked from the bottom to the centre of the steel.

Deflection formulas for homogeneous beams imply that

the material of the beam obeys Hooke's law ("stress is proportional to strain"), up to working stresses at least, and that the moduli of elasticity of the material for tension and compression are equal. While it is true that concrete does not obey the law strictly, still its stress-strain relation for compression is nearly linear up to working stresses. But the stress-strain relation for tension is far from linear, and the assumption that it is, herein made for simplicity in formulas, must be regarded as a rough approximation. It is true that the "initial moduli" (Art. 24) of concrete for compression and tension are nearly equal, but the deflection of a beam depends on the elongations and shortenings of all the fibres, and hence not upon initial modulus but on some sort of a mean value. This is not the modulus corresponding to the mean unit fibre stress, but certainly the average or secant modulus is nearer correct than the initial or the modulus at the maximum unit stress.

The formulas also imply that the moments of inertia of the cross-sections of the beam are equal. This condition is not fulfilled in most reinforced concrete-beams, due account being taken of the steel, because of presence of bent-up rods and stirrups. Still the amount of steel in, and hence the moments of inertia of, sections in the middle third or middle half are commonly constant; and since the middle half contributes nearly 85% of the maximum deflection in the case of a simple beam constant in section and uniformly loaded, and 82% when the beam is loaded at the two outer points, it must be that a small change in the moments of inertia of end sections of a simple beam would produce a much smaller change in the maximum deflection. In fact, if a simple beam is uniformly loaded, for example, and the moment of inertia of sections in its middle half is I_1, and that of sections in its outer quarters is I_2, then its maximum deflection is $Wl^3(67I_2+13I_1)/6144EI_1I_2$; and if the sections are uniform and the common moment of inertia is I_1, then the maximum deflection is $5Wl^3/EI_1\,384$; hence the ratio of the deflections is

$(67I_2+13I_1)/80I_2$, and if I_1 and I_2 differ by 10%, say, the maximum deflections differ by less than 2%.

For the reasons stated above, the deflection formulas for homogeneous beams will be used for reinforced-concrete beams, but modified in accordance with the following assumptions:

1. That the representative or mean section has a depth equal to the distance from the top of the beam to the centre of the steel;
2. That it sustains tension as well as compression, both following the linear law;
3. That the proper mean modulus of elasticity of the concrete equals the average or secant modulus up to the working compressive stress; and
4. That the allowance for steel in computing the moment of inertia of the mean section should be based on the amount of steel in the mid-sections.

92*d*. *Deflection of Rectangular Beams.*—For homogeneous beams the deflection formulas commonly involve the load or the maximum unit fibre stress. The following are corresponding formulas for rectangular reinforced concrete beams:

$$D=\frac{c_1}{E_s}\frac{Wl^3}{bd^3}\frac{n}{\alpha}, \quad . \quad . \quad . \quad . \quad . \quad . \quad . \quad . \quad . \quad (1)$$

and

$$D=\frac{1}{2}\frac{c_1}{c_2E_s}\frac{f_c l^2}{d}(kj)\frac{n}{\alpha}, \quad . \quad . \quad . \quad . \quad . \quad . \quad (2)$$

or

$$D=\frac{c_1}{c_2E_s}\frac{f_s l^2}{d}(pj)\frac{n}{\alpha}. \quad . \quad . \quad . \quad . \quad . \quad . \quad . \quad (3)$$

In these the notation is as follows:

D=maximum deflection (if desired in inches, the units specified below should be used);

b=breadth of the section (in.);

d=depth of the section to the centre of the steel (in.);

c_1 = the numerical coefficient in the formula for deflection of homogeneous beams, c_1Wl^3/EI, depending on the loading and support (see page 126);

c_2 = the numerical coefficient in the formula for maximum bending moment, c_2Wl, also depending on the loading and support (see page 126);

E_s = modulus of elasticity of the reinforcing steel (lbs/in^2);

E = modulus of elasticity of the concrete (lbs/in^2);

n = ratio of the moduli of elasticity of steel and concrete;

p = steel ratio (area of steel section $\div bd$);

α = a numerical coefficient depending on p and n;

f_c = greatest unit compressive stress in the concrete (lbs/in^2);

f_s = greatest unit tensile stress in the steel (lbs/in^2);

k = proportionate depth of the neutral axis (see Fig. 16);

j = proportionate distance of the centroid of the compressive stress from the steel (see Fig. 16).

The schedule (page 126) gives values of c_1/E_s for use in formula (1) and values of c_1/c_2E_s for formulas (2) and (3) for certain standard cases more or less close approximations to which are met in practice; and the diagrams (Figs. 37*f* and 37*g*) furnish values of n/α, kj, and pj. It is recommended that 8 or 10 be used for n in the first diagram (see Art. 112*c*); in the second that value of n is to be used which the computer prefers in his own formulas, tables, or diagrams for the strength of beams. These two values of n will probably be unlike; the apparent inconsistency is discussed at the close of this article.

Example 1.—A concrete beam rests on end supports 16 ft. apart, the breadth of its section is 10 in., the depth (to the steel) is 15 in., the reinforcement consists of four ¾-in. rods extending along the whole length (and stirrups). What is its probable deflection when sustaining a uniform load of 10,000 lbs., including its own weight?

The amount of steel is 1.767 in^2, hence $p = 1.767 \div 150 = .012$. Entering the diagram (page 124) at percentage 1.2, tracing upward to the $n=8$ curve say, and then horizontally, it is found that $n/\alpha = 76$.

From the schedule (page 126) it is found that $c_1/E_s=0.000434/1,000,000$. hence from eq. (1)

$$D=\frac{.000434\times 10,000\times 192^3\times 76}{1,000,000\times 10\times 15^3}=.07 \text{ in.}$$

Example 2.—The deflection of the beam described in the preceding example is desired, (1) when it is loaded so that the working compressive fibre stress is 500 lbs/in², and (2) when the working stress in the steel is 14,000 lbs/in².

(1) From the schedule it is seen that $c_1/c_2E_s=.00347/1,000,000$, and, as in example 1, $n/\alpha=76$. Entering the diagram (page 125) at $p=1.2\%$ and tracing upwards to the $n=15kj$ curve (a value of n much used in strength formulas), and then horizontally to the left, we find that kj is .38; hence from eq. (2),

$$D=\frac{.00347\times 500\times 192^2\times .38\times 76}{2\times 1,000,000\times 15}=.06 \text{ in.}$$

(2) Entering the diagram at $p=1.2\%$ and tracing upward to the $n=15pj$ curve and then horizontally to the right we find that $pj=.0102$; hence from eq. (3)

$$D=\frac{.00347\times 14,000\times 192^2\times .0102\times 76}{1,000,000\times 15}=.09 \text{ in.}$$

Analysis for Formulas and Diagrams.—Since the total tension (in concrete and steel) and the total compression are equal (see Fig. 37*d*), at any section,

$$\frac{1}{2}\frac{f_s}{n}b(d-kd)+f_sA=\tfrac{1}{2}f_cbkd.$$

Also $f_s/n=f_c(1-k)/k$ and $A=pbd$, and these values substituted in the first equation yield one from which it follows that

$$k=\frac{1+2np}{2+2np}. \qquad (4)$$

The moment of inertia, with respect to the neutral axis, of the part of section in compression is $\frac{1}{3}bk^3d^3$, that of the concrete section in tension is $\frac{1}{3}b(1-k)^3d^3$, and that of the weighted steel sections is practically $nA(1-k)^2d^2$; hence

$$I=\tfrac{1}{3}[k^3+(1-k)^3+3np(1-k)^2]bd^3,$$

or

$$\alpha=\tfrac{1}{3}[k^3+(1-k)^3+3np(1-k)^2], \quad . \quad . \quad . \quad . \quad (5)$$

and

$$D=c_1Wl^3/E_cI=c_1Wl^3n/E_sbd^3\alpha,$$

which is eq. (1).

From eqs. (4) and (5), the value of α for any values of p

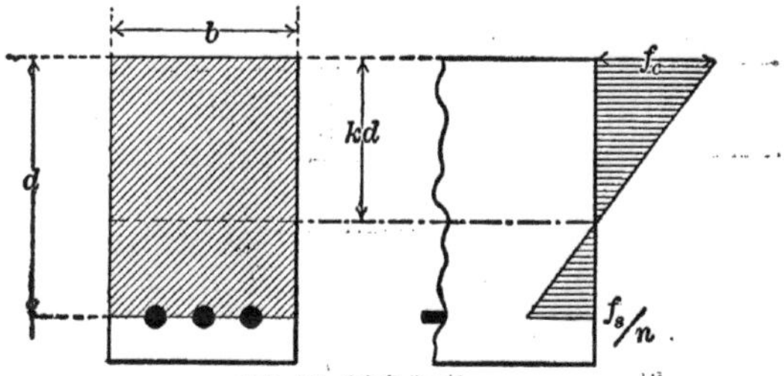

FIG. 37*d*.

and n may be computed; a sufficient number of these were thus computed to determine the n/α curves in Fig. 37*f*.

The transformation of the deflection formula (1) (in terms of the load) into (2) and (3) (in terms of the working unit stresses f_c and f_s respectively) will now be made. For this purpose, strength formulas based on cracked sections (see Fig. 16) and a linear variation of compression are used. These well-known strength formulas based on concrete and steel are respectively, $M=\frac{1}{2}f_ckjbd^2$ and $M=f_spjbd^2$ (see page 56). Since $M=c_2Wl$ also, $W=\frac{1}{2}f_ckjbd^2/c_2l=f_spjbd^2/c_2l$. These two values of W substituted in eq. (1) yield eqs. (2) and (3) respectively.

The formulas for k and j of Fig. 16 are also well known; they are (see Art. 55)

$$k=\sqrt{2pn+(pn)^2}-pn \quad . \quad . \quad . \quad . \quad . \quad (6)$$

and

$$j=1-\tfrac{1}{3}k. \qquad (7)$$

By means of these the values of kj and pj can be computed for any values of p and n; a sufficient number of these were thus computed to determine the kj and pj curves of Fig. 37*g*.

Choice of different values of n in n/α and kj or pj for use in any particular case is not an inconsistency. The first value depends on the unit fibre stresses at all points of the beam, and when the numerical value is chosen from experiments on deflection, then n becomes also a sort of empirical coefficient-

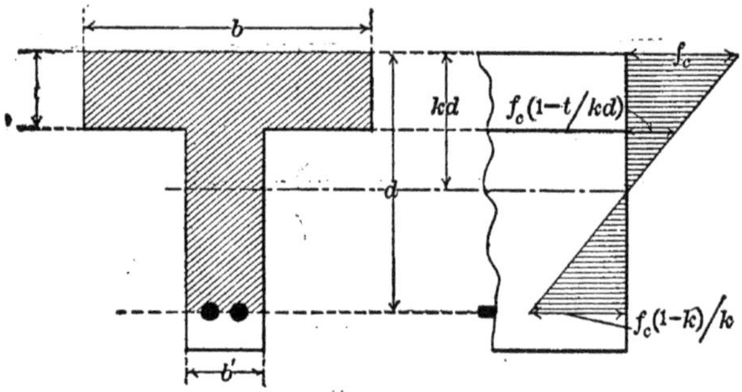

FIG. 37*e*.

making correction for various errors in the deduction of the deflection formula, whereas the second depends on the unit stresses in the cracked section and when its numerical value is chosen from experiments on the strength of beams, then it also becomes in part an empirical coefficient correcting errors of approximation in the strength formulas used.

92*e*. *Deflection of T-beams.*—Under the four assumptions stated in Art. 92*c*, the deflection formula for T-beams in terms of the load becomes

$$D=\frac{c_1}{E_s}\frac{Wl^3}{bd^3}\frac{n}{\beta}, \qquad (1)$$

in which β is a coefficient depending upon the steel ratio and n, b width of flange, d depth to steel (see Fig. 27); other symbols are explained in Art. 92*d*.

In accordance with assumption 2, the neutral axis of the representative section will be in the web, or stem, generally, as is implied in Fig. 37*e*. Then the total tension and compression at the section are given by

$$T=b'(1-k)d\,\tfrac{1}{2}f_c(1-k)/k+pbdnf_c(1-k)/k$$

and

$$C=bt\tfrac{1}{2}[f_c+f_c(1-t/kd)]+b'(kd-t)\tfrac{1}{2}f_c(1-t/kd).$$

Since $T=C$, their values may be equated; the resulting equation leads to

$$k=\frac{np+\frac{1}{2}\left[\frac{b'}{b}-\frac{b'}{b}\left(\frac{t}{d}\right)^2+\left(\frac{t}{d}\right)^2\right]}{np+\frac{b'}{b}-\frac{b'}{b}\frac{t}{d}+\frac{t}{d}}. \quad . \quad . \quad . \quad . \quad (2)$$

The moment of inertia of the concrete-steel section, the steel area being weighted n-fold, is given by

$$I=bd^3\left[k^3-\left(1-\frac{b'}{b}\right)\left(k-\frac{t}{d}\right)^3+\frac{b'}{b}(1-k)^3+3pn(1-k)^2\right]\tfrac{1}{3},$$

and if β be used to denote this coefficient of bd^3, then

$$\beta=\tfrac{1}{3}\left[k^3-\left(1-\frac{b'}{b}\right)\left(k-\frac{t}{d}\right)^3+\frac{b'}{b}(1-k)^3+3pn(1-k)^2\right]. \quad (3)$$

Example.—A T-beam rests on end supports 10 ft. apart and sustains loads of 5000 lbs. at its third points. The dimensions of the section are $b=16$ in., $b'=8$ in., $d=10$ in., and $t=3\tfrac{1}{4}$ in.; and the reinforcement consists of three $\tfrac{3}{4}$-in. square bars. What is the probable deflection due to the load?

Solution. The steel ratio is .011; and with $n=8$, eq. (2) gives $k=.485$, and eq. (3) gives $\beta=.0835$. Now for loads at third points, $c_1=23/1296$; hence

$$D=\frac{23}{1296}\,\frac{10{,}000\times120^3}{30{,}000{,}000\times16\times10^3}\,\frac{8}{0.0835}=.06\text{ in.}$$

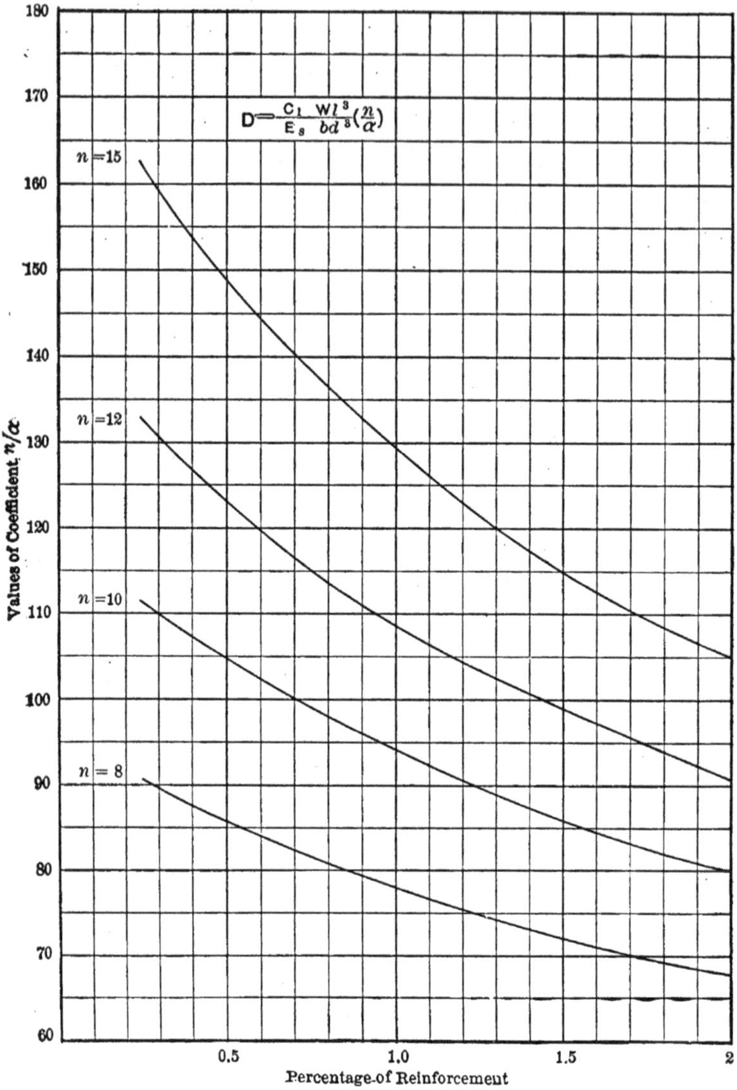
$D = \frac{C_1}{E_s} \frac{Wl^3}{bd^3} \left(\frac{n}{\alpha}\right)$
n =15
n =12
n =10
n = 8
Values of Coefficient n/α
180
170
160
150
140
130
120
110
100
90
80
70
60
0.5
1.0
1.5
2
Percentage of Reinforcement

FIG. 37f.

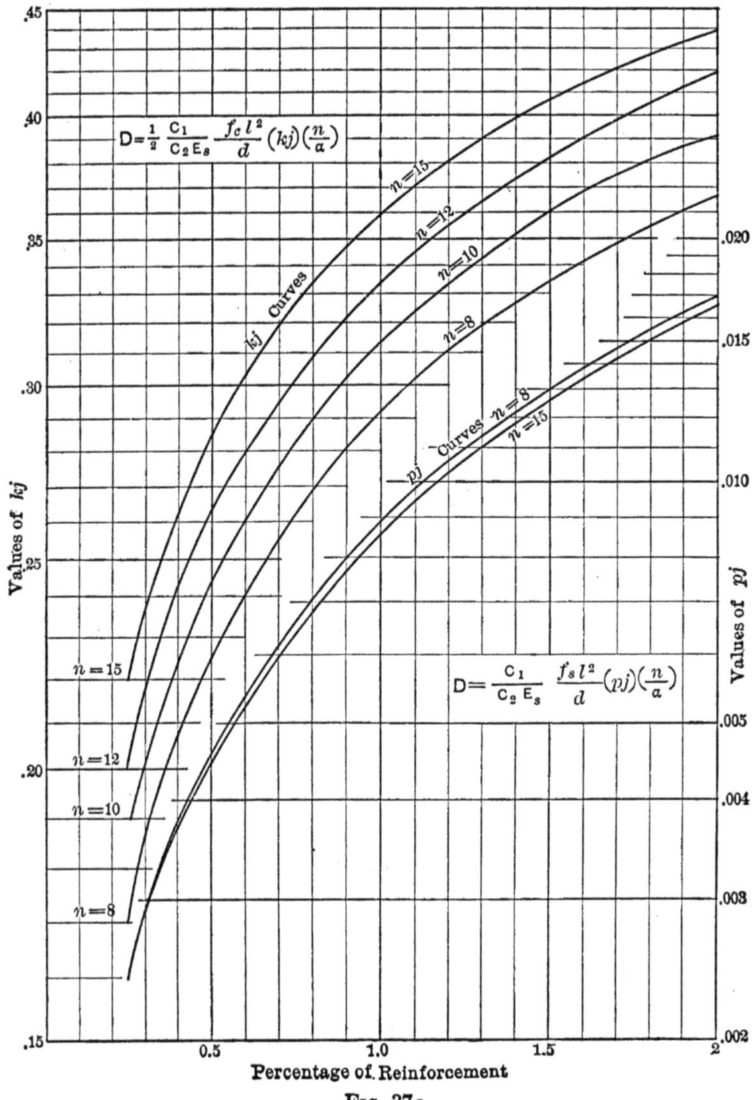
.45
.40
.35
.30
.25
.20
.15
D=½ C1/C2Es · fcl²/d (kj)(n/a)
n=15
n=12
n=10
n=8
kj Curves
pj Curves n=8
n=15
n=15
n=12
n=10
n=8
D= C1/C2Es · f's l²/d (pj)(n/a)
.020
.015
.010
.005
.004
.003
.002
0.5
1.0
1.5
2
Percentage of Reinforcement
Values of kj
Values of pj

Fig. 37g.

SCHEDULE OF COEFFICIENTS.

$D=c_1Wl^2/EI$ $M=c_2Wl$	c_1	c_2	In millionths.	
			$\frac{c_1}{*E_s}$	$\frac{c_1}{*c_2E_s}$
	$\frac{1}{3}$	-1	.0111	.0111
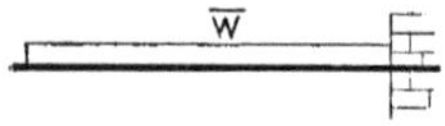	$\frac{1}{8}$	$-\frac{1}{2}$	.00417	.00834
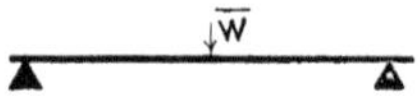	$\frac{1}{48}$	$\frac{1}{4}$	.000694	.00278
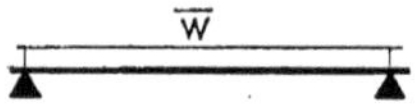	[illegible]	$\frac{1}{8}$	.000434	.00347
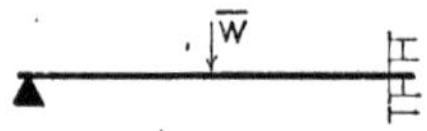	.00932	$-\frac{3}{16}$	.000301	.00160
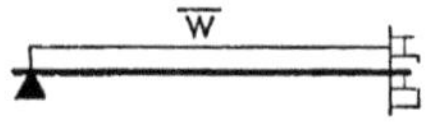	.0054	$-\frac{1}{8}$	.000180	.00144
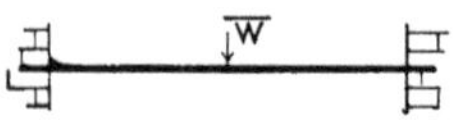	$\frac{1}{192}$	$\pm\frac{1}{8}$	.000173	.00139
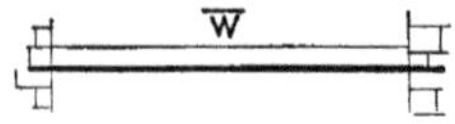	$\frac{1}{384}$	$-\frac{1}{12}$	.000087	.00210

* For $E_s=30{,}000{,}000$ lbs/in² in this schedule.

93. Strength of Columns.—Concrete columns need rarely be calculated as long columns. In ordinary construction the ratio of length to least width will seldom exceed 12 or 15, while the results of tests indicate little or no difference in strength for ratios up to 20 or 25. It will be desirable then to determine first the strength of a reinforced column considered as a short column. If the conditions require it a general column formula may then be applied to provide for cases where the length is excessive.

94. *Methods of Reinforcement.*—Columns are reinforced in two ways: (1) by means of longitudinal rods extending the full length of the column, and (2) by means of bands or spirally wound metal. In the first case the steel aids by carrying a part of the load directly, the stresses in the two materials being proportional to their moduli of elasticity. In the other case the steel supports the concrete laterally, preventing lateral expansion to a greater or less degree, and thus strengthening the concrete. Usually both methods are more or less combined, the longitudinal rods being frequently bound together at intervals by circumferential bands of some sort, and on the other hand hoops or spiral wire being conveniently held in place by longitudinal rods. Experiments show that both types of reinforcement are effective in raising the ultimate strength of a column, but conclusive results have not been reached as to the true relative effect of different types and amounts of reinforcement.

95. *Columns with Longitudinal Reinforcement.*—As long as the steel and concrete adhere the relative intensities of stress in the two materials will be as their moduli of elasticity, using the modulus as explained in Art. 24.

Let A denote total cross-section of column;
- A_c " cross-section of concrete;
- A_s " cross-section of steel;
- p " ratio of steel area to total area $=A_s/A$;
- f_c " stress in concrete;
- n " ratio of moduli of steel and concrete at the given stress $f_c, =E_s/E_c$;
- P " total strength of a plain column for the stress f_c;
- P' " total strength of a reinforced column for the stress f_c.

Then $$P=f_cA \quad \text{(a)}$$

and $$P'=f_cA_c+f_sA_s=f_c(A-pA)+f_cnpA,$$

whence $$P'=f_cA[1+(n-1)p], \quad (1)$$

from which also

$$\frac{P'}{P}=1+(n-1)p. \quad (2)$$

The relative increase in strength caused by the reinforcement is

$$\frac{P'-P}{P}=(n-1)p. \quad (3)$$

The elastic limit of the steel, if low, may affect the ultimate strength of the column. The value of P' is then

$$P'=f_cA_c+f_{el}A_s, \quad (4)$$

in which f_{el} is the elastic limit strength of the steel. (See Chapter IV for further discussion of this question.)

Eq. (2) is convenient to use in determining the relative strength of a reinforced as compared to a plain concrete column for a given percentage of steel. Thus if $p=1\%$ and $n=15$, we have $\frac{P'}{P}=1+0.14=1.14$. Thus a reinforcement of 1% increases the strength by 14%.

From these relations it is seen that the relative increase in strength caused by a given amount of reinforcement depends on the value of n and is greater the larger n is.

The economy of steel reinforcement is also dependent upon the working stresses permissible in the concrete since

$f_s = nf_c$. The following table shows the various working stresses in the steel corresponding to various values of working stress in the concrete and to various values of the modulus E_c; there is given also the percentage increase in strength for each one per cent of steel.

Table No. 6.

LONGITUDINAL REINFORCEMENT OF COLUMNS.

f_c, lbs./in.²	E_c, lbs/in²	Ratio of Moduli, n	f_s, lbs/in²	Percentage Increase in Strength for each 1% Reinforcement.
300	750,000	40	12,000	39
	1,000,000	30	9,000	29
	1,500,000	20	6,000	19
	2,000,000	15	4,500	14
400	1,000,000	30	12,000	29
	1,500,000	20	18,000	19
	2,000,000	15	6,000	14
	2,500,000	12	4,800	11
500	1,000,000	30	15,000	29
	1,500,000	20	10,000	19
	2,000,000	15	7,500	14
	2,500,000	12	6,000	11
600	1,500,000	20	15,000	19
	2,000,000	15	10,000	14
	2,500,000	12	7,200	11
	3,000,000	10	6,000	9
800	2,000,000	15	12,000	14
	2,500,000	12	9,600	11
	3,000,000	10	8,000	9
	3,500,000	8.6	6,900	7.6

From this table the relation among the various quantities may be clearly appreciated. It is to be noted that the working stresses in the steel must be relatively low except in the unusual combination of high working stresses in the concrete with low modulus. High-grade concrete, permitting high working stresses, will have a high modulus. For further discussion of the relations of working stresses, see Chapter V.

Examples.—(1) What will be the safe strength of a column 15″×15″ in cross-section which is reinforced with 1.5% of steel, the working stress in the concrete being 400 lbs/in². Take $n=15$.

From eq. (1) we have

$$P'=400\times15\times15\times\left(1+14\times\frac{1.5}{100}\right)=90,000(1+0.21)=108,900 \text{ lbs.}$$

The strength of the plain concrete column would be 90,000 lbs., and the relative increase in strength is 21%. The stress in the steel would be $15\times400=6000$ lbs/in².

(2) The area of a column is 120 sq. in., load to be carried is 60,000 lbs., and working stress on the concrete is 400 lbs/in². What percentage of steel will be required? Take $n=15$.

The safe strength of a plain concrete column would be $120\times400=48,000$ lbs. Hence, from eq. (2), $\frac{P'}{P}=\frac{60}{48}=1+(15-1)p$. Hence

$$p=\left(\frac{60}{48}-1\right)\div14=1.8\%.$$

96. *Columns with Hooped Reinforcement.*—Whenever a material subjected to compression in one direction is restrained laterally, then lateral compressive stresses are developed which tend to neutralize the effect of the principal compressive stresses and thus to increase the resistance to rupture. Were the compressive stresses equal in all directions there would be no rupture (as there would be no shear). The strengthening effect of lateral banding depends then upon the rigidity of the bands, that is, upon the amount of steel used and its closeness of spacing. Its elastic limit may also affect the ultimate strength of the column.

On the basis of the relative lateral and horizontal deformation of the concrete (Poisson's ratio) it is possible to deduce a theoretical relation between the lateral and the longitudinal stresses, and thence the portion of the longitudinal stress remaining unbalanced. Let μ=Poisson's ratio, f_c=unbalanced or excess of longitudinal over lateral compressive unit stress, f_c'=total longitudinal unit stress, f_s=unit tensile stress in steel, p = steel ratio (reinforcement to be closely spaced). We find approximately

$$f_c' = f_c\left(1 + \frac{\mu n p}{2}\right) \quad . \quad . \quad . \quad . \quad . \quad . \quad . \quad . \quad (1)$$

and

$$f_s = \mu n f_c.^* \quad . \quad . \quad . \quad . \quad . \quad . \quad . \quad . \quad . \quad . \quad (2)$$

Recent experiments by Talbot indicate that Poisson's ratio for concrete is quite small, probably not greater than $\frac{1}{10}$ or $\frac{1}{8}$.

---

* *Demonstration.* (See Johnson's "Materials of Construction".)—Let μ = Poisson's ratio; p = steel ratio considered as a thin cylinder of equivalent area surrounding the concrete; A_s = cross-section of this steel cylinder; r = radius. Then

$$A_s = p\pi r^2 \quad \text{and} \quad \text{thickness of cylinder} = \frac{p\pi r^2}{2\pi r} = p\frac{r}{2}.$$

With no steel banding the stress f_c' would cause a proportionate lateral swelling of $\frac{f_c'}{E_c}\mu$. If the actual stress in the steel is f_s then the compression per sq. in. developed in the concrete by the steel reinforcement $= f_s p\frac{r}{2} \div r = \frac{f_s p}{2}$. This compression caused by the banding is equal in all horizontal directions, and has the same effect on distortion as two pairs of equal compressive forces acting on two sets of faces of a cube. The resultant lateral compression due to these horizontal forces is equal to $\frac{f_s p}{2E_c}(1-\mu)$. Combining this compression with the lateral swelling caused by f_c' we have the net lateral deformation equal to $\frac{f_c'}{E_c}\mu - \frac{f_s p}{2E_c}(1-\mu)$. This net deformation must equal the actual deformation in the steel under the stress f_s, which is $\frac{f_s}{E_s}$ or $\frac{f_s}{nE_c}$. Hence we have

$$\frac{f_c'}{E_c}\mu - \frac{f_s p}{2E_c}(1-\mu) = \frac{f_s}{nE_c}.$$

A part of f_c' may be considered to be balanced by the lateral compression of $\frac{f_s p}{2}$; it is the unbalanced portion only which is significant. Call this unbalanced portion f_c; then $f_c' = f_c + \frac{f_s p}{2}$. Then eliminating f_s from these two equations we find for f_c' the value

$$f_c' = f_c\left(1 + \frac{np\mu}{np(1-2\mu)+2}\right) \quad . \quad . \quad . \quad . \quad . \quad . \quad . \quad . \quad (a)$$

At the latter value eqs. (1) and (2) would become $f_c'=f_c(1+np/16)$, and $f_s=\frac{1}{8}nf_c$. Comparing these equations with those of Art. 95 it would appear that within the limit of elasticity the hooped reinforcement is much less effective than longitudinal reinforcement; in fact it would seem that very little stress can be developed in the steel under elastic conditions as here assumed. Such reinforcement may, however, be quite effective in increasing the ultimate strength of a column.

Results of tests appear to accord in a general way with these theoretical relations. Hooped columns have a relatively large deformation, reaching at an early stage a deformation equal to the maximum for plain concrete. Under further loading the concrete is prevented by the banding from actual failure, but continues to compress and to expand laterally, increasing the tension in the bands, the elasticity of the bands rendering the column in large degree still elastic. Final failure occurs upon the breakage of the bands or their excessive stretching. Banded columns thus exhibit a toughness or ductility much greater than other forms, but without a corresponding increase in stiffness under lower loads. Ultimate failure is likely to be long postponed after the first signs of rupture, and the column will sustain greatly increased loads even after the entire failure of the shell of concrete outside the bands.

Considère has made extensive theoretical and experimental investigations of hooped columns, from which he concludes that

---

We also have

$$f_s=\frac{2\mu n}{np(1-2\mu)+2}f_c. \qquad (b)$$

For ordinary values of p eqs. (a) and (b) are reduced approximately to

$$f_c'=f_c\left(1+\frac{\mu np}{2}\right) \qquad (1)$$

and

$$f_s=\mu nf_c. \qquad (2)$$

the ultimate strength is given by the formula

$$P' = f_c A + 2.4 f_{el} p A, \quad \ldots \ldots \quad (5)$$

in which f_c is the strength of concrete and f_{el} is the elastic-limit strength of the steel. This formula virtually counts the steel worth 2.4 times as much as in longitudinal reinforcement. (For further discussion see Chapter IV.)

96a. *Columns with both Longitudinal and Hooped Reinforcement.*—From the theoretical consideration of the preceding article it would appear that the addition of bands or hoops to columns having longitudinal reinforcement would not have a large effect upon the deformation of such columns up to the point of failure of the columns without hooping. The effect of such hooping would be rather to increase the ultimate possible deformation and, to a less extent, the ultimate strength. It would thus insure the integrity of the concrete up to a deformation corresponding to the elastic limit of the longitudinal steel, but below such limit it can hardly come much into action. Results of tests discussed in Chapter IV bear out these conclusions.

96b. *Long Columns.*—For columns of such a length that flexural strength and stiffness become of importance (more than about 20 diameters) the working stress should be reduced by the use of a long column formula. Until more data are available from tests, the authors would propose the use of the theoretical form of Rankine's formula which is, for pivoted ends,

$$P'' = \frac{P'}{1 + \frac{f}{\pi^2 E}\left(\frac{l}{r}\right)^2}, \quad \ldots \ldots \quad (1)$$

in which P' is the strength of a short column, and f and E are the ultimate strength and the modulus of elasticity of the concrete. This formula gives results materially too low when applied to steel columns, but it is believed that it is not too conservative for material like concrete. The value of f/E should, for conservative design, be taken at its maximum rather than minimum value, say $^1/_{1000}$, giving finally the formula

$$P''=\frac{P'}{1+\frac{1}{10,000}\left(\frac{l}{r}\right)^2}. \quad \ldots \ldots \quad (2)$$

For fixed ends the constant in the denominator may be made $^1/_{20000}$ giving

$$P''=\frac{P'}{1+\frac{1}{20,000}\left(\frac{l}{r}\right)^2}. \quad \ldots \ldots \quad (3)$$

It may be observed that if formulas be derived from the Rankine-Gordon formulas for steel columns, by taking account of the difference in ultimate strength and modulus of elasticity of the materials, the resulting formulas would contain constants of very nearly the same value as for steel, namely, $^1/_{18000}$ and $^1/_{36000}$. The low values above given represent a larger degree of safety, which is to be desired. For a value of l/r of 100, or a length of about 30 diameters, the formula for pivoted ends gives an ultimate strength of two-thirds that of the short column. Because of the fact that it is difficult to secure thoroughly homogeneous concrete, and that variations in quality will affect the strength of long columns more seriously than any other structural form long columns should generally be avoided.

CHAPTER IV.

TESTS OF BEAMS AND COLUMNS.

BEAMS.

97. Methods of Failure of a Reniforced-concrete Beam.—A reinforced-concrete beam tested to destruction will usually fail in one of three ways:

(*a*) By the yielding of the steel at or near the section of maximum bending moment.

(*b*) By the crushing of the concrete at the same place.

(*c*) By a diagonal tension failure of the concrete at a place where the shear is large.

Methods (*a*) and (*b*) may be called "moment" failures. Method (*c*) is sometimes called a shear failure, but this term is somewhat misleading, as the concrete in such cases does not fail by shearing.

(*a*) As a beam is progressively loaded and the steel has reached its yield point any further load will rapidly increase the deformation. The effect of this is to open up large cracks in the tension side and to raise the neutral axis. This causes a rapid increase in the compressive stress in the concrete and ultimate failure soon occurs by the concrete crushing. Such yielding may also result in final failure by diagonal tension if large shear exists near the place of maximum moment. In this case the primary cause of failure is the yielding of the steel and such failure may properly be called a tension failure. The additional load carried after the yield point is reached depends on the excess strength of the concrete, position of loads, and other causes, but it is usually not large and cannot be safely considered. The yield point of the steel may therefore be considered its ultimate strength for reinforcing purposes.

(*b*) If the beam is relatively long and the amount of steel is sufficient so that the crushing strength of the concrete is reached before the yield point of the steel, a failure by crushing is likely to result. In this case tension cracks may appear, but will not become large. Fig. 38, (*a*) and (*b*), illustrates methods of failure (*a*) and (*b*) respectively.

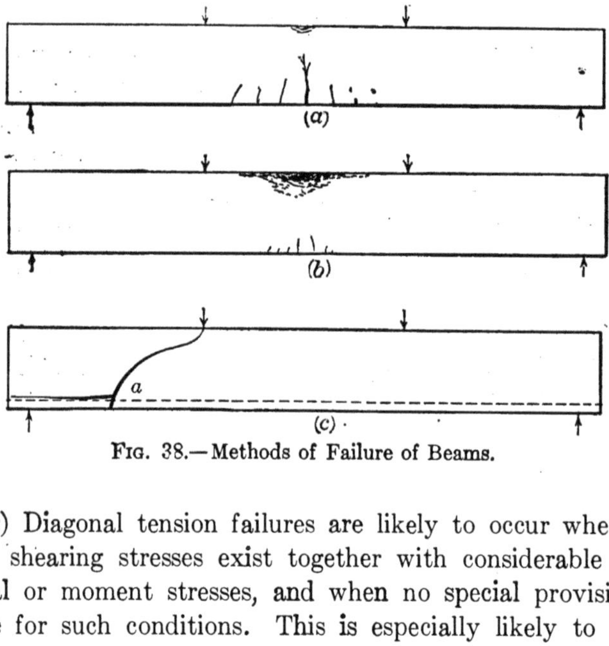

FIG. 38.—Methods of Failure of Beams.

(*c*) Diagonal tension failures are likely to occur whenever large shearing stresses exist together with considerable horizontal or moment stresses, and when no special provision is made for such conditions. This is especially likely to occur in beams of relatively great depth, beams having a ratio of depth to length of more than about 1:10 being likely to fail in this way if no special provision is made for web reinforcement.

Fig. 38, (*c*), illustrates the typical diagonal tension failure where only horizontal bars are used. The initial crack forms at *a*. This gradually extends upwards in an inclined line and

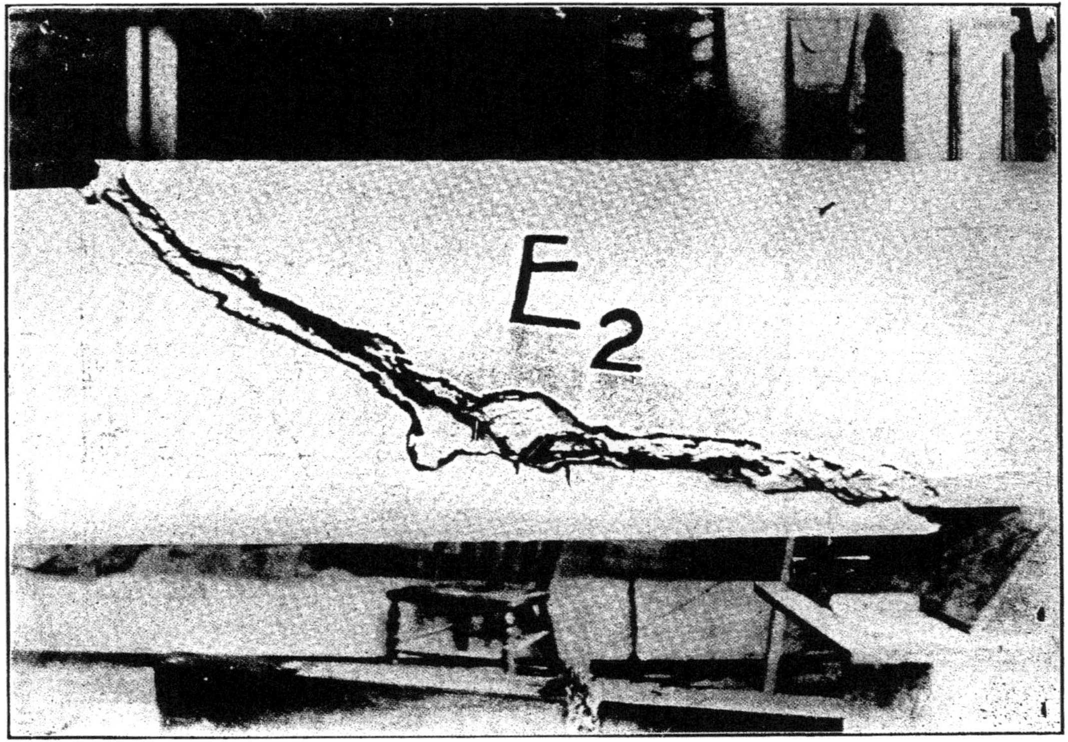

FIG. 39.

FIG. 40.

a little later the concrete begins to fail in a horizontal tension crack just above the rods, progressing from *a* towards the end of the beam. Tension along this line is brought about by the new conditions existing after the concrete has become cracked along the diagonal line and the normal diagonal tension has thus ceased to act. Usually this horizontal crack rapidly extends to the end of the beam and the failure is complete. In other cases the diagonal crack may extend to the top of the beam, allowing the part on the right to drop down and causing final failure. In such a case the concrete on the left may remain intact. Figs. 39 and 40 are photographs representing "diagonal-tension" failures.

A rupture of the concrete on a diagonal line also causes an increase in the stress on the rod at *a*, as shown more fully in Art. 108. This may result in a failure of bond, especially if the support is too near the end of the beam.

Final failure thus often results from stresses which are developed after initial failure has occurred, and while the cause of final failure is important from the standpoint of ultimate strength, yet of more importance in design is the initial failure and its cause. Other conditions besides those already mentioned may influence final failure so as often to mislead the observer as to the cause of the initial failure.

98. *Minor Causes of Failure.*—Slipping of the bars may cause failure, but under usual conditions it will not occur; and as it can readily be obviated by proper construction it need not be considered as limiting the strength of the beam. Failure by the shearing of the concrete near the support is possible where the load is very close thereto, but as the shearing strength of concrete is about one-half the crushing strength, such failures are exceedingly unlikely and need rarely be considered. The usual so-called "shear" failures are in reality diagonal-tension failures.

99. **Tests of Beams Giving Steel-tension Failures.**—The diagrams of Figs. 41 and 42 present in a roughly classified form results of the most important tests on reinforced-concrete beams

in which the failure appears to have been caused primarily by the yielding of the steel. In such a case the strength of the beam is directly proportional to the elastic-limit strength of the steel, and hence the tests have been classified as nearly as practicable with respect to this limit. The tests are thus divided into four groups according to values for the elastic limit as given in the diagrams. On each of the diagrams are drawn theoretical curves of strength using values for the steel stress corresponding to the elastic limit for the group. The full line is based upon the parabolic law of stress variation, the full parabola being used; the dotted line is based upon the straight-line law of stress variation. The value of n was taken at 15.*

Considering the nature of the material and of the tests the agreement between theory and experimental results is very satisfactory. It is to be expected that the theoretical values should represent minimum rather than average results, since the strength of a beam as determined by the elastic limit of the steel should be at least equaled, and generally slightly exceeded, in a test if failure does not occur in some other way. If the conditions are favorable the strength may considerably exceed that corresponding to the elastic limit of the steel, and in a few tests the steel has been pulled apart before complete collapse has taken place. Such excess of strength cannot be counted upon, however, as is well indicated in the diagrams.

---

* The sources of information are as follows:

1. Boston Transit Commission, Fourth Annual Report, 1904.
2. Bulletins Nos. 1 and 4, University of Illinois, Engineering Experiment Station
3. Jour. West. Soc. Eng., Vol. X, 1905, p. 705 (C., M. & St. P. R'y Co.'s tests).
4. Jour. West. Soc. Eng., Vol. IX, 1904, p. 239 (tests of M. A. Howe).
5. Bulletins No. 4, Vol. 3, and No. 1, Vol. 4, Engineering Series, University of Wisconsin, 1907.
6. Proc. Am. Soc. Test. Materials, Vol. IV, 1904, p. 508 (Univ. of Penn. tests).
7. Eng. Record, Vol. LI, 1905, p. 545 (Purdue Univ. tests).

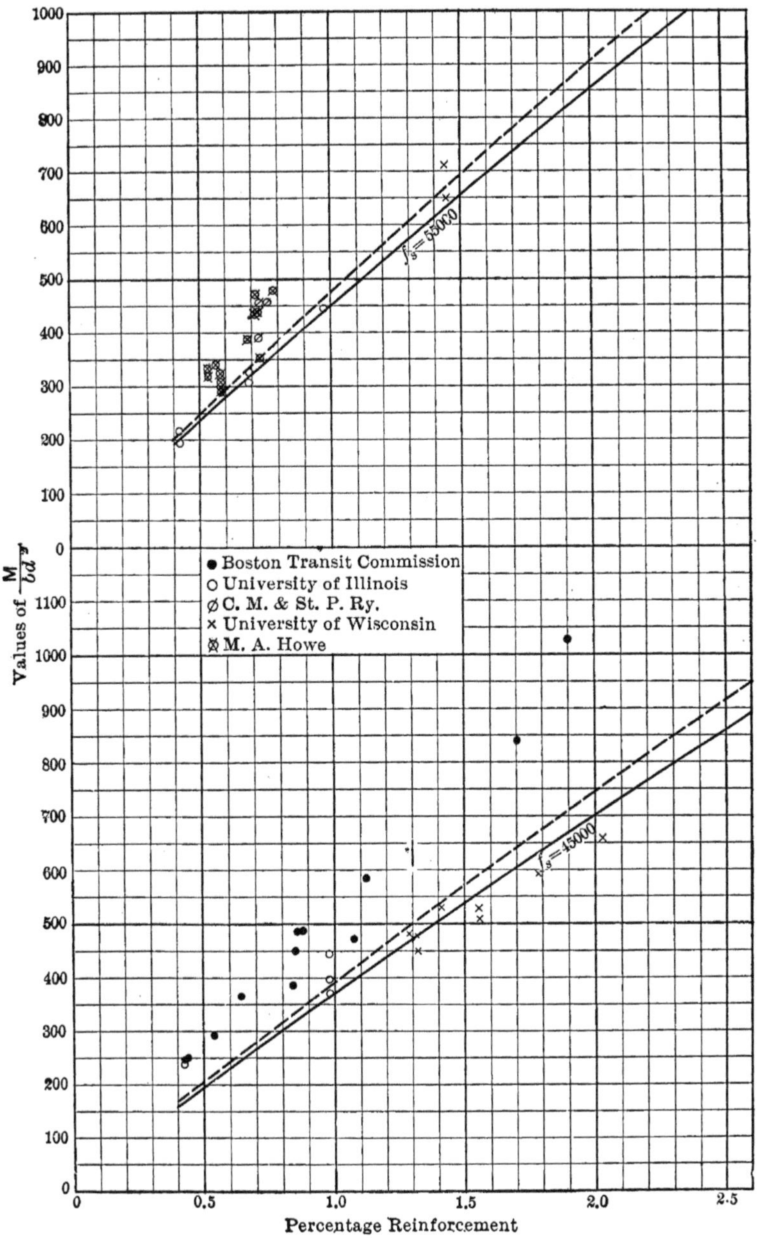

FIG. 41.—Steel-tenison Failures.

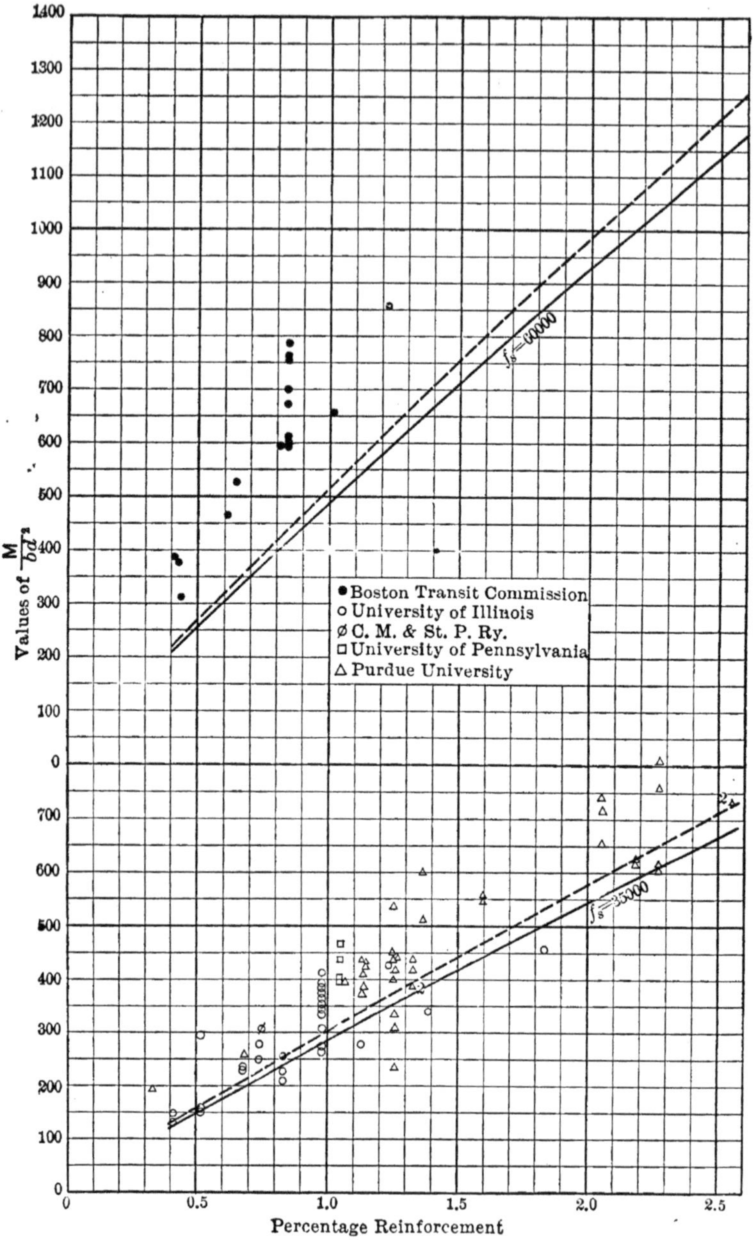

FIG. 42.—Steel-tension Failures.

In the tests of the Boston Transit Commission, which range uniformly high, the conditions were favorable, inasmuch as the beams were tested with center load. The concrete was also of very high grade, having a crushing strength of about 4000 lbs/in^2, thus enabling the steel to elongate very considerably before final failure occurred through the crushing of the concrete.

No distinction has been made in these diagrams between the different grades of concrete employed. Variations in concrete will affect the results only by slightly affecting the position of the neutral axis, and hence the resisting moment of the steel, and by postponing somewhat the final failure, as noted above.

Later tests by Talbot * gave tension failures in which the calculated steel stress closely agreed with the yield-point of the metal. His tests also showed a somewhat higher strength for a center load than for loads at two or more points.

100. *Results from Individual Tests.*—Numerous tests of beams have been made in which extensometers have been used to measure distortions so that the deformation of the steel and of the extreme fiber of the concrete could be calculated and the neutral axis determined. Results of such measurements of deformations and also of center deflections are shown in Figs. 43 and 44 for two typical beams. In Fig. 43 the proportions were such that the failure occurred by diagonal tension; neither the steel nor the concrete was stressed to the limit of failure. During the first stage of the test, up to a load of about 2500 pounds, the deformations in both steel and concrete are proportional to the loads. Up to this point the tension deformation has not been great enough to begin to rupture the concrete, but with increasing loads and deformations the concrete begins to fail, as shown by the appearance of minute cracks (the "water-marks" discussed in Art. 42), indicated on the diagram by the letters *W M*. The deformation

---

* Bull. No. 14, Univ. of Ill.

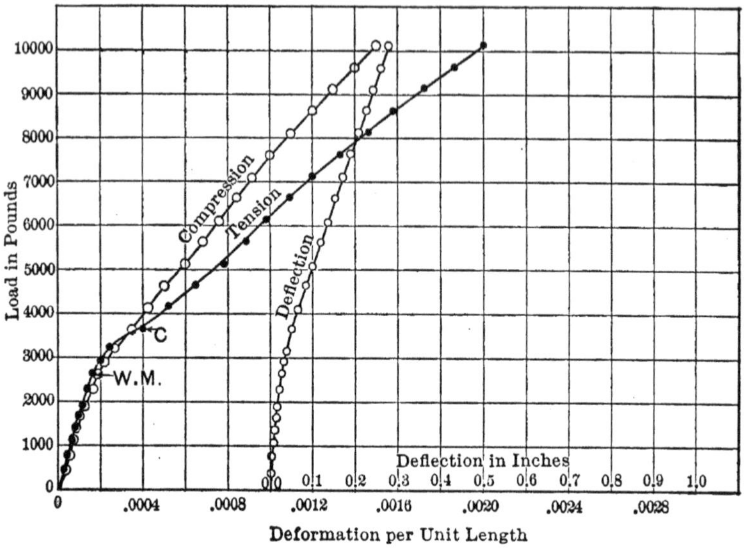
Load in Pounds
10000
9000
8000
7000
6000
5000
4000
3000
2000
1000
0
Compression
Tension
Deflection
C
W.M.
Deflection in Inches
0.0 0.1 0.2 0.3 0.4 0.5 0.6 0.7 0.8 0.9 1.0
0 .0004 .0008 .0012 .0016 .0020 .0024 .0028
Deformation per Unit Length

FIG. 43.

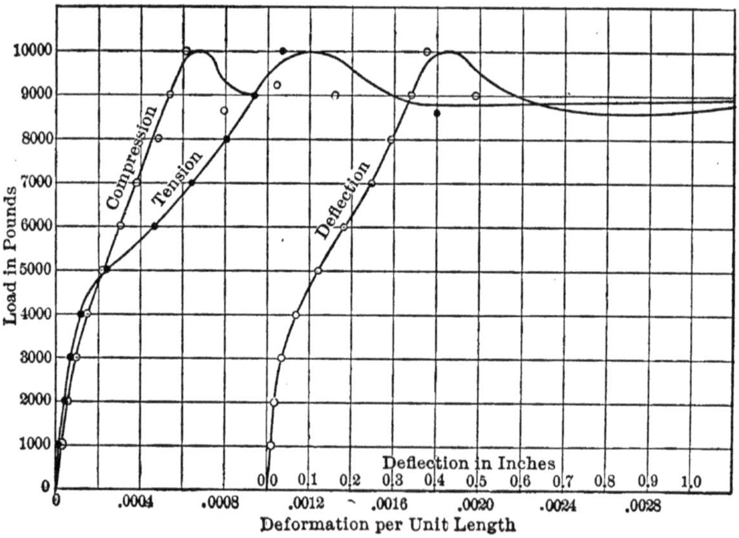
Load in Pounds
10000
9000
8000
7000
6000
5000
4000
3000
2000
1000
0
Compression
Tension
Deflection
Deflection in Inches
0.0 0.1 0.2 0.3 0.4 0.5 0.6 0.7 0.8 0.9 1.0
0 .0004 .0008 .0012 .0016 .0020 .0024 .0028
Deformation per Unit Length

FIG. 44.

at the first "water-mark" in this case was about .00018, corresponding to a stress of 270 lbs/in^2, assuming a modulus of elasticity of 1,500,000. The first visible crack appeared at the point marked C.

The failure of the concrete in tension takes place somewhat gradually and causes a gradual increase in the rate of deformation as indicated by the curved part of the diagram between loads of 2500 and 4000 pounds. After the concrete has ceased to offer any considerable resistance in tension the deformations again become nearly proportional to the loads, but at a different ratio from that obtaining previously, giving nearly straight lines for both steel and concrete—in this case to the end of the test.

In Fig. 44 the amount of steel was small and a tension failure occurred. This is indicated by the great deformations at the end of the test. The curves in the early stages of the test are very similar, in general form, to those in Fig. 43.

In the case of a compressive failure the curve for compression shows an increased rate of deformation towards the end, somewhat similar to the diagram for simple compression.

101. Position of Neutral Axis and Value of n.—Reference to the analysis of Arts. 55 and 56 show that in the calculation of the strength of reinforced beams the determination of the position of the neutral axis is of prime importance. This being known, the strength can be determined with little uncertainty. In determining the position of the neutral axis eq. (1) of Art. 55 shows it to depend only upon the amount of steel used and upon the ratio E_s/E_c or n. The only element of uncertainty is the value of E_c. This is the modulus of elasticity of the concrete in compression and it might be considered sufficient to take the value as determined in the ordinary compression test. However, the variation of E_c for different stresses, and the effect of the tensile stresses in the concrete below the neutral axis (a stress which is properly not allowed for in the resisting moment), make it desirable to compare experimental determinations of the neutral axis with theoretical position for various assumed values of E_c or of n.

Many experiments have been made in which the position of the neutral axis has been determined. Among the best are those by Bach,* made on 1:4 gravel concrete, 6 to 7 months old. The beams were 2 m. long and 30 cm. deep and were loaded at quarter points. The observed positions of the neutral axis (values of k), at various loads are given in the following table:

POSITION OF NEUTRAL AXIS.

(VALUES OF k.) (BACH.)

No. of Beams.	Percent Reinf.	Values of k for Various Proportions of Ultimate Load.					Theoretical Values.	
		Initial	¼ Load.	½ Load.	¾ Load.	Full Load	for $n=12$	for $n=15$
5	0.54	.56	.53	.43	.33	.31	.30	.33
3	0.43	.59	.55	.47	.31	.28	.27	.30
5	1.32	.59	.55	.45	.44	.46	.43	.46

The theoretical positions are also given for $n=12$ and $n=15$. The value of E for this concrete, at a load of 600 lbs/in^2, as determined by compression tests, was 3,300,000 lbs/in^2.

Similar tests have been made on T-beams by Bach and also by Withey.† All of Bach's tests and those on T-beams by Withey are plotted in Fig. 44*a*.

In Figs. 45 and 46 are plotted in a different form results of various tests on rectangular beams. On the diagrams are also plotted the theoretical positions of the neutral axis for various values of n. The full lines are based on the straight-line stress variation assumption, and the dotted lines on the assumption of a parabolic law in accordance with Professor Talbot's method (see Art. 65). The dotted lines have been drawn only for a single value of 15 for n. For a three-quarter load the dotted line for $n=15$ would coincide very closely with the full line for $n=20$. The value of q has been taken at $\frac{1}{4}$, $\frac{1}{2}$, and $\frac{3}{4}$, respectively.

* Mit. über Forsch. a. d. Gebiet des Ing.. 1907, 45–47.

† Bull. Univ. of Wis., 1908, Vol. 4, No. 2.

It will be noted that for the lower loads and the small percentages of steel the neutral axis is more uncertain and generally lower than for the higher loads and larger percentages. This is due to the relatively large influence of the tensile strength of the concrete in such cases. The T-beam

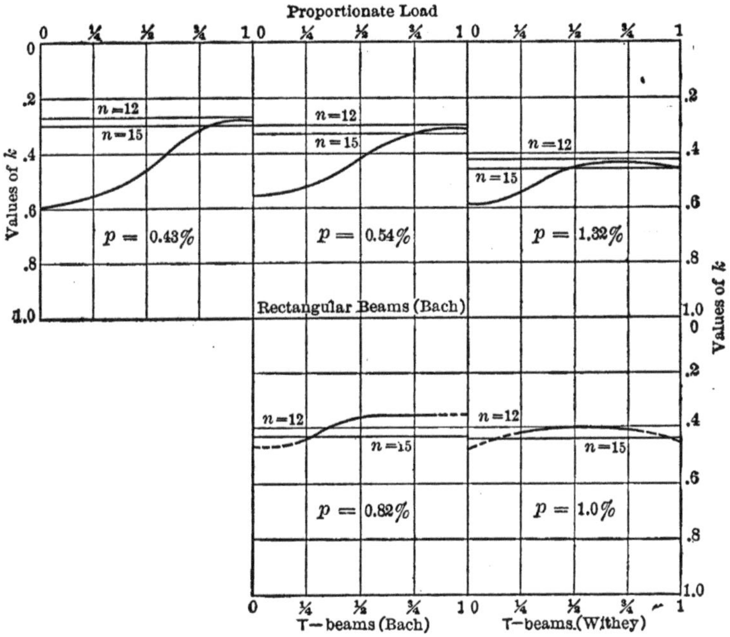

FIG. 44*a*.—Position of Neutral Axis.

tests show relatively little of this effect owing to the small area of concrete in tension.

From these results it appears that a value of 15 for n is not too large for calculations of strength of beams under the usual assumptions. This value is the one most generally used, but a value of 12 is also frequently employed. The value of 15 corresponds to a value of E_c of 2,000,000, which is somewhat low as determined by compressive tests. A value of $n=10$, corresponding to $E=3,000,000$, does not

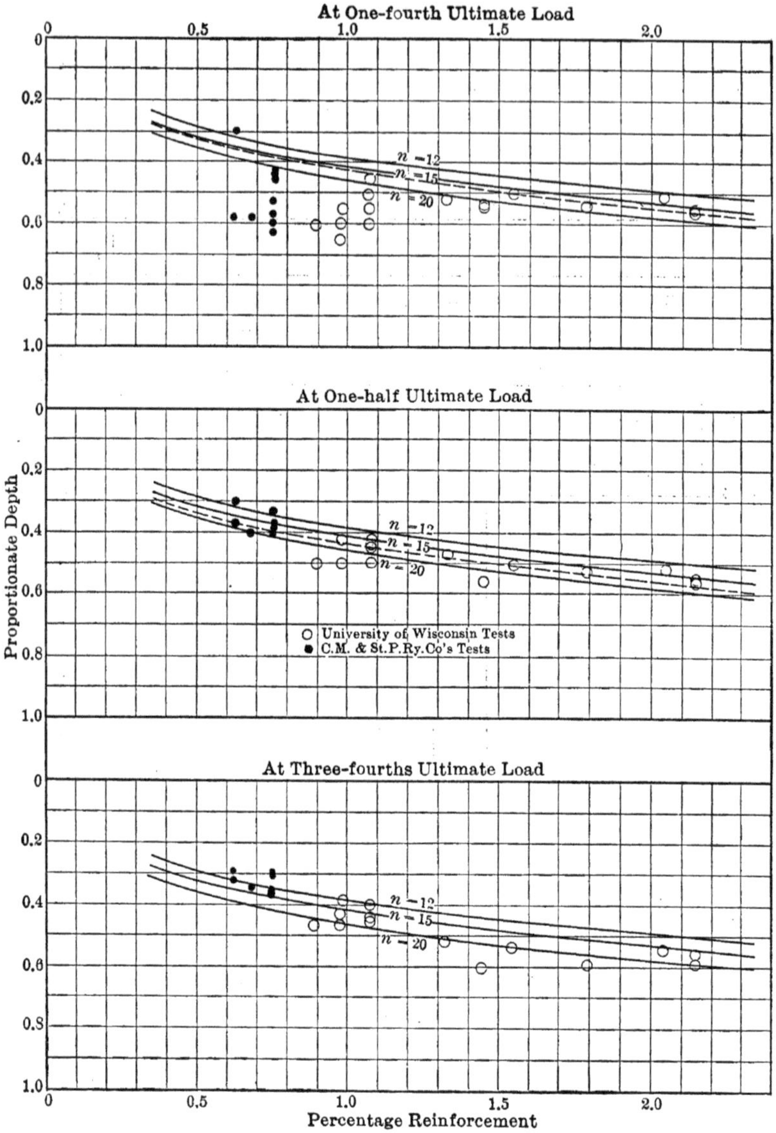

FIG. 45.—Position of Neutral Axis. (1:2:5 Concrete.)

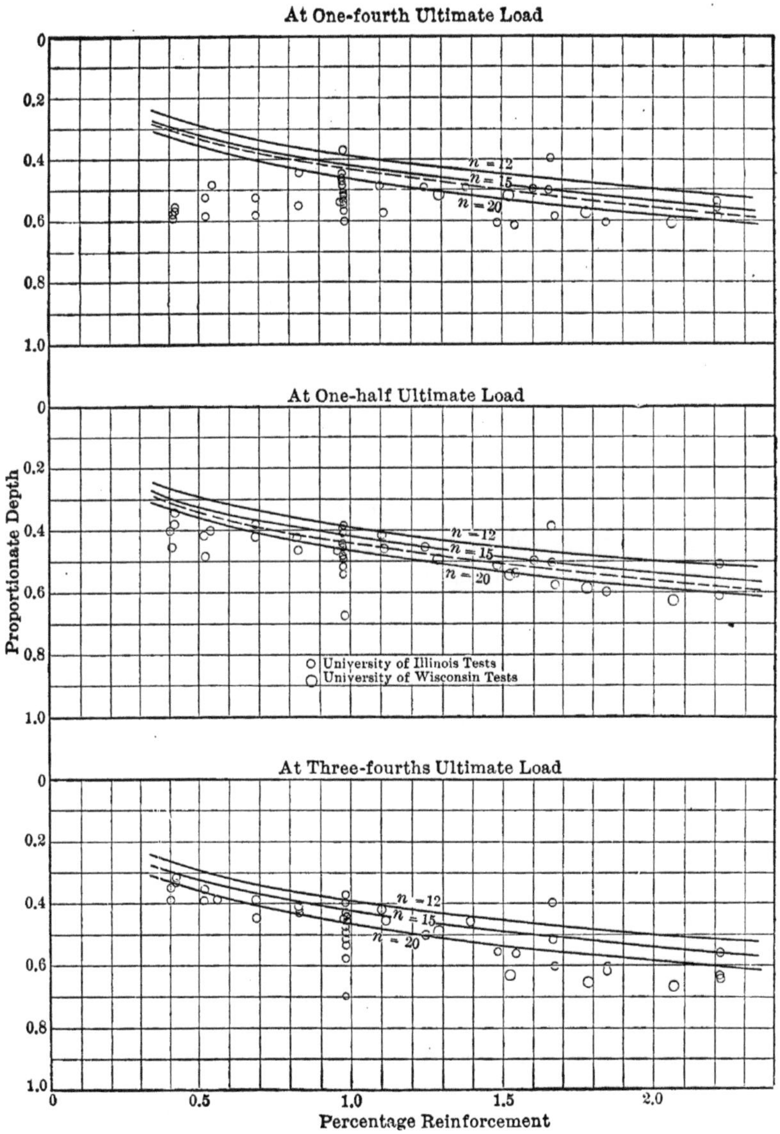

FIG. 46.—Position of Neutral Axis. (1:3:6 Concrete.)

accord with results from bending tests. If the comparison between measured and calculated positions of the neutral axis be made on the basis of the parabolic law of stress variation the results will differ considerably in the latter stages of the tests, but very slightly at the quarter load. It should be noted, however, that it is only with the high percentages of steel that the concrete stress reaches nearly to its ultimate value, and hence is the only condition where the full parabolic law can be expected to give consistent and rational results.

In some of the tests whose results are plotted here the concrete was cut away from the steel for the measured distance, leaving it exposed. The position of the neutral axis was very slightly affected.

102. Observed and Calculated Stresses in Steel.—Where the neutral axis is determined by extensometer measurements a check upon theoretical results can be obtained by calculating the stress in the steel in two ways: (1) from the observed deformations at the plane of the steel, and (2) from the known bending moment and known position of the neutral axis. In the first calculation the tensile strength of the concrete, which is neglected, causes some error, especially under light loads, and in the second calculation the exact position of the centroid of pressure in the concrete, especially in the later stages of the test, is to a small degree uncertain, but as the variation in steel stress is only about 2%, using the two extreme assumptions of stress variation, this source of error is not great. Table No. 7 presents several representative results derived from such calculations. The stresses calculated from moments are based on the assumption that the concrete takes no tension.*

Tests have been made at the University of Illinois and at the University of Wisconsin in which the rods have been exposed for a considerable distance along the center of the beam, and thus have been much less affected by any possible tensile stress in the concrete. Measurements of extension made in such

* For further data see Table No. 12, Art. 112.

cases show little variation from those made on the ordinary beam.

TABLE No. 7.

STRESSES IN STEEL REINFORCEMENT.

Authority	Per Cent Reinforcement.	Observed Position of Neutral Axis, *k*.	Calculated Stress in Steel, lbs/in².	
			From Moments.	From Extensions in Steel.
Talbot; *Bull. Univ. of Ill., 1906.*	.74	.410	33,100	36,000
	1.23	.470	35,000	36,000
	1.60	.501	29,500	35,400
	1.66	.505	30,600	30,000
	1.84	.606	25,600	27,200
	1.84	.552	28,300	30,000
Withey; *Bull. Univ. of Wis., 1907.*	2.9	.670	35,200	36,000
	2.9	.6[illegible]0	31,600	33,000

Considering the nature of such experiments the results obtained may be considered as according with theory very satisfactorily.

103. Compressive Stresses in Concrete in Beams and in Compression Specimens.—An important question relating to proper working stresses is whether the ultimate compressive strength of concrete in a beam is the same as determined by a direct compression test.

The results of certain tests indicate that the compressive strength and ultimate deformation in a beam may be somewhat greater than in a prismatic compressive piece; and it would seem that the differences in condition are sufficient to make such a difference possible. In a compressive specimen the material is free to shear in any direction, thus limiting the strength of the specimen to its weakest shearing plane. In a beam the (shear) failure is practically confined to planes perpendicular to the side of the beam. Furthermore, in a beam the material is not subjected to the secondary stresses due to possible poor bedding of the test specimen or nonparallel motion of the testing machine, as is the case in compression tests.

In most of the tests reported both the beams and the accompanying compression specimens have been hardened in air. Under these conditions there is usually some drying-out effect resulting in a weaker concrete than if hardened in water, and owing to the smaller dimensions of the compressive specimens the effect will be greater with them than with the relatively large beams. Many tests have therefore shown a compressive strength of concrete in the beam considerably greater than results obtained on cubes. When both beam and cube are hardened in water the results do not differ greatly. The following are some results obtained on tests made relative to this point.* The beams were $5\frac{3}{4}'' \times 6''$ net section and 5 ft. span. They were reinforced with $2\frac{1}{2}\%$ of steel and gave compressive failures. The cubes were 4 inches in dimension and the cylinders 6″ in diameter by 18″ high.

		Stress in Concrete at Rupture, lbs/in².		
		Beam.	Cube.	Cylinder.
Hardened in air	1	1770	1187	1380
	2	1460	1350	1295
Hardened in water	3	1810	1450	1265
	4	1850	1750	1680

The stresses in the beams were calculated on the basis of the parabolic variation of stress, the neutral axis being determined by extensometers.

It will be seen that in case of the specimens hardened in air there is a marked difference in strength, but where hardened in water the difference is much less. The difference is hardly sufficient to warrant much consideration in the determination of working stresses.†

* Bulletin No. 1, Vol. 4, Engineering Series, University of Wisconsin, 1907.

† For further data see Table No. 12, Art. 112.

104. Conclusions Regarding Moment Calculations.—The comparison of experimental results with theoretical analysis herein given shows that the simple beam theory as generally employed, neglecting the tension in the concrete, can be used with confidence. In particular, the results appear to show that calculated on the basis of such theory the yield point (commonly called the elastic limit) of the steel may safely be taken as its ultimate strength in reinforced beams; that the crushing strength of concrete as determined by tests on cubes hardened under exactly similar conditions as the beams will be fully realized in the beam; that for working loads the straight-line law of stress variation is sufficiently exact; that the value of n may be taken at about 15, but that great accuracy in this respect is unnecessary; that for ultimate values, especially where the concrete is near failure, the parabolic assumption of stress variation may well be used.

105. Tests in which Failure Occurred by Diagonal Tension. *Influences Affecting Failure by Diagonal Tension.*—The strength of a beam in diagonal tension is not a simple function of the shear, but as shown in Art. 90 it depends also upon the horizontal tension or bending-moment stresses in the concrete. These will in turn depend upon the actual bending moment at the section of failure and the amount of horizontal reinforcement, a large percentage of reinforcement reducing the horizontal deformation and therefore the tension in the concrete and tending to strengthen the beam as regards failure in diagonal tension. The strength of the beam therefore depends upon the relation between shear and bending moment and upon the amount of reinforcement. The chief factor is, however, the shearing stress.

From the preceding considerations it is evident that the nature of the loading will influence the strength of the beam. Most structures are calculated for uniform or approximately uniform loading, and in experimental work two concentrated loads applied at the third points are commonly used as representing roughly the conditions which exist in the uniformly loaded beam. Fig. 47 represents the variation in moment and shear in a beam loaded at the third points, while Fig. 48 shows similar curves for a uniformly loaded beam. It is to be noted that in the first case maximum shear occurs where maximum moment exists, while in the latter case maximum shear occurs at the point of zero moment. In the former case

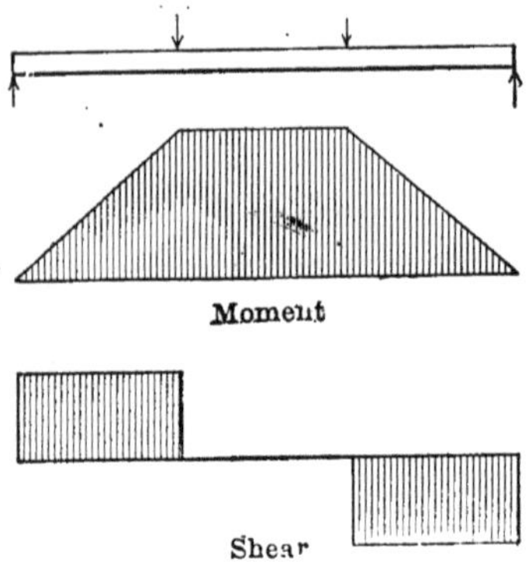

Fig. 47.

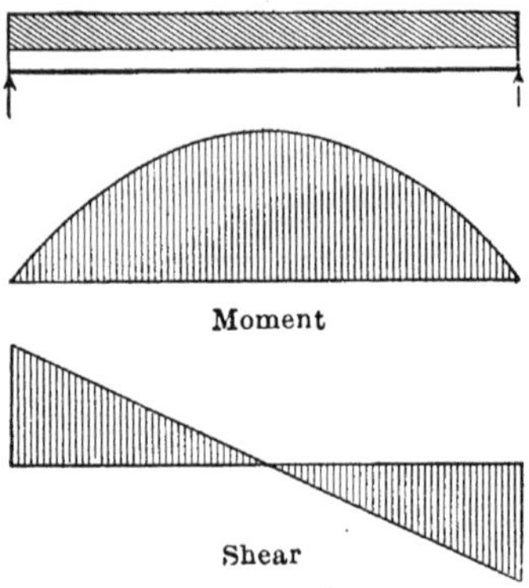

Fig. 48.

diagonal-tension failure will occur just outside the loads, while in the latter case it will occur nearer to the support where the moment is considerably less than the maximum. Conditions as to shear are therefore somewhat more favorable in the continuously loaded beam. A single concentrated load causes less shear for a given moment than the double load, and is therefore more favorable as regards shear.

As continuous beams are commonly used in building construction it will be useful to note here the variation in shear and moment in such a beam. This is shown in Fig. 49, and it will be seen that the conditions here are quite unfavorable, large shear occurring near the supports where the negative bending moment is large.

Whether a beam will fail from moment stresses or shearing stresses will depend largely upon its relative length and depth. For any given distribution of loads and given stresses there is a definite ratio of length to depth for equal strength as given in Chap. III, Art. 91, but by reason of the variation in shearing strength due to the direct effect of moment and amount of steel, these formulas can be considered as only roughly approximate.

106. Methods of Web Reinforcement.—There are in use many methods of placing steel in the web so as to reinforce it against inclined tension failure. The various methods may, for convenience, be divided into three groups: (1) Reinforcing metal placed at an inclination; (2) Reinforcing metal placed vertically; (3) Miscellaneous methods.

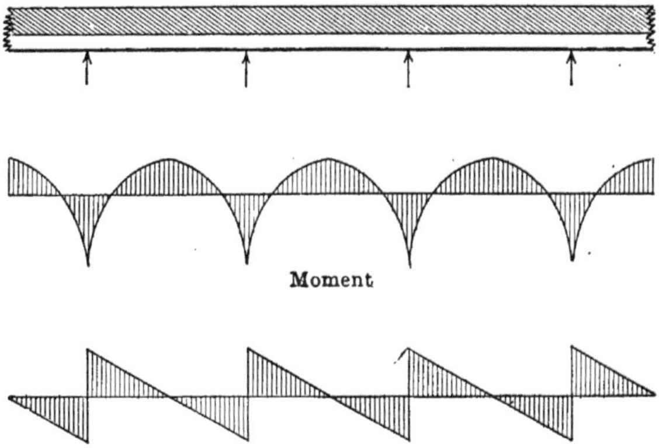

FIG. 49.

(1) Theoretically the most effective way to reinforce against tension failure in any direction is to place reinforcement across the lines of rupture, or in the direction of the maximum tensile stresses. In the case of web tension the lines of maximum stress vary in direction, but it is not practicable or necessary to have the inclination of the reinforcing rods exactly the same as the lines of maximum tension, and various arrangements will serve to accomplish the purpose. The most common method is to use several rods for the horizontal reinforcement and then to bend a part of these upwards as they approach the end, where they are not needed to resist bending stresses. Such an arrangement is shown in Fig. 50, (*a*) and (*b*). Separate

inclined rods may also be used, attached or not to the horizontal bars. The "stirrups" commonly placed in a vertical position may thus be inclined. Special forms of bars may be used, as the Kahn bar, Fig. 7, p. 31, in which strips are sheared from the main bar and bent up.

(2) Vertical reinforcement has long been the established practice in European work where the experience has extended over many years. It has proven its effectiveness and in connection with bent rods has many practical advantages. Vertical reinforcement usually consists of some form of bent rod or band styled a "stirrup", placed as shown in Fig. 50, (*c*) and (*d*). The Hennebique system, widely and successfully used, employs both the inclined rods and the vertical stirrup (see Fig. 85, Art. 162). Combined with bent rods many arrangements of stirrups are possible, especially in continuous-girder constructions, the chief object being to secure good connection of stirrup to top and bottom steel.

(3) Some form of web of woven wire or expanded metal may be used for web reinforcement, and still other arrangements of wire or rods employed as illustrated in Fig. 50, (*e*), (*f*), and (*g*). In (*g*), representing the Visintini system, the beam is made into a truss in which the chords and the tension diagonals are reinforced.

107. *Action of Web Reinforcement.*—To aid in appreciating the action of steel placed in various ways, consider the typical diagonal tension failure, Fig. 51, as it occurs where only horizontal rods are used. The inclined crack at *a* usually appears first, due to rupture of the concrete in tension. To assist in preventing this rupture in its initial stage the most efficient reinforcement would be such as supplied by the inclined rod 1, fastened to or looped about the horizontal bar, or by the bent end of one of the horizontal bars. Reinforcement in this direction is in a position to take stress immediately. The vertical rod 2 can hardly be as effective as the inclined rod in preventing initial rupture, for so long as the concrete is intact the deformation on a vertical line is practically zero, owing to the combined

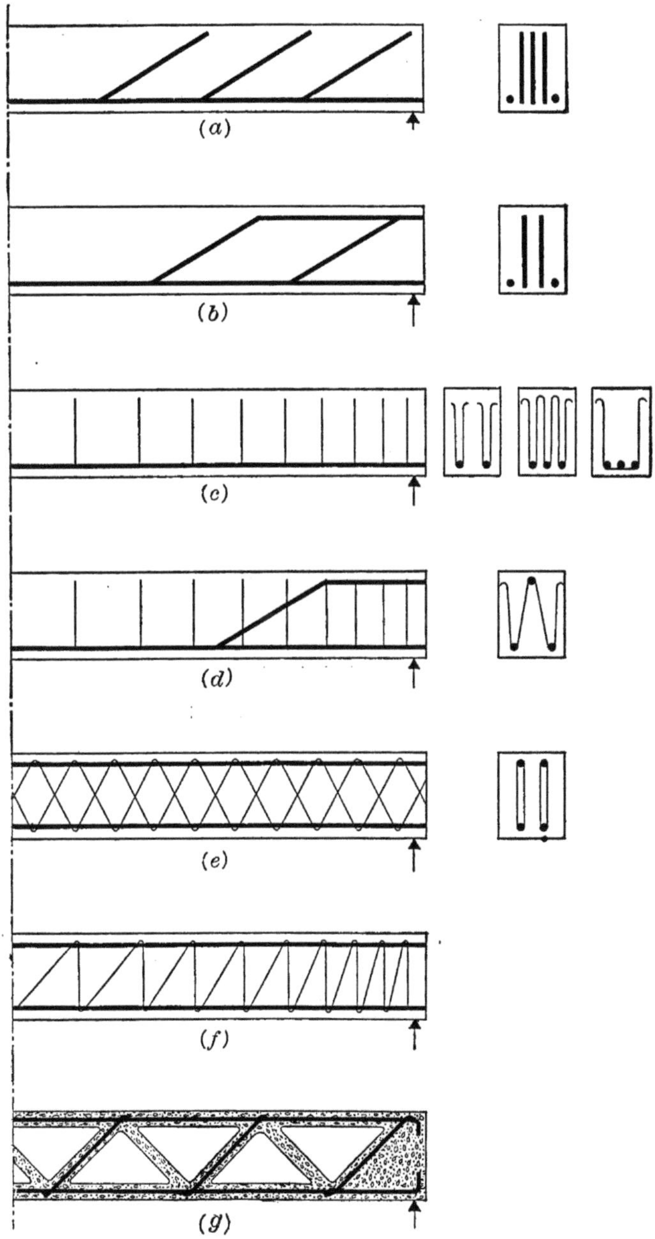

FIG. 50.—Some Methods of Web Reinforcement.

action of web tension and web compression at right angles to each other. Unless the unit stresses in the steel be made very low, however, it is likely that the concrete has received excessive tensile stress even under working conditions, and may be assumed to be ruptured more or less, in the same manner as on the tension face of the beam at points of maximum moment. At least the distortion in tension will be greater than in compression, and there will be a vertical movement of the concrete on the left of the crack, *a*, *downwards* with respect to the part of the right, and the vertical rod 2 will be brought into direct action if looped around or attached to the horizontal bars. Such a rod may then be more effective (allow of less vertical movement) than the inclined rod. Practically, there is no great difference in the effectiveness of the two forms of reinforcement if closely spaced so as to be in position to prevent excessive deformation all along the lower portion of the beam. To secure thoroughly effective reinforcement in this respect requires very careful

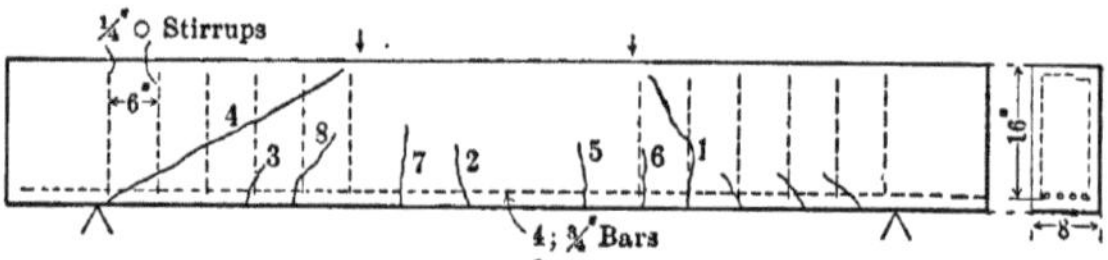

Fig 50*a*.

arrangement of the rods and faithful execution of the work. The action of vertical stirrups in resisting vertical deformation and the extension of inclined cracks is very well illustrated in Fig. 50*a*, which represents the conditions which developed in a test of a beam. The cracks are numbered in the order of their appearance, final failure occurring at crack No. 4 and being due to inadequate web reinforcement. The stirrups were stressed beyond their yield-point.

Vertical stirrups spaced a distance apart equal to or greater than the depth of the beam will give little aid in the prevention of diagonal cracks between successive stirrups although they may prevent final failure by the extension of a crack horizontally along the reinforcing rods. Stirrups should be looped around the horizontal rods so as to be firmly anchored at their lower end (or upper end at points of negative moment), where the stress is a maximum, but attachment to the rod is not necessary, as the office of the stirrup is to prevent vertical, or nearly vertical, distortion. The value of a stirrup unless anchored or looped at the top is limited by its strength of bond, and as its length is not great this point may need consideration. In some tests at the University of Wisconsin final failure has resulted from slipping of the stirrups. Stirrups made of small sections or bent in loops are advantageous in this respect. Where separate inclined reinforcement is used there is danger of its slipping along the horizontal rods if the inclination is too great. If attached to the horizontal rods, however, such reinforcement is very effective, not only with respect to shear but also in increasing the bond strength of the main bars.

Bent rods alone are apt to be of limited value, owing to the difficulty of providing rods close enough together. Convenience of horizontal reinforcement calls for comparatively few rods of large size, which provides too few for effective diagonal reinforcement. Where large rods are bent up the length of the bent end should be made sufficient, by bending at a small angle, to develop the requisite bond strength. Some tests of beams show failure of bond in the case of short bent rods. In

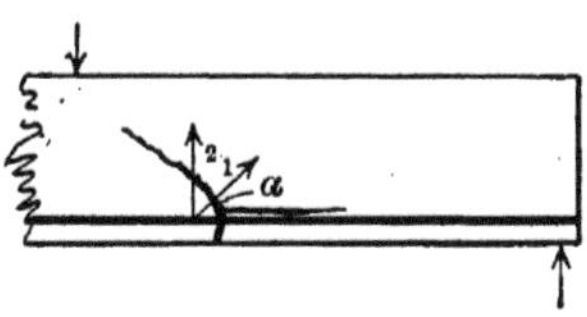

FIG. 51.

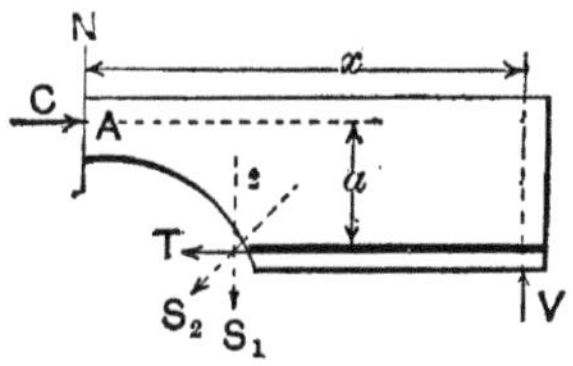

FIG. 52.

the case of continuous girders it is convenient to extend the bent rod horizontally at the top over the support to furnish

TABLE No. 8.

TESTS GIVING SHEAR OR "DIAGONAL TENSION" FAILURES.

Authority and Kind of Concrete.	Net Cross-section $b \times d$	Span.	Method of Loading.	Percentage of Reinforcement.	Arrangement of Reinforcement.	Kind of Bars.	Average Shearing Stress at Failure, Lbs/in²	No. of Tests	Kind of Failure.
(2) Talbot: 1:3:6	8″×10″	12′	⅓ points	1.66–2.21	Straight	Plain round	102	7	Shear
(6) Marburg: 1:2:4 crush. strength = 1700 lbs/in²	7″×8″	5′ 8″	Center	.96–1.36	Straight	Sq. corrugated and twisted	95	13	Shear
(3) Harding: 1:2:5	12″×20″	12′	⅓ points	.75	Straight	Plain round	108	3	Shear
crush. strength = 1530 lbs/in²	12″×21 7/16″ } 12″×23¼″ }	12′	"	.72–.75	"	Corrugated	128	6	"
	12″×18¾″	12′	"	.75	"	Twisted	99	3	"
	12″×20″	12′	"	.75	1 bent, 2 straight (a)	Plain round	132	3	Tension
	12″×18¾″	12′	"	.75	" " (b)	Twisted	155	3	Shear
	12″×19⅜″	12′	"	.75	5 bent, 2 straight (c)	"	184	3	Shear & tens.
	12″×21 [illegible]/16″	12′	"	.72	3 bent, 2 straight (d)	Corrugated	186	3	" "
	12″×23¼″	12′	"	.75	1 bent, 2 straight (b)	"	200	3	Shear

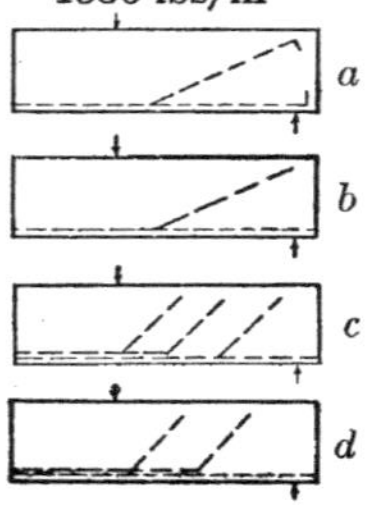

(4) **Carson:** 1:2:3 about crush. strength= 4100 lbs/in²	6″×7½″ about	5′	Center	1.22–2.26	Straight	Corrugated	201	6	Shear
	"	5′	"	1.51–2.26	"	Twisted	213	2	"
	"	5′	"	1.21–1.71	"	Pl. round & sq.	182	2	Tension
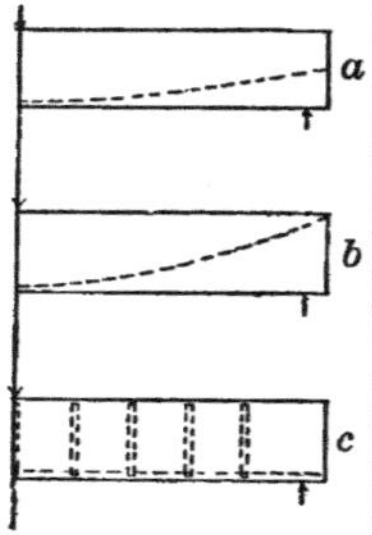	"	5′	"	1.22–1.62	Curved (*a*) and (*b*)	Corrugated	227	8	Shear & tens.
		5′		1.22–1.62	Straight with Stirrups (*c*)	"	220	5	Tension & comp.
Double reinforced beams				2.44 total	"	"	259	2	" "
				3.25 total	"	"	338	2	" "
(5) Withey: 1:2:4 crush. strength= 1780 lbs/in²	7¾″×9″ and 7¾″×11″		Overhanging beams. *a*	2.04–3.01 (top) 1.22–1.63(bot.)	Straight	Plain round	161	4	Shear
			a	1.28–2.00 (top) .65–1.00 (bot.)	Str. and bent	" "	258	8	Shear & tens.
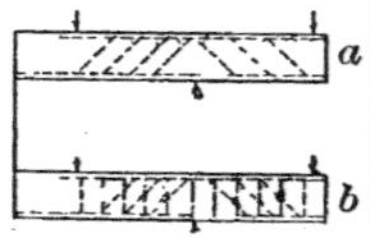			*a*	2.00 at top 1.00 at bottom	Stght with Stirrups	" "	240	10	" "
			b	"	Str. and bent with stirrups	" "	334	3	Tension
			b	"	Expanded metal	" "	240	8	Shear & ts.
			b	"	Straight with expanded met	" "	239	8	Shear
Double reinforced beams, simple span	8″×10″	12′	⅓ points	2.9 bot.; .0 top	2 str., 3 bent	" "	155	2	Compression
				2.9 bot.; .48 top	"	" "	189	2	Shear
				2.9 bot.; .96 top	"	" "	184	2	"
				2.9 bot.;1.44 top	"	" "	194	2	"

tension reinforcement. A very satisfactory arrangement of web reinforcement is a combination of bent rods and vertical stirrups, and especially is this the case in continuous-beam construction. Tests of various arrangements, so far as the authors have been able to find, show the best results from this method under the ordinary conditions and proportions. Web reinforcement of woven wire or expanded metal should give good results.

108. *Effect of Stirrups on Stress in the Horizontal rods.*—A careful study of the distribution of stress which exists after a beam begins to rupture on a diagonal line will show the fact that a stirrup, whether vertical or inclined, will relieve the stress in the horizontal rods at the point of rupture. Thus in Fig. 52, if the concrete no longer has tensile strength, the value of the tension T in the horizontal rods at the line of rupture, if unaided by the stirrup stress S_1 or S_2, is equal to Vx/a, the same as its value was at section N before rupture began. The moment of the stress in the stirrup about the point A, whether the stirrup be vertical or inclined, serves to reduce the value of T. Without the stirrup there is therefore more danger of failure of *bond* near the ends of the beam.

109. *Results of Tests.*—In Table No. 8 are given in a classified form the most important tests of rectangular beams which lend information on web stresses and web reinforcement. The reference number in the first column refers to the list of authorities on p. 120. In this table are given the significant facts as far as practicable, although a detailed inspection of the reports referred to is necessary for a thorough study of the tests. The kind of failure denoted as a "shear failure" is so called for convenience; they are diagonal-tension failures brought about by large shearing stresses and hence may be measured by the shearing forces present. The average shearing stress on a vertical section at failure is given. While the maximum shearing stress is somewhat greater than this (Art. 89) the average stress is practically as good a standard of measure and is much more readily calculated. Where the failure was not a shear failure the figures for shearing stress are valuable as indicating what the maximum stresses were, although the beam may have withstood still larger stresses if failure had not occurred in some other way.

Straight Reinforcement Only.—The tests of Professor Talbot, Professor Marburg, and Mr. Harding give values of from 95 to 128 lbs/in² under quite a variety of conditions. Mr. Carson, with specially good concrete, secured values of about 200 in the case of high-elastic-limit deformed bars and 182 for plain bars, which, however, failed in tension. In the University of Wisconsin tests on overhanging beams, which represented beams of great depth, those with straight bars gave a value of 161 lbs/in² and double reinforced beams values from 155 to 194 lbs/in², depending upon the per cent of steel used. Other tests by Talbot * in which straight bars only were used gave results for average shear as follows:

No. of Tests.	Kind of Concrete.	Average Shearing Stress (v') at Rupture, lbs/in².	
		Minimum and Maximum.	Average of Group.
2	1:2:4	117–122	120
11	1:3:5½	71–124	90
4	1:3:6	83–92	88
1	1:4:7½	46	46
4	1:5:10	54–64	58

The amount of steel varied from 1 to 2.2%, but this had no apparent effect upon the strength, all beams failing in diagonal tension. Stirrups spaced 12 in. apart on 10-in. beams showed little or no strengthening effect. Professor Talbot estimates the *maximum* shearing strength (which is about 15% greater than the average values above given) at ½ to ⅓ of the tensile strength of the concrete. The very low values obtained in the case of the poorer concretes should be noted.

* Bull. No. 14, Univ. of Ill., 1907.

As stated in Art. 105 the amount of horizontal steel has a direct bearing on shear failures for the reason that large areas of steel with low unit stresses permit less extension of the concrete than small areas with high working stresses. This effect is shown in a marked manner in a series of tests made at the University of Wisconsin on small mortar beams of 1·3 mixture. The beams were $3'' \times 4\frac{1}{2}''$ in cross-section and 4 ft. span length. Loads were applied at two points a varying distance apart. Only straight reinforcement was used, amounting to 1.41%. The tensile strength of the material was high, being 490 lbs/in². The results were as follows:

Distance Apart of Loads. Inches.	Average Shearing Stress. Lbs/in².
Centre Load.	177
12	200
24	220
32	316
36	512
40	850
44	1035

The increase in strength as the loads approached the supports must be due largely to the decrease in moment stress and consequent distortion, which is essentially what occurs when large areas of steel and low working stresses are used.

Beams with Web Reinforcement.—Mr. Harding's tests included only bent rods, and with these very considerably higher ultimate values were obtained than for straight rods, averaging for the three groups 190 lbs/in². Plain bars, bent, gave tension failures, these bars being of lower elastic limit than the deformed bars. These results are therefore of negative value. In some of Mr. Harding's tests the inclined bars pulled out, the bent ends being relatively short, as indicated in the sketches. An inspection of the deflection curves of these beams will show that those in which the rods were not bent were the stiffer beams, owing to the greater average amount of steel carrying the bending moment. Mr. Carson's results average 227 lbs/in² for curved bars and from 220 to 338 lbs/in² for straight bars with stirrups, the strength increasing with increasing per cent of metal. The stirrups were 1″×⅛″ straps spaced about 7 in. apart. Tests by Talbot in which curved and inclined rods were used, but in which no rods continued straight for the entire length of the beam, showed results very little better than for straight rods.*

A reference to Table No. 13 will show the effect of stirrups on the ultimate strength and method of failure of beams reinforced for compression. In the tests of Mr. Withey the bent rods alone gave 258 lbs/in² and stirrups alone average 240, while the combination gave 334, with a tension failure, showing still greater web stresses possible. Expanded metal, as used, proved too weak, as it pulled apart at a shearing value of 240 lbs/in². T-beam tests described in Art. 110 indicate that a value of 350 lbs/in² may readily be reached with stirrups and bent rods even with a relatively poor concrete.

The importance of tensile strength in the concrete should be noted in this connection, as the diagonal tension or shear failure is the one to be most feared and therefore most carefully guarded against.

110. Tests on T-Beams.—The reinforcing of T-beams requires special care in providing against shearing stresses. Where a floor slab forms the upper part of a beam there will usually be ample strength in compression for any depth likely to be selected. The design of the stem of the T, or the beam below the slab, is therefore largely a question of providing sufficient concrete and reinforcement to take care of the shearing stresses. In this case, therefore, it is important to provide a strong web for shearing stresses, as the strength in this respect will commonly determine its size. In Tables Nos. 9 to 11A are given the most important tests on T-beams known to the authors. The percentage of steel is calculated with reference to a rectan-

* Bull. No. 14, Univ. of Ill., 1907.

TABLE No. 9.

T-BEAM TESTS MADE BY F. VON EMPERGER.*

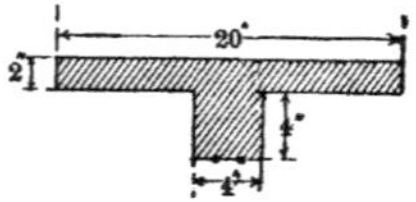

Concrete, 1:4; age, about 100 days.
Plain round steel used except where noted.
Beams loaded with two loads at ¼ points. Span = $78\frac{3}{4}''$.

Number.	Percentage Reinforcement.	Number and Approximate Size of Rods.	Arrangement of Rods.	Number and Spacing of Stirrups between Load and Support.	Total Breaking Load, Pounds.	Average Shearing Stress on Beam 4″×6″, Lbs/in².
1	.84	Two; .8-inch	Straight	None	5700	119
2	.64	Two Thacher bars, .5-inch	"	"	5700	119
3	.84	Eight; .4-inch	"	"	9470	198
4	.84	Two; .8-inch	"	3 double, spaced 8″	9800	205
5	.84	" "	"	5, spaced 4″	16000	334
6	.84	" "	"	6, spaced 2″ to 6″	16500	345
7	.64	Two Thacher bars, .7-inch	"	" " " " "	17200	360
8	.84	Eight; .4-inch	"	6 double, spaced $3\frac{1}{2}''$	15600	325
9	.84	" "	4 straight, 4 bent	None	12100	252
10	.84	" "	" "	4, spaced $5\frac{1}{2}''$	16000	334
11	.84	Two; .8-inch	Straight	4 inclined, spaced 6″	13800	288
12	.84	" "	"	4 double, inclined, spaced 6″	21800	455
13	.84	" "	"	6 double, inclined, spaced 4″	23100	482
14	.75	Two; .32″×1.4″	"	" " " " "	23700	495

* Forscherarbeiten auf dem Gebiete des Eisenbetons, Heft V, 1906.

gular beam having a cross-section equal to the circumscribing rectangle.

The yield point of the plain steel in the tests of Table No. 9 was about 37,000 lbs/in², and its ultimate strength 51,000 lbs/in². A load of about 19,000 lbs. would stress the steel in the beams having .84% reinforcement to the yield point. This limit is exceeded only in the last three of the list. In these beams inclined stirrups were used, placed in a notch in the bar, in all other series the stirrups were placed vertically.

Reviewing these experiments we note, first, the results with straight bars and no stirrups. The beams having the .8-in. rods and the Thacher bars developed a value of 119 lbs/in² average shearing stress, while the .4-in. rods developed 198, the difference being due doubtless to the difference in bond strength, although the previous experiments cited would indicate that not much greater value than the latter figure could be expected from straight bars only.

Noting the next five beams, all have straight rods and vertical stirrups, No. 4 having stirrups spaced 8″ apart, while the others have a spacing of 4″ or less near the support. For the former a value of 205 lbs/in² was reached, while the three others averaged 341, all being nearly the same despite the variety of bars used. No. 9 had bent bars and no stirrups, giving a strength of 252, while No. 10 had bent rods and stirrups rather widely spaced, developing 334. Nos. 11–14 had inclined stirrups attached to the bars and all but the first gave high values of over 450 for the shear.

In these tests it should be noted that a load of 16,000 lbs. would develop in the rods a theoretical stress of $(8000\times 20)/4.2 =$ 38,000 lbs/in². For the .8-inch rods this would require an average bond strength of about $38,000/(2\times\frac{3}{4}\times\pi\times 25\frac{1}{2}) =$ 320 lbs/in², about all that could be expected. The .4-inch rods would be stressed one-half as much in bond. The spacing of stirrups in No. 10 was too great to be entirely efficient. The inclined attached stirrups gave the best results in these tests, but whether similar results would be obtained where strength

TABLE No. 10.

T-BEAM TESTS AT THE UNIVERSITY OF WISCONSIN.*

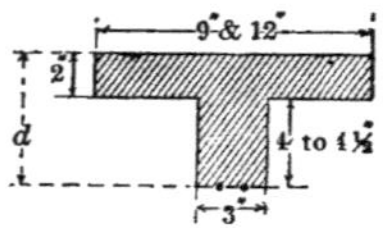

Concrete, 1:2:4; age, 30 days
Compression strength of cubes = 1120 lbs/in².
All rods plain round. Length of beams, 6' 8".
Span length: Nos. 1–4 = 6 feet; all others = 5 feet.
Loaded at third points. Width of flange, 12" for Nos. 9–12; all others 9".
Rods bent up in all beams, two rods in those containing three rods and four rods (in pairs) in those containing six rods. All stirrups placed about $3\frac{1}{2}$" apart.

Number.	Percentage Reinforcement.	Number and Size of Rods.	Web Reinforcement in Addition to Bent Rods.	Total Breaking Load, Pounds.	Average Shearing Stress on Section, $3'' \times d$.	Kind of Failure.
1	.94	Three; $\frac{1}{2}$-inch	None	4000	107	Bond
2	.94	" "	"	5100	136	"
3	.52	Three; $\frac{3}{8}$-inch	$\frac{1}{4}$" stirrups	4750	127	"
4	.52	" "	" "	4000	107	"
5	.52	" "	$\frac{3}{16}$" stirrups	9380	238	Tension and shear
6	.52	" "	" "	9400	246	" " "
7	1.05	Six; $\frac{3}{8}$-inch	" "	12850	330	Compression and shear
8	1.05	" "	" "	13550	349	Compression
9	.39	Three; $\frac{3}{8}$-inch	$\frac{1}{4}$" stirrups	8400	216	Shear
10	.39	" "	" "	8000	205	"
11	.78	Six; $\frac{3}{8}$-inch	" "	11400	304	"
12	.78	" "	" "	12700	346	"
13	.52	Three; $\frac{3}{8}$-inch	Expanded metal	7800	217	"
14	.52	" "	" "	7000	190	"
15	1.05	Six; $\frac{3}{8}$-inch	" "	13800	384	Tension and shear
16	1.05	" "	" "	12600	350	Compression

* Bulletin No. 1, Vol. 4, 1907.

Table No. 11.

T-BEAM TESTS AT THE UNIVERSITY OF ILLINOIS.*

Concrete, 1:2:4; age about 60 days; comp. strength of cubes=1820 lbs/in².

Steel: yield point of plain round=38300 lbs/in²; of Johnson bars= 53800 lbs/in².

Size of beams: thickness of flange=3¼ in.; thickness of web=8 in.; depth to center of steel=10 in.; total length=11 ft.; span length=10 ft.; width of flange varied.

Stirrups: made of ½-in. Johnson bars; five stirrups at each end spaced 6 in. apart.

Loads applied at third points. All failures were steel tension failures.

Number.	Width of Flange. Inches.	Percentage Reinforcement.	Number and Size of Rods.	Total Breaking Load. Pounds.	Average Shearing Stress on Section 8″×10″ lbs/in².	Stress in Steel. lbs/in².
1	16	1.05	3 ¾″ Johnson	46700	293	64300
4		1.10	4 ¾″ Plain round	32410	203	41500
7		1.10	4 ¾″ " "	30100	188	38100
3	24	0.93	4 ¾″ Johnson	55700	347	57500
6		0.92	5 ¾″ Plain round (2 bars bent up)	39300	246	40700
8		0.92	5 ¾″ Plain round (2 bars bent up)	40100	250	41200
2	32	1.05	6 ¾″ Johnson	80500	503	55700
5		1.05	6 ¾″ Johnson (2 bars bent up)	83300	521	57400
9		0.97	7 ¾″ Plain round (3 bars bent up)	50900	318	37600

of bond was not in question cannot be stated. In case of weak bond an attached inclined stirrup virtually adds much to the bond strength of the bar.

In Table No. 10 are given further results of tests. In the first four tests the supports were placed too near the ends of the beam (4 inches) with the result that after the initial cracking the bars soon pulled out. After reducing the span length to 5 feet no further trouble in this respect was experienced. The results correspond closely with those given in the other tables.

* Bulletin No. 12, Eng. Exp. Station, Univ. of Ill , 1907.

Table No. 11 contains results of recent tests by Professor Talbot. The maximum values of shearing stress are unusually high and indicate very effective web reinforcement. As no shear failures occurred the possible limit of strength of web was not determined. The excess of stress in the steel as compared to the yield points should be noted, due in large measure no doubt to the excessive compressive strength of the flange and the thorough web reinforcement.

Table No. 11A contains results of additional tests on T-beams made at the University of Wisconsin. In these tests the yield point of the corrugated bars was about 48,000 lbs/in² and that of the ¾-in. rods was 41,000 lbs/in². These limits correspond closely to the stresses in the steel at failure, excepting in the case of beams G_1 and G_2 which failed by shear. The table contains results of value with respect to shearing stresses and the use of stirrups and bent rods for shear reinforcement. In the progress of the tests the occurrence of the first diagonal crack was carefully noted and the maximum shearing stress at this load is calculated and given in the table. It will be noted that there is a fairly close agreement between this value and the tensile strength of the concrete as given in the next column. The average value for the maximum shearing stress is 179 lbs/in² whereas the average tensile strength is 187 lbs/in². This would indicate that in spite of stirrups the concrete is likely to open up at a diagonal tensile stress about equal to its usual tensile strength. The table also gives the shearing stress at ultimate load, both the average and maximum stress (at neutral axis and below) being given. The web reinforcement was effective in preventing shear failures, excepting in the case of beams G_1 and G_2 where no bent rods were used. In all other cases the web reinforcement was ample and hence no conclusions can be drawn as to relative value of the different kinds of reinforcement. The wire mesh gave good reinforcement and was convenient to use. In beams G_1 and G_2 failure by diagonal tension occurred, the stirrups breaking at the maximum load. These tests indicate that by the use of bent

Table No. 11a.

T-BEAM TESTS, UNIVERSITY OF WISCONSIN.*

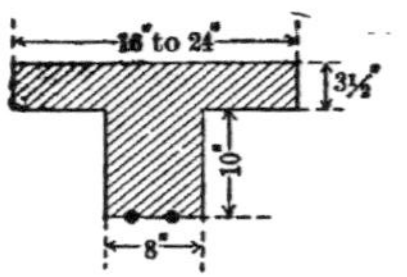

Concrete 1:2:4; age, 28 days; compressive strength = 1940 lbs/in².
Beams D had 24-in. flanges, all others 16-in. Span length = 10 ft.; loaded at third points.
Stirrups uniformly spaced between loads and supports.
All beams failed in tension except G_1 and G_2, which failed by breaking of stirrups.

No. of Beam.	Kind of Reinforcement.	At First Diagonal Crack.		Tensile Strength of Concrete, lbs/in².	At Maximum Load.			$\frac{M}{bd^2}$
		Load, Lbs.	Maximum Shearing Stress, lbs/in².		Load, Lbs.	Average Shearing Stress, lbs/in².	Maximum Shearing Stress, lbs/in².	
A_1	Four $\frac{3}{4}''$ cor. bars (2 bent).....	24000	128	256	66600	308	361	452
A_2	Fourteen $\frac{1}{4}''$ cor. stirrups......	32000	171	167	66200	306	357	450
B_1	Four $\frac{3}{4}''$ cor. bars (2 bent).....	42000	226	260	65600	304	357	444
B_2	Sixteen $\frac{1}{4}''$ round stirrups......	26000	150	157	62400	289	354	425
C_1	Five $\frac{3}{4}''$ round rods (3 bent)....	34000	181	197	60000	278	322	407
C_2	Sixteen $\frac{1}{4}''$ round stirrups......	34000	185	216	57400	265	316	390
D_1	Six $\frac{3}{4}''$ cor. bars (3 bent)......	46000	247	182	96200	445	526	437
D_2	Fourteen $\frac{3}{8}''$ cor. stirrups......	58000	306	...	101400	470	544	460
E_1	Six $\frac{3}{4}''$ cor. bars (3 bent).......	46000	244	148	92800	429	505	420
E_2	Twenty-four $\frac{3}{8}''$ cor. stirrups...	40000	215	184	88000	407	485	400
F_1	Four $\frac{3}{4}''$ cor. bars (2 bent).....	30000	158	142	67600	313	368	459
F_2	No. 11 wire mesh, 1 inch.......	28000	150	174	65400	302	352	445
G_1	Four $\frac{3}{4}''$ cor. bars (straight)....	24000	126	184	48000	222	259	323
G_2	Sixteen $\frac{1}{4}''$ round stirrups......	28000	149	164	48200	223	260	324

* Bulletins No. 2, Vol. 4, 1908.

rods and stirrups an average shearing strength of 400 lbs/in² can readily be developed with good concrete.

111. Conclusions as to Shearing Strength.—From the available data it would appear that with ordinary concrete and no web reinforcement the ultimate average shearing strength is about 100 lbs/in² and that this strength can readily be increased by the use of web reinforcement to 300 to 400 lbs/in². The latter figure may, from our present knowledge, be taken as about the maximum value with ordinary, closely spaced web reinforcement. It appears also that the shearing strength of a T-beam is about the same as that of a rectangular beam of the same depth and a width equal to the width of the stem of the T. It is to be understood that the shearing stress is here used merely as a convenient measure of the diagonal tensile stress, which is really the stress involved. This being the case it would be incorrect to take any account of the shearing strength of the steel in designing the reinforcement, as is sometimes done.

112. Beams Reinforced for Compression. — Generally speaking, it is more economical to carry compressive stresses by concrete than by steel, but limitations as to size sometimes makes it desirable to strengthen the compressive side of a beam. In cases, also, where both positive and negative moments exist in the same beam, either as alternating stresses or simultaneously at different points, steel reinforcement will be used on both sides and its value on the compressive side needs to be known. Obviously, steel reinforcement on the compression side will have little effect in beams that would otherwise fail in tension or shear, although there would be some gain owing to increased distance between centers of tensile and compressive forces. The effectiveness of steel in compression has sometimes been questioned, but the results of tests on beams and columns indicate that, in ordinary ratios at least, the steel does its share of work.

Table No. 12 gives results of tests on double reinforced beams made at the University of Wisconsin. No stirrups were used in the first eight beams and, as a consequence, all

Table No. 12.

TESTS OF BEAMS REINFORCED FOR COMPRESSION.

University of Wisconsin, 1906, 1907.*

Size of beams, 8″×10″ net section; 12′ span Concrete, 1:2:4; age, 28 days. Plain round rods used. Elastic limit of ¾″ rods=about 42000 lbs/in²; elastic limit of ½″ rods for first 8 beams=38600 lbs/in², and for last 4 beams=46100 lbs/in².

No.	Reinforcement, Tension Side.		Reinforcement, Compression Side.		Ultimate Load, Lbs.	$\frac{M}{bd^2}$	Average Shearing Str'gth, lbs/in².	Load Considered.	Neutral Axis, k.	Stress in Tension Steel, Lbs/in².		Stress in Compression Steel by Deformations, Lbs/in².	Maximum Compressive Stress in Concrete, Lbs/in². B	Crushing Strength by Compression Test, Lbs/in². S	Ratio B/S.	Ratio of Comp. Stress in Steel to Max. Stress in Concrete.	Kind of Failure.
	Bars.	Per Cent.	Bars.	Per Cent.						By Moments.	By Deformation.						
P_1	Five ¾″ round bars.	2.9	None	0.0	24780	794	165	22160	.670	35200	36000	....	2260	1290	1.75	....	Comp.
P_3		"	None	0.0	21380	683	141	20160	.675	31600	33000	....	2050	1275	1.61	....	Comp. &shear
R_1		"	2 ½″ R	0.49	28740	922	190	26160	.590	39000	39000	40000	2370	1715	1.43	16.9	Shear
R_2		"	"	0.49	27840	895	184	26160	.600	39100	39000	45000	2260	1580	1.46	19.9	"
T_1		"	4 ½″ R	0.98	24180	774	159	22160	.570	32800	33000	33000	1610	1715	0.94	20.5	"
T_2		"	"	0.98	31260	1003	206	24160	.565	35700	33800	36000	1760	1385	1.27	20.5	"
V_1		"	6 ½″ R	1.47	27580	884	182	25160	.510	36100	36000	28500	1780	2240	0.79	16.0	"
V_2		"	"	1.47	31380	1008	206	29160	.555	42500	42000	39000	1670	1770	0.94	23.4	"
W_1		"	4 ½″ R	0.98	34000	1060	218	34000	.503	46100	43500	34000	2880	2600	1.11	11.8	Tension
W_2		"	"	0.98	34800	1085	223	34000	.503	46100	45000	33900	2880	2600	1.11	11.8	"
X_1		"	6 ½″ R	1.47	33500	1043	218	33000	.511	45000	48000	39000	2000	1500	1.33	19.5	"
X_2		"	"	1.47	34800	1085	223	34000	.550	47100	46500	42000	1830	1530	1.20	22.9	"

* Bulletins No. 1 and 2, Vol. 4, Engineering Series, University of Wisconsin, 1907, 1908.

these beams, except the first two, failed by diagonal tension. The last four beams were supplied with stirrups looped around the lower bars and consisting of ¼-in. round steel, ten stirrups being used at each end between the load and support. In all beams two of the bottom rods were bent up at about 2 feet from the end and one bar at 4 feet from the end. Fig. 52*a* shows arrangement of reinforcement and location of cracks in beam W_1. They are numbered in the order of occurrence, crack "F" being the point of final failure, the steel passing its yield point.

The neutral axis was found by the use of extensometers, after which the stresses in steel and concrete at "load considered" were found, assuming the compression in the con-

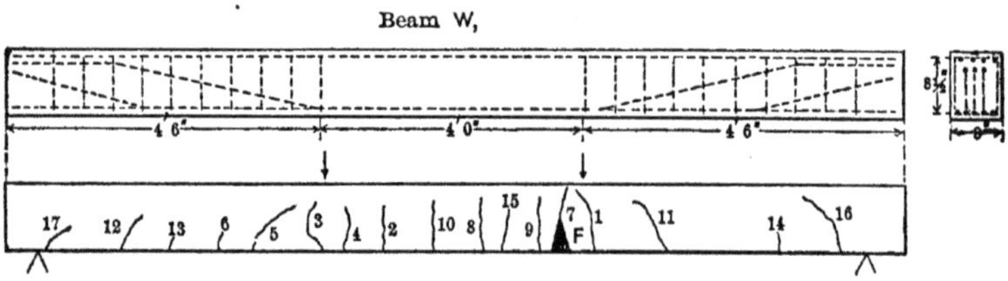

Fig. 52*a*.

crete to follow the parabolic law. The tensile stresses in the steel as calculated by the two methods agree very closely. The compression in the concrete is determined by subtracting from the total compression the compressive stress in the steel.

The crushing strength of corresponding compression specimens is also given in column *S*, and in the following column the ratio of the calculated stress in the beam to the crushing strength. Except in the case of the first two beams the ultimate compressive strength of the concrete was not reached, although in the last four beams crushing took place very soon after the maximum load was reached. In the next column are given the ratios of calculated stresses in compression steel and in concrete. These are in fair agreement with the value of *n* for concrete at rupture. The concrete in beams *W* happened to be especially good and in beams *X* it was poor. The stirrups used in the last four beams proved effective in preventing shear failures. The results of these tests indicate that the steel is taking its share of stress and that the compression side of the beam is strengthened in accordance with the usual theory. Obviously, in order to secure full benefit of the steel up to rupture, a fairly high elastic limit material should be used. The tests show, further, that the compressive strength of concrete in a beam is fully equal to and probably somewhat greater than the strength as determined from the usual compression test.

Fig. 53 gives a typical set of curves for double reinforced beams. Beams T_2 and W_2 had equal amounts of compressive reinforcement, but beam W_2 was provided with stirrups while T_2 was not. These conditions resulted in a shear failure in the former case and a tension failure in the latter. Comparing with those shown in Art. 100, it will be seen that these beams are much stiffer and apparently more perfectly elastic, as would be expected from the nature of the reinforcement.

Table No. 13.

TESTS OF BEAMS REINFORCED FOR COMPRESSION.

Boston Transit Commission.*

Beams and material as described in Table No. 8. All beams reinforced with ½″ corrugated bars, with same number top and bottom. Stirrups 1″×⅛″, spaced about 7″. Centre loads.

Number.	Total Reinforcement. Number of Bars.	Total Reinforcement. Percentage	Use of Stirrups.	Load at First Sign of Failure. Pounds	Ultimate Load. Pounds.	$\frac{M}{bd_2}$	Average Shearing Stress, v'. Lbs/in².	Kind of Failure.
72	4	1.62	No	9920	10980	513	126	Tension
78	4			11424	14148	660	162	"
71	4		Yes	11000	16506	766	188	Shear & tens.
77	4			11224	15072	701	172	" " "
70	6	2.44	No	14992	16096	740	182	Shear
76	6			16716	17300	796	195	"
69	6		Yes	17724	23972	1106	272	Tension
75	6			14476	21284	990	244	"
68	8	3.25	No	19044	19044	880	215	Shear
74	8			17200	18584	854	210	"
67	8		Yes	21200	30168	1400	344	Tension
73	8			22132	29178	1347	332	"

* Tenth Annual Report, 1904.

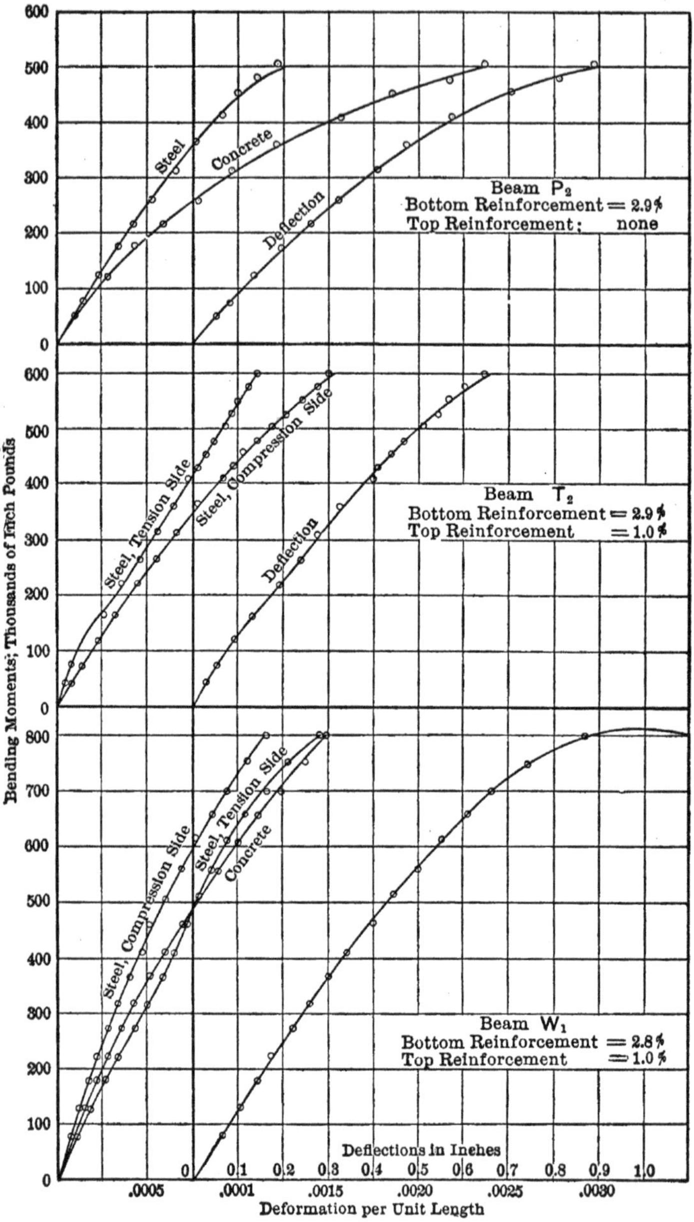
Bending Moments; Thousands of Inch Pounds
Steel
Concrete
Deflection
Beam P_2
Bottom Reinforcement = 2.9%
Top Reinforcement; none
Steel, Tension Side
Steel, Compression Side
Deflection
Beam T_2
Bottom Reinforcement = 2.9%
Top Reinforcement = 1.0%
Steel, Compression Side
Steel, Tension Side
Concrete
Beam W_1
Bottom Reinforcement = 2.8%
Top Reinforcement = 1.0%
Deflections in Inches
0 0.1 0.2 0.3 0.4 0.5 0.6 0.7 0.8 0.9 1.0
.0005 .0001 .0015 .0020 .0025 .0030
Deformation per Unit Length

FIG. 53.

Table No. 13 gives results of tests on double-reinforced beams by the Boston Transit Commission. The table is of value mainly in showing the benefit of stirrups. Crushing failures were obtained in but few cases in this series of tests, even where no compressive reinforcement was used, so that little advantage could be expected. It should be noted that where stirrups are not used the results shown in this table are very nearly the same as those of Table No. 12, although the quality of the concrete in the latter case was much inferior. Conditions were such that the full strength of the concrete was not developed in the tests of Table No. 13.

112*a*. Experiments on Deflections of Beams.—Probably the most complete and accurate deflection measurements ever made are those by Bach. About 50 rectangular and 20 T-beams are reported on by him.* The *rectangular beams* were 2 m. long, 30 cm. deep and 15, 20, or 30 cm. wide. They were reinforced with a single straight rod, several straight rods, straight rods with stirrups, or several rods, some bent up; the percentages varied from about .4 to 1.35. The T-beams were 3 m. long, 45 cm. wide, 48 cm. deep, flange 10 and web 20 cm. thick. They were reinforced with straight rods, with or without stirrups, or rods, some bent up, with and without stirrups; the percentage of steel was about .8 in all of them. The beams rested on end supports, and were loaded at the quarter or third points. They were made in sets of three (as nearly alike as possible) and the load-deflection curves for any set are in remarkably good agreement. Deflections were measured at 5 or 7 different points along the beam and to the nearest .005 mm.

Fig. 53*a* shows the load-deflection curves for 4 sets of beams. Groups 1 and 2 relate to rectangular beams; in (1) the beams were 15 cm. wide, reinforced with .5% of steel in 3 rods, 2 bent up at each end; in (2) the beams were 20 cm. wide and

---

* Mitteilungen über Forschungsarbeiten auf dem Gebiete des Ingenieurwesens, Hefte 39, 45, 46, 47 (1907).

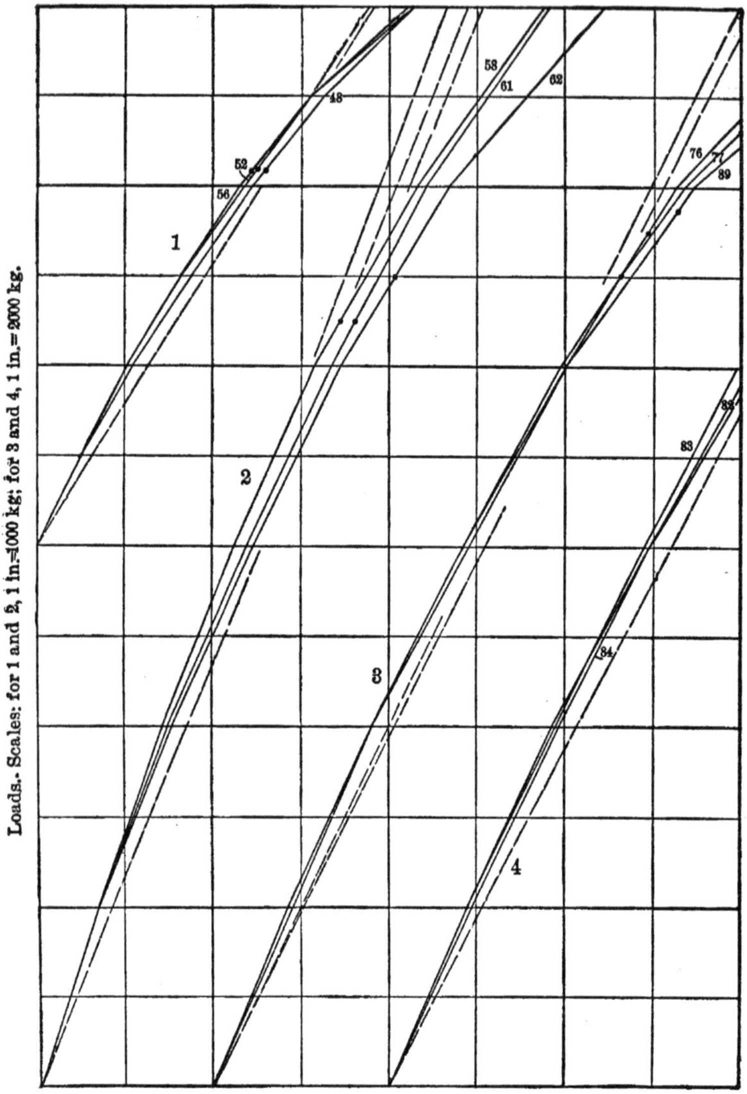
Loads.- Scales: for 1 and 2, 1 in.=1000 kg; for 3 and 4, 1 in.=2000 kg.
Deflections.- Scale: 1 in.=0.2 mm.
1
2
3
4
48
52
56
58
61
62
76
77
89
83
84

FIG. 53a

reinforced with 1.35% of steel in 3 rods, 2 bent up at each end. Groups 3 and 4 relate to T-beams; in the beams of (3) there were 3 straight rods (.8%) and 24 stirrups, and in those of (4) there were 5 rods, 4 bent up at each end, (.87%) and 24 stirrups. Only a part of each curve is given. The dot on each corresponds to one-quarter ul timateload; dots on extensions of group 4 would be a trifle higher than in group 3.

The dashed lines are graphs of the deflection formulas (see Arts. 92*d* and 92*e*) corresponding to the various beams, *n* having been taken as equal to 8 for reasons given in the next article. The deflection formula agrees as well with other sets in Bach's tests except in a few cases in which the reinforcement consisted of a single straight rod and stirrups.

Deflection measurements on beams tested in America seem not to have been made with special care, as there is considerable discordance in the published results. Among the best are some reported by Talbot, a few * of which are represented in Fig. 53*b*. Groups 1 and 2 relate to two sets of T-beams, 12 in. deep (over all), flange 3¼ and web 8 in. thick; the span was 10 ft., and loads at third points. The three in group 1 were 16 in. wide, reinforced with straight bars (about 1%) and stirrups, the three in group 2 were 24 in. wide, reinforced as others except some rods were bent up. Curve 3 is for a very large rectangular beam; its breadth was 25 in., depth to steel 30.5 in., span 23.5 ft., and percentage of steel 1.25. Only a portion of each curve is shown; the dot on each corresponds to one-fourth the ultimate applied load. The dashed lines are the graphs of the deflection formulas (Arts. 92*d* and 92*e*) for the corresponding beams; 8 is the value of *n* used in groups 1 and 2.

112*b*. Stirrups and Bent-up Rods do not affect the stiffness of the beam materially for working loads; but they do increase the ultimate deflection as well as strength. Bach's tests clearly show this to be true, for example:

---

* Bulletin Univ. of Ill. Eng. Exp. Station, No. 12 (1907); Eng. News, Vol. LX, p 145 (1908).

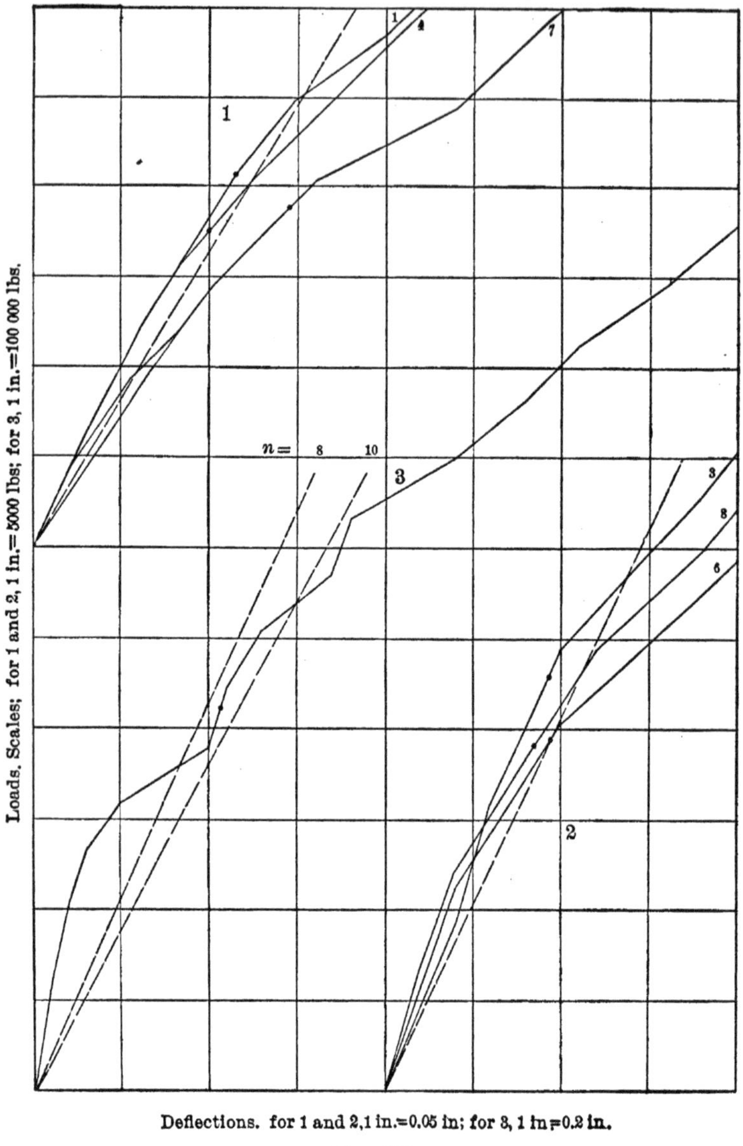
Loads. Scales; for 1 and 2, 1 in.=5000 lbs; for 3, 1 in.=100 000 lbs.
1
1
4
7
n= 8 10
3
3
8
6
2
Deflections. for 1 and 2,1 in.=0.05 in; for 3, 1 in=0.2 in.

Fig. 53b.

(1) Column *a* of the adjoining table (No. 13A) gives the average deflections for three beams (numbers 7, 13, and 14) corresponding to the loads tabulated; the beams were reinforced with a single straight rod (p about .9%). Column *b* gives the average deflections for another set of three (29, 32, and 37); these were reinforced like the first set but with sixteen stirrups added. The fourth column gives the percentage differences between the deflections of the two sets of beams up to 4000 kg. The average ultimate deflections of the two sets were 1.78 and 2.3 mm., and the ultimate loads 18,900 and 23,250 kg. respectively.

TABLE No. 13A.

DEFLECTIONS OF RECTANGULAR BEAMS.

Load, (Kilos).	Deflection (millimeters).					
	a.	*b.*	Diff.	*A.*	*B.*	Diff.
500	.052	.052	0%	.048	.050	+4.0%
1000	.110	.107	−2.7	.107	.110	+2.8
1500	.175	.165	−5.7	.167	.173	+3.6
2000	.245	.232	−4.9	.232	.248	+6.8
2500	.322	.307	−4.6	.308	.330	+6.9
3000	.428	.417	−2.6	.403	.403	+6.7
3500	.608	.580	−4.6	.538	.585	+8.7
4000	.793	.767	−3.3	.775	.902	+14.0

(2) Column *A* of the same table gives the average deflections for a set of beams (40, 43, and 45) which were reinforced with three straight rods (p=.55%); and *B* the average deflections for a set (49, 51, and 53) reinforced like the first, but two of the rods were bent up at each end. The last column gives the percentage differences between the average deflections of the two sets of beams. The average ultimate deflection of sets *A* and *B* were 3.38 and 3.45 mm. and their average ultimate loads 8250 and 8600 kg. respectively.

(3) The numbered columns in the adjoining table (No. 13B) give the average deflections of six sets of T-beams, three in

TABLE No. 13B.

DEFLECTIONS OF T-BEAMS.

Loads (kilos).	Deflections (millimeters).						
	1	2	3	4	5	6	Diff.
2000	.090	.087	.088	.093	.092	.095	9.2%
4000	.193	.183	.180	.190	.190	.192	7.2
6000	.307	.290	.287	.303	.293	.303	10.5
8000	.435	.400	.398	.422	.407	.420	6.0
10000	.577	.535	.568	.553	.553	.563	1.8

each set, for the loads tabulated. The beams were alike except as to reinforcement. Beams of set 1 were reinforced with three straight rods ($p = .8\%$); set 2 like 1 and 24 stirrups; set 3 like 1 and 48 stirrups; set 4 with five rods ($p = .87\%$),

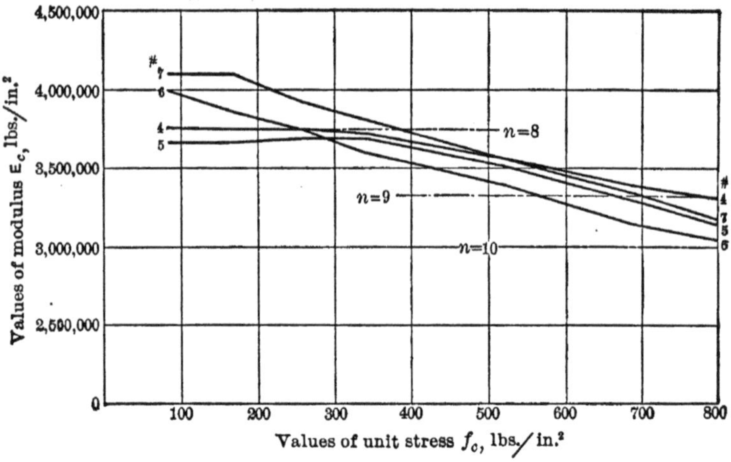

FIG. 53c.

four bent up at each end; set 5 like 4 and 24 stirrups; and set 6 like 5 except that a hook was formed at each end of the fifth rod. The horizontal lines in the table are drawn to correspond to one-quarter ultimate loads. The last column of the table gives the greatest percentage difference for the various working loads. The average ultimate deflections were 2.4, 3.2, 3.8, 6.0, 5.8, and 9.4 mm.; the average ultimate loads 23,000, 30,500, 37,800, 33,300, 41,000, 46,000 kg. respectively. All average ultimate deflections are not reliable.

112c. On the Value of n for Deflection Formulas.—As explained in Arts. 92c and b, the value of the modulus of elasticity to be used in deflection formulas should correspond not to the greatest unit stress in the concrete but to a fair average of the unit stresses at all points in the beam. Fig. 53c shows how the secant modulus varied in four compression specimens representing the concrete of the Bach beams referred to in Art. 112a. It was a 1:4 gravel concrete and the specimens were about eight months old when tested. The curves show that for unit stresses as high as 600 lbs/in^2 the moduli averaged over $3\frac{1}{3}$ million ($n=9$), and for the fair average unit-stress in Bach's beams under working loads n would be about 8.

COLUMNS.

113. Tests of Plain Concrete Columns.—An important series of tests on columns are those made at the Watertown Arsenal, and reported in *Tests of Metals*, 1904, and subsequent volumes. The principal results on plain concrete are given in Table No. 14.

These tests indicate an average strength for 1:2:4 concrete of 1600 to 1700 lbs/in^2, with no excessive variation in individual tests. For the weaker mixture, 1:3:6, the individual tests are much more at variance, indicating greater unreliability. The great strength of very rich mortars is noteworthy, and this fact is borne out by experiments on columns slightly reinforced. Considering relative cost, a rich mortar may often be the more advtantageous.

Table No. 14A gives results of tests by Professor Talbot on plain concrete.

TABLE No. 14.

TESTS OF PLAIN CONCRETE COLUMNS.

WATERTOWN ARSENAL, 1903–1905.

All columns were 8 ft. high and ranged from 10 in. in diameter to 12 in. square. The age of the concrete ranged from 5 to 8 months.

Kind of Concrete.	Crushing Strength, Lbs/in².	
	Results of Individual Tests.*	Average Crushing Strength.
1:1 mortar	5011+ 4320	4665
1:2 "	3652 2488	3070
1:3 "	2062 2692	2377
1:4 "	1564 1471 1050	1362
1:5 "	1038 1082	1060
1:1:2 (pebbles)	1525 1720	1622
1:1:2 (trap rock)	3900	3900
1:2:4 (pebbles)	1506 1710	1608
1:2:4 (trap-rock)	1750 1990 1413	1718
1:3:6 (pebbles)	462 700 1260	807
1:3:6 (trap-rock)	1350 750 1446	1182
1:2:4 (cinders)	871	871
1:3:6 (cinders)	1060 698	879

* Where two lines of values are given, those in the first line are results obtained in the 1904 series, those in the second line are from the 1905 series.

In general, it was found that the richer mixtures tended to fail by true shear failures, while the poorer mixtures generally failed by gradual crushing. The very superior results obtained on the 1:1½:3 mixture as compared to the 1:2:4 mixture, or poorer, should be noted. It shows the value of the use of rich mixtures for columns, the increase in strength over the 1:2:4 concrete being about 32% while the increase in cost would not be over 10 or 15%. Compared to the results of Table No. 14 these results agree as closely as could be expected. The great variation in individual tests in Table No. 14A should be noted, the results for group 2 varying from 33% below to 27% above the average. Results of comparative

Table No. 14a.

TESTS OF PLAIN CONCRETE COLUMNS.

University of Illinois, 1907.*

All columns were 12 in. in diameter by 10 ft. long.

Group.	Col. No.	Kind of Concrete.	Crushing Strength, Lbs/in².		Age of Specimen, Days.
			Individual Tests.	Average for Group.	
1	111	1:1½:3	2120	2300	66
	112		2480		62
2	101	1:2:4	1165	1740	58
	102		2000		69
	103		2210		65
	104		1590		64
	105		1945		62
	108		1460		72
	109		1810		64
3	116	1:3:6	955	1030	61
	117		1110		62
4	121	1:4:8	575	575	63
	122		575		63
5	110	1:2:4	1925	2025	203
	128		1845		194
	129		1770		181
	163		2680		187
	164		2160		187
	168		1770		201
6	21	1:2:3¾	2650	2710	12 mo.
	22		2770		16 mo.

tests on short cylinders of 1:2:4 concrete, stored in damp sand for 9 to 11 months, gave an average crushing strength of 2650 lbs/in^2; tests on 12-inch cubes stored in air at age of 60 days gave an average value of about 1950 lbs/in^2, and at age of about 200 days, of 2350 lbs/in^2.

Other tests on plain concrete are given in Tables Nos. 17 and 18a. The results in Table No. 17 appear to be unusually

* From Bulletin No. 20, Eng. Exp. Sta., University of Illinois.

low. Tests made at the University of Wisconsin in 1908 indicate that with careful workmanship and testing an average value of about 2000 lbs/in² can be obtained in 60 days on 1:2:4 concrete, the results obtained being very uniform.

114. Tests of Columns with Longitudinal Reinforcement only.—The results of a valuable series of experiments made at the Massachusetts Institute of Technology are given in Table No. 15.* The concrete was 1:3:6 broken stone concrete; the rods were partly plain square rods and partly twisted rods, the strength of the plain rods being 56,000–60,000 lbs/in², and of the twisted rods about 80,000 lbs/in². Where single rods were used they were placed in the centre, and where four rods were used they were placed in the form of a square one-half the dimensions of the column. The columns were approximately thirty days old.

TABLE NO. 15.

TESTS OF REINFORCED COLUMNS.

MASSACHUSETTS INSTITUTE OF TECHNOLOGY.

Number.	Cross-section.	Ratio: Length / Diam.	Number of Rods and Size (Square).	Plain or Twisted.	Area of Steel, Sq. In.	Percentage of Reinforcement.	Crushing Strength, Lbs/in².
1	8″×8″	25.5	1 1″	P	1	1.56	1670
2	"	25.5	1 1″	T	1	1.56	1985
3	"	18.0	1 1″	P	1	1.56	1560
4	"	18.0	1 1″	T	1	1.56	1970
5	"	9.0	1 1″	P	1	1.56	2160
6	"	9.0	1 1″	T	1	1.56	2080
7	"	25.5	1 1¼″	P	1.56	2.44	2125
8	"	25.5	1 1¼″	T	1.56	2.44	2410
9	"	25.5	4 ¾″	P	2.25	3.51	2840
10	"	25.5	4 ¾″	T	2.25	3.51	2610
11	"	18.0	4 ¾″	T	2.25	3.51	2300
12	"	18.0	4 ¾″	P	2.25	3.51	2390
13	"	9.0	4 1″	T	4.0	6.25	2470
14	"	9.0	4 1″	P	4.0	6.25	3810
15	10″×10″	20.4	1 1″	P	1	1	2150
16	"	7.2	1 1″	P	1	1	2000
17	"	7.2	1 1″	T	1	1	2284
18	"	14.4	1 1¼″	T	1.56	1.56	2620
19	"	14.4	1 1¼″	P	1.56	1.56	2570
20	"	14.4	4 ¾″	T	2.25	2.25	3000
21	"	14.4	4 ¾″	P	2.25	2.25	2740

* Trans. Am. Soc. C. E., Vol. L, 1903, p. 487.

Grouping these tests in accordance with the amount of reinforcement we have the following average values:

	Per Cent Reinforcement.	Average Strength, Lbs/in².	Calculated Strength, $f=1470(1+19p)$. Lbs/in².
8″×8″ columns. Average length = 12.4 ft.	1.56	1904	1904
	2.44	2267	2170
	3.51	2535	2450
	6.25	3140	3250
			$f=1800(1+19p)$.
10″×10″ columns. Average length = 11.0 ft.	1.0	2145	2145
	1.56	2452	2320
	2.25	2870	2600

It is evident that the larger columns are, for like reinforcement, stronger than the smaller columns, showing an effect either of ratio of length to diameter or of diameter directly. Little difference is observed between plain and twisted bars. The effect of amount of reinforcement can be observed by considering each size separately. The results have been studied on the basis of the theoretical formula of Art. 95, Chapter III,

$$\frac{P'}{P}=1+(n-1)p, \quad . \quad . \quad . \quad . \quad . \quad . \quad . \quad (1)$$

in which P'/P represents the ratio of the strength of the reinforced to that of the plain concrete column.

No results are given for plain concrete columns, but assuming that the column with the lowest percentage of steel follows the theoretical law the strength of the ideal plain concrete column is calculated to be 1470 lbs/in² for the first group and 1800 lbs/in² for the second group, making $n=20$. Taking these values then as a basis the results are plotted in Fig. 54. Abscissas represent per cent of reinforcement and ordinates the relative strengths, that of the ideal plain concrete being 100. The theoretical relation is shown by the straight line drawn for $n=20$. This value of n corresponds to a value of E_c of 1,500,000, which would be a reasonable value at rupture on the basis of total deformation, as explained in Art. 24. While the results are not sufficiently numerous to be at all conclusive, they do indicate that the relative strength

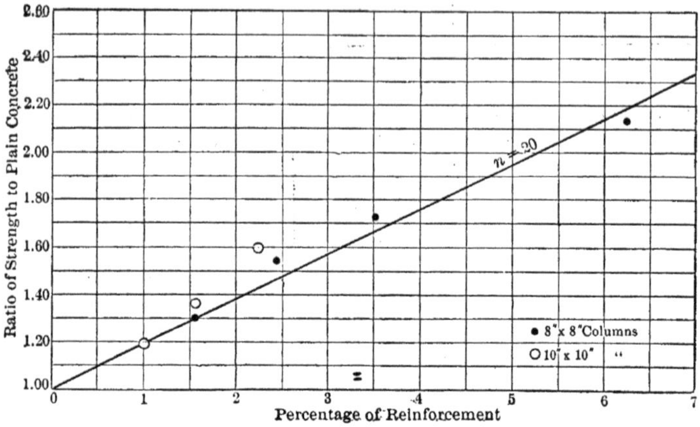

FIG. 54.—Tests of Reinforced Columns. (Mass. Inst. of Technology.)

of such columns is fairly represented by the theoretical law. Calculated values corresponding to the theoretical lines of the diagram are given by the formulas

$$f=1470(1+19p)$$

and

$$f=1800(1+19p).$$

These values are given in the table on p. 153. Eliminating the longest columns of the first group a fairly correct value for the ultimate strength of all would be given by $f=1600(1+19p)$ (n is assumed equal to 20).

The following table gives results of tests made at the Watertown Arsenal on concrete columns reinforced with longitudinal bars only. All columns were 8 ft. long and approximately $12'' \times 12''$ square; age, $3\frac{1}{3}$ to 8 months.

TABLE NO. 16.

TESTS OF REINFORCED COLUMNS.

WATERTOWN ARSENAL, 1904–1905.

Kind of Concrete.	Reinforcement. Description.	Reinforcement. Per Cent.	Compressive Strength, Lbs/in².	Strength of Plain Concrete. (See Table No. 14.)	Ratio of Strength of Reinforced Concrete to Plain Concrete.
1:2 mortar.......	8 ¾″ bars	2.85	4200	3070	1.37
1:3 "	" "	2.87	3841	2377	1.61
1:4 "	" "	2.86	3377	1518	2.22
1:5 "	" "	2.86	2813	1060	2.65
1:5 "	13 ¾″ "	4.63	3905	1060	3 68
1:1:2 (pebbles)...	4 ¾″ twisted	1.46	2890	1720	1.68
1:2:4 " ...	" "	1.43	1990		1.17
" " ...	4 ¾″ Thacher	1.03	1990		1.17
" " ...	4 ⅝″ corrugated	.97	2180		1.28
" " ...	4 ¾″ twisted	1.45	1820	1710	1.06
" " ...	8 ¾″ "	2.86	3160		1.84
" " ...	8 ¾″ Thacher	2.09	2760		1.62
" " ...	8 ⅝″ corrugated	1.94	2830		1.66
1:3:6 " ...	4 ¾″ twisted	1.44	1370	462	2.96
1:3:6 (trap-rock)..	8 ⅝″ corrugated	1.94	2290	1350	1.82
" "	" "	1.93	2650		
1:2:4 (cinders)....	4 ¾″ twisted	1.45	2095	871	2.40
1:3:6 " ...	4 ¾″ bars	1.42	1932	1060	1.82
" "	8 ¾″ "	2.83	3100		2.92

On Fig. 55 are plotted the results of the mortar tests and the 1:2:4 concrete in the same manner as the values in Fig. 54, using as a standard the results on plain concrete given in Table No. 14. Average values have been plotted for the columns with percentages of .97 and 1.03 and of 1.43 and 1.45. Lines have also been drawn representing the theoretical relations for different values of n. In the mortar tests the results show that for the poorer mortars the relative effect of the steel is high, corresponding to what would be obtained theoretically by using a value of $n=40$ to 50. In the 1:2:4 concretes the results do not vary widely from the theoretical results for $n=30$, or a value of E_c at rupture of 1,000,000.

It is assumed in the theoretical discussion that the steel is not stressed beyond its elastic limit. It is to be noted that in these tests the stress on the steel bars must have been as high as 45,000 to 50,000 lbs/in^2, showing the usefulness of a fairly high elastic-limit steel in this case. (See further discussion in Chapter V.)

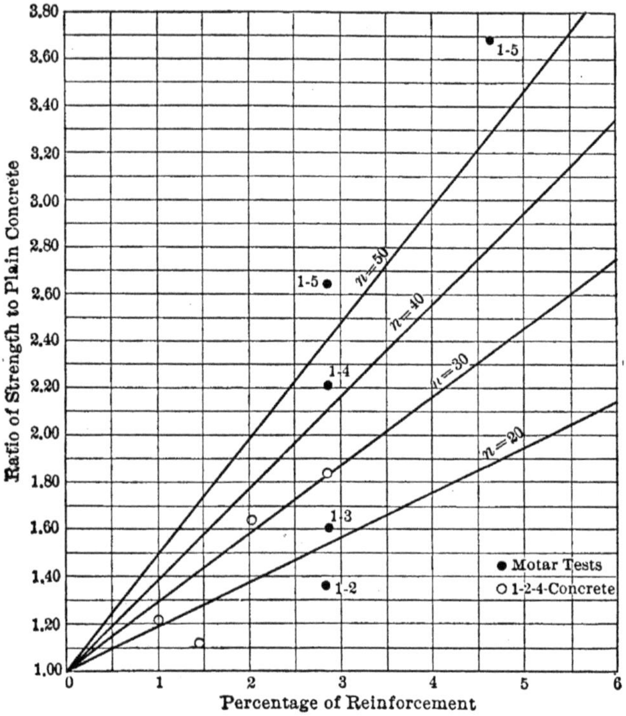

FIG. 55.—Tests of Reinforced Columns. (Watertown Arsenal.)

Table No. 17 contains the results of tests made by Professor A. N. Talbot at the University of Illinois. The columns were made of 1:2:3–3/4 concrete and plain steel of 39,800 pounds per square inch elastic limit. The age was from 59 to 71 days. Comparing the reinforced with the plain concrete,

TABLE No. 17.

TESTS OF REINFORCED COLUMNS.

UNIVERSITY OF ILLINOIS, 1906.*

No.	Length.	Cross-section.	Reinforcement.		Crushing Strength. Pounds per sq. in.	
			Kind.	Per cent.	Individual Test.	Average of Group.
1			4 $\frac{3}{4}$-in. rods	1.20	1587	
3			4 $\frac{3}{4}$-in. rods 12 $\frac{1}{4}$-in. ties	1.21	1862	
7	12 ft.	12″×12″	4 $\frac{3}{4}$-in. rods	1.21	1850	1809
11			4 $\frac{3}{4}$-in. rods 12 $\frac{1}{4}$-in. ties	1.21	1936	
2	12 ft.		4 $\frac{5}{8}$-in. rods	1.52	1577	
6	"		4 $\frac{5}{8}$-in. rods	1.52	1600	
10	"		4 $\frac{5}{8}$-in. rods 12 $\frac{1}{4}$-in. ties	1.50	1280	
12	9 ft.	9″×9″	4 $\frac{5}{8}$-in. rods 9 $\frac{1}{4}$-in. ties	1.48	2335	1710
14	12 ft.		4 $\frac{5}{8}$-in. rods 12 $\frac{1}{4}$-in. ties	1.50	1367	
16	9 ft.		4 $\frac{5}{8}$-in. rods 9 $\frac{1}{4}$-in. ties	1.49	1607	
17	6 ft.		4 $\frac{5}{8}$-in. rods	1.47	2206	
5	12 ft.	12″×12″			1710	
8	"	9″×9″			2004	
9	"	12″×12″			1610	
13	"	" "	Plain	0	1709	1550
15	6 ft.	" "			1189	
18	"	9″×9″			1079	

the average strength of the 12″×12″ columns with 1.2 per cent reinforcement is about 1.17 times as great, and the 9″×9″ columns with 1.5 per cent reinforcement is about 1.10 times as great. These tests indicate a less effect of reinforcement than some of the other tests quoted. The smaller cross-section of the columns containing the larger amount of reinforcement may have been the cause of the lower strength of this group. It is important to note the wide variation in the individual results of these and other tests; they indicate what may be expected in practice, and show clearly the necessity of adopting conservative values of working stress. Careful measurement

* Bulletin No. 10, Engineering Exp. Sta., 1907.

of distortions showed that tbe ratio of stress in steel to stress in concrete varied from about 14 at the beginning to about 27 at rupture, taking average values. The low values for ultimate strength of the reinforced columns appeared to be due to a lower actual crushing strength of the concrete in these columns than in the plain columns.

115. Tests of **Hooped Columns.**—If a compression member be reinforced by bands or hoops closely spaced, such reinforcement will raise the ultimate strength by preventing lateral expansion under the compressive forces. It was shown in Art. 96 that under this system of reinforcement the steel cannot be stressed to any considerable extent under loads below the usual elastic limit strength of the concrete. This limit being exceeded, however, the banding becomes very effective in holding the concrete together so that it will endure large deformations without rupture, thus increasing greatly its ultimate strength. Longitudinal reinforcement is also used with hoops or bands. This part of the reinforcement will receive stress in proportion to the longitudinal deformations and will thus be more effective at low loads than the bands. Results on both forms of columns are here given.

In 1902 and 1903 Considère * published certain tests made on columns reinforced by spirally wound wire and by longitudinal rods or wire. His most important results were those obtained upon a number of octagonal columns 5.9 in. short diameter. As a result of these and other tests, as well as from a theoretical basis, he came to the conclusion that steel in the form of spiral reinforcement was 2.4 times as efficient as in the form of longitudinal reinforcement, presuming the spacing of the wire to be not great ($\frac{1}{4}$ to $\frac{1}{10}$ of the diameter of the spiral) and that ordinary mild steel be used. It was found also desirable to use a small amount of steel in the form of longitudinal reinforcement. Tests on the elastic properties showed considerable deformation and set, but after the first application of

* Génie Civil, 1902.

a load the column is relatively rigid, with greatly increased value of E.

Table No. 17A gives the results of an important series of tests on hooped columns made by Professor Talbot, in 1907.

TABLE No. 17A.

TESTS OF HOOPED COLUMNS.

UNIVERSITY OF ILLINOIS, 1907.*

Concrete 1:2:4; age, from 56 to 69 days; length, 10 ft.; diam., 12 ins.

Group.	Col. No.	Reinforcement.			Crushing Strength, Lbs/in.²	
		Kind.	Size and Spacing.	Per Cent.	Individual Tests.	Average of Group.
1	131	Electrically welded bands.	No. 16, 2 in. c.-c.	1.08	2384	2239
	132			1.08	2150	
	133			1.05	2182	
2	136		No. 12, 2 in. c.-c.	2.08	2860	2877
	137			2.07	2660	
	138			2.12	3110	
3	146		No. 8, 2 in. c.-c.	.22	3000	3202
	147			.20	3715	
	148			3.20	2890	
4	143		No. 12, 3 in. c.-c.	1.39	2735	2735
5	141		No. 12, 4 in. c.-c.	1.02	2275	2226
	142			1.02	2178	
6	171	High carbon wire spiral.	No. 7	0.85	2503	2505
	172			0.85	2506	
7	181		$\frac{1}{4}$ in.	1.73	2718	3437
	182			1.67	3800	
	183			1.68	3793	
8	176	Mild steel wire spiral	No. 7	0.84	2080	2168
	177			0.85	2203	
	178			0.84	2220	
9	186		$\frac{1}{4}$ in.	1.64	2068	2736
	187			1.71	3404	
	188			1.61		

* Bull. No. 20, Eng. Exp. Sta., Univ of Ill.

Two forms of hooping were used, electrically welded bands 1 in. wide and of various gage thickness, and spirally wound wire at a pitch of 1 in. The steel used in the bands had a yield point of about 48,000 lbs/in^2. The wire was of two kinds, high carbon and mild steel. The former had a yield point of 115,000 lbs/in^2 for the $\frac{1}{4}$-in. size and 60,000 lbs/in^2 for the No. 7; the latter had yield points for the same sizes of 54,000 and 38,500 lbs/in^2, respectively. The columns were 10 ft. long by 12 in. in diameter. A thin film of mortar covered the hooping.

In a study of these tests it is desirable to keep in mind the two more or less independent elements, namely, the ultimate strength, and the behavior of the specimen previous to rupture, which is shown best by the stress deformation diagram. The latter element may be of greater importance than the former.

As to ultimate strength, the results may be compared by groups with those for plain concrete given in Table No. 14A, group 2. The figures are here brought together:

COMPARISON OF HOOPED AND PLAIN CONCRETE COLUMNS.

Group.	Kind of Reinforcement.	Average Amount of Reinforcement, Per Cent.	Average Ultimate Strength, Lbs/in^2.	Excess over Plain Concrete, Lbs/in^2.	
				Total.	Per 1% Reinforcement.
2	Plain concrete	0	1740	....	....
1	Bands	1.07	2239	599	560
2	Bands	2.09	2877	1137	540
3	Bands	3.21	3202	1462	450
4	Bands	1.39	2735	995	710
5	Bands	1.02	2226	486	480
6	High carbon wire	0.83	2505	765	920
7	High carbon wire	1.69	3437	1697	1000
8	Mild steel wire	0.84	2168	428	510
9	Mild steel wire	1.65	2736	996	600

The average increase per 1% of steel for the banded columns is about 570 lbs/in^2; for the high carbon wire, 960 lbs/in^2: and for the mild steel wire, 555 lbs/in^2. As shown in Art. 114, the effect of longitudinal reinforcement may be taken, in accordance with theory, as equal to $f_c(n-1)1/100$ for each 1% of steel. For the plain concrete columns in these tests the value of n at rupture was about 17, hence for 1% longituinal reinforcement the strength should be equal to $1740 \times (17-1)1/100$ or about 280 lbs/in^2. Comparing this with the results in the table it is seen that the 1% of steel in the form of bands added about twice as much to the strength of the column, and in the form of spirally wound wire, from twice to three and one-half times as much. Furthermore, it would appear that the increase in strength within the limits of the tests is about proportional to the amount of steel used. It should be said that in Professor Talbot's analysis the strength of the plain concrete is estimated for each column by a study of its deformation and not from the test on plain columns. He thus arrives at values for plain concrete averaging about 1600 lbs/in^2, resulting in a still better showing for the reinforcement.

The effect of hooping upon the deformation of the columns and their general behavior before rupture is of perhaps greater importance than its effect upon ultimate strength. The results in general are in accordance with the discussion of Art. 96. For loads below that corresponding to the ultimate strength of a plain concrete column there is no strengthening effect of the hooping apparent, but beyond this load the column shortens rapidly and the hooping comes into action. The hooped columns in fact seem to be somewhat less stiff at low loads than the plain concrete, due, possibly, to the interference of the bands in the fabrication. The total deformation at rupture is very great, amounting to from 6 to 12 times that for plain concrete; and at maximum load it is about 5 times as great. Scaling of the exterior shell occurred at loads corresponding to the ultimate strength of plain concrete.

The deformation of plain concrete and of hooped concrete

with different amounts of reinforcement is well shown in Fig. 55a, in which have been reproduced typical stress-strain curves for some of the columns tested. These curves should be carefully studied. They bring out in a striking manner the fact that the chief effect of hooping is to increase the "toughness" or "ductility" of the concrete, to increase its ultimate strength, though to a less extent, but to cause little or no change in the deformation of the column in the early part of the test. It is seen that up to a stress of 1200 to 1500 lbs/in^2 the deformations are about the same in all columns, but that beyond this the deformation of the hooped columns rapidly increase. As usually interpreted, it would appear that the elastic limit of all the columns is in this region of stress and is about the same for all. Doubtless a small set would occur at still smaller loads, but up to 1200–1500 lbs/in^2 the set could not be great. That the hooped column would act in this respect much like mild steel is indicated by the test on column No. 173 in which the load was removed and reapplied with results shown in the figure.

Tests made at the Watertown Arsenal in 1905 * showed results very similar to those quoted above. The reinforcement consisted of riveted bands 1.5×0.12 in. and longitudinal angle bars 1×1×⅛ in. The columns were 1:2:4 concrete 5 and 6 months old and were 10½ in. in diameter by 8 ft. long. The entire column was inclosed by the bands. The results are given in Table No. 17B.

The additional strength of the hooped columns over the plain concrete for each 1% of reinforcement was 819 lbs/in^2 for 13 hoops, 1120 lbs/in^2 for 25 hoops and 1140 lbs/in^2 for 47 hoops. The additional strength due to the angle bars over the same column without the bars was 797 lbs/in^2 for the one with 13 hoops and 761 lbs/in^2 for that with 25 hoops. These values are about equal to the highest obtained by Talbot.

Of much significance in these tests also is the character

* Tests of Metals, 1906.

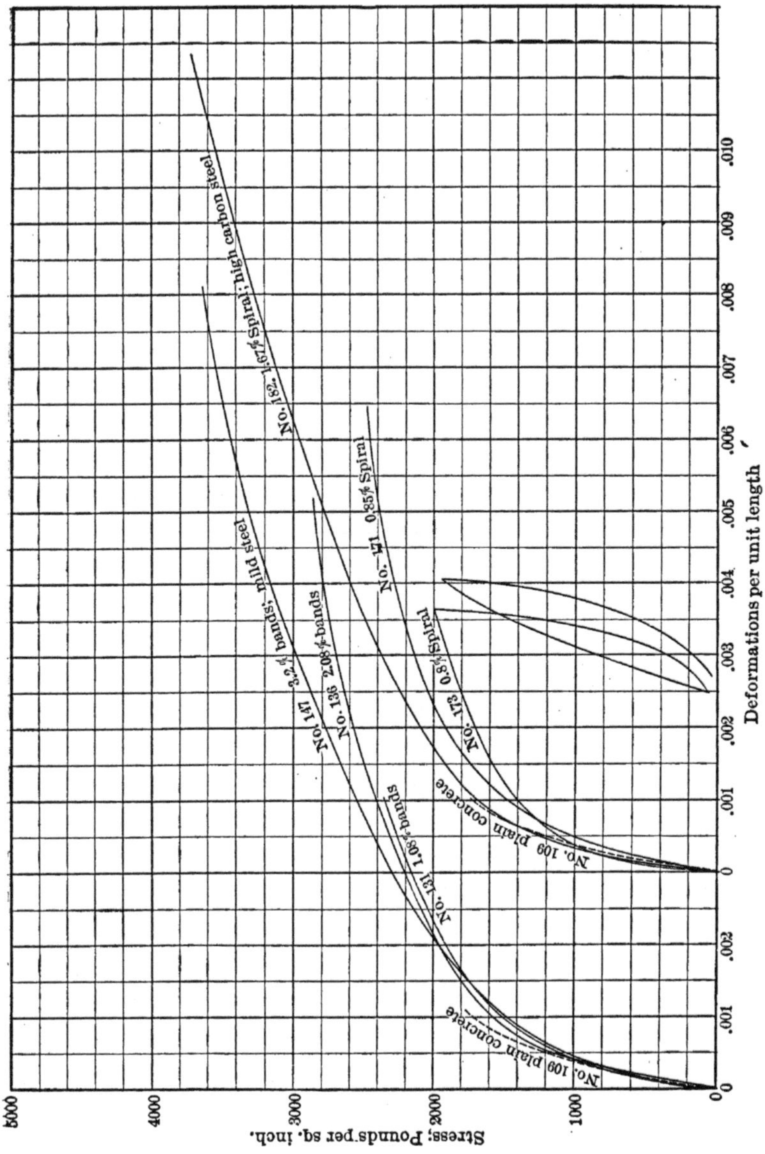

FIG. 55a.—Tests of Hooped Columns (Talbot.)

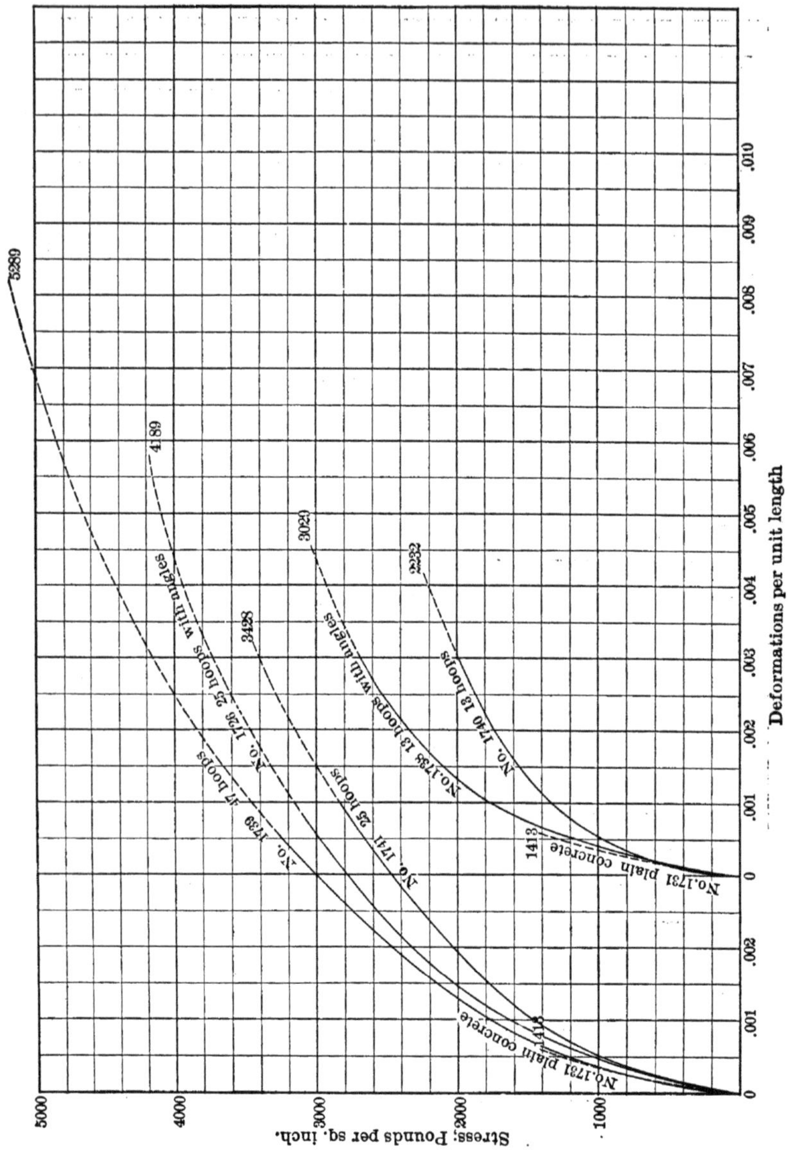

FIG. 55b.—Tests of Hooped Columns (Watertown Arsenal).

TABLE No. 17B.

TESTS OF HOOPED COLUMNS.

WATERTOWN ARSENAL, 1905.

Reinforcement.			Strength, Lbs/in².
Kind.	Per Cent Hoops.	Per Cent Longitudinal.	
Plain concrete	....	....	1413
13 hoops	1.0	.0	2232
13 hoops, 4 L's	1.0	1.0	3029
25 hoops	1.8	.0	3428
25 hoops, 4 L's	1.8	1.0	4189
47 hoops	3.4	.0	5289

of the stress deformation curves, as shown in Fig. 55*b*. The general effect of banding is the same as indicated in Talbot's tests; it increases greatly the ultimate strength but does not stiffen the column up to the ultimate strength of plain concrete. The effect of longitudinal reinforcement on the deformation curve is marked. In stiffening the column at early stages, or in raising its elastic limit, it has a much greater effect than an equal amount of band reinforcement.

Thus at the deformation of .00015, corresponding to a stress in the longitudinal steel of 45,000 lbs/in², 1% of longitudinal steel increases the resistance for the column of 13 hoops, from 1520 to 2080, or about 560 lbs/in², whereas adding .8% in hoops increases it to 1780, or 260 lbs/in². Even 47 hoops, or 2.4% of added steel, adds no more strength at this deformation than 1% of longitudinal steel. If the elastic limit of the longitudinal steel is 45,000 lbs/in², then its ultimate resistance would be reached at a deformation of .0015, and at this deformation 1% of such reinforcement would, theoretically, add $45{,}000 \times .01 = 450$ lbs/in². This figure appears to have been exceeded in these tests.

Table No. 18 gives results of a series of experiments on hooped columns conducted by Bach.* The columns were of

* Quoted from Mörsch, Eisenbetonbau, p. 70.

TABLE No. 18.
TESTS ON HOOPED COLUMNS.
(BACH.)

Group.	No.	Spiral Reinforcement.		Longitudinal Reinforcement.		Per Cent Spiral Reinforcement.	Per Cent Longitudinal Reinforcement.	Per Cent Total Reinforcement.	Crushing Strength in Lbs./Sq. In.	Ratio of Strength to Plain Concrete.
		Pitch, Inches.	Diameter, Inches.	Number of Rods.	Diameter, Inches.					
	1		No reinforcement			....	....	....	1890	1.00
I	2	1.496	.197	4	.276	0.64	0.25	0.89	2260	1.20
	3	1.457	.276	4	.276	1.29	0.25	1.54	2530	1.34
	4	1.654	.394	4	.276	2.30	0.25	2.55	3410	1.80
II	5	1.496	.197	8	.433	0.64	1.22	1.86	3220	1.70
	6	1.457	.276	8	.433	1.29	1.22	2.51	3270	1.73
	7	1.694	.394	8	.433	2.25	1.22	3.47	4000	2.12
III	8	1.222	.276	4	.276	1.53	0.25	1.78	2850	1.51
	9	1.576	.394	4	.276	2.42	0.25	2.67	3000	1.59
	10	1.615	.473	4	.276	3.41	0.25	3.66	3640	1.93
	11	1.457	.551	4	.276	5.12	0.25	5.37	3500	1.85
IV	12′	1.575	.276	8	.197	1.19	0.25	1.44	2320	1.23
	12²	1.575	.394	8	.276	2.42	0.49	2.91	3270	1.73
	12³	1.575	.551	8	.394	4.75	1.01	5.76	4300	2.28
V	13′	3.150	.276	8	.276	0.60	0.49	1.08	2310	1.22
	13²	3.150	.394	8	.394	1.22	1.01	2.23	2580	1.36
	13³	3.150	.551	8	.473	2.38	1.45	3.83	2[illegible]30	1.50
VI	14′	4.725	.276	8	.394	0.40	1.01	1.41	2205	1.17
	14²	4.725	.394	8	.473	0.82	1.45	2.27	2600	1.38
	14³	4.725	.551	8	.551	1.60	1.96	3.56	2950	1.46

octagonal form with short diameter equal to 275 mm. and height of 1 m. The diameter of the spirals was 245 mm. The concrete was 1:4 gravel concrete 5–6 months old. Each result is the average of three tests, except in the case of the unreinforced concrete, where four tests were made. The steel was mild steel. The strength is calculated with reference to the gross section of the column.

It is difficult to draw any definite conclusions from these

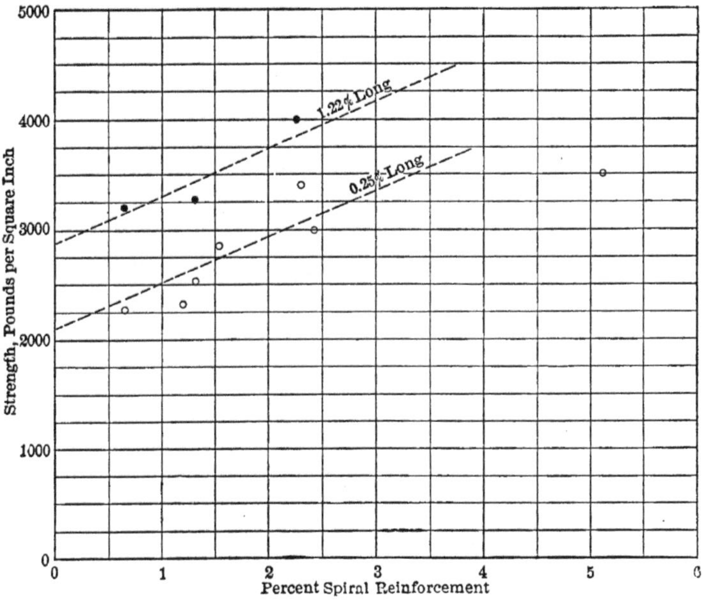

FIG. 56.—Tests of Hooped Columns (Bach).

tests as to the relative value of longitudinal and hoop reinforcement. However, some suggestions of value may be obtained from a consideration of those tests where the longitudinal steel was .25 and 1.22%. These include all of the first three groups and the first of the fourth group. These results are plotted in Fig. 56 in two groups, corresponding to the two percentages of longitudinal steel, using the percentages of spiral steel as abscissæ. Approximate average straight lines are drawn through each group. They indicate, roughly, that the strength is increased by 425 lbs/in² for each percentage of spiral reinforcement for both groups and that the strength of the upper group is about 800 lbs/in² greater than the lower group, showing a strengthening effect of about 800 lbs/in² for 1% additional longitudinal reinforcement when used with spirals. Thus, so far as these tests go, the longitudinal steel is fully as effective as the spiral steel in the proportions here used. Notice that the strength of groups V and VI is relatively less than the others. This appears to be due to the wider spacing of the spirals in these groups. These results, as compared to those already quoted, are doubtless modified by the fact that there was a considerable thickness of concrete outside of the spirals and generally such outer shell will crack and fall away sometime before final failure. The values given refer to gross-section.

The results of an extensive series of tests of reinforced columns made by Mr. M. O. Withey at the University of Wisconsin are given in Table No. 13A and in Figs. 56*a* and 56*b*.* Columns *A* were unreinforced; *E* were reinforced by longitudinal bars only with wire ties widely spaced; *B* were reinforced by latticed angles forming a square 8″× 8″ in section; the remaining columns by spiral wire and varying proportions of longitudinal reinforcement wired to the spirals on the inner side. The diameter of the columns was made equal to the outside diameter of the spirals, but the calculations have been made with reference to the inside diameter of 10 inches. The yield-point of the steel reinforcement was 38,000 to 45,000 lbs/in² for the longitudinal steel and 80,000 to 105,000 lbs/in² for the spiral steel.

Noteworthy are the characteristics shown by the curves of Fig. 56*a*. The "toughening" effect of the hooping is well shown; it increases greatly the ultimate strength of the column

* *Eng. Record*, July 10, 1909.

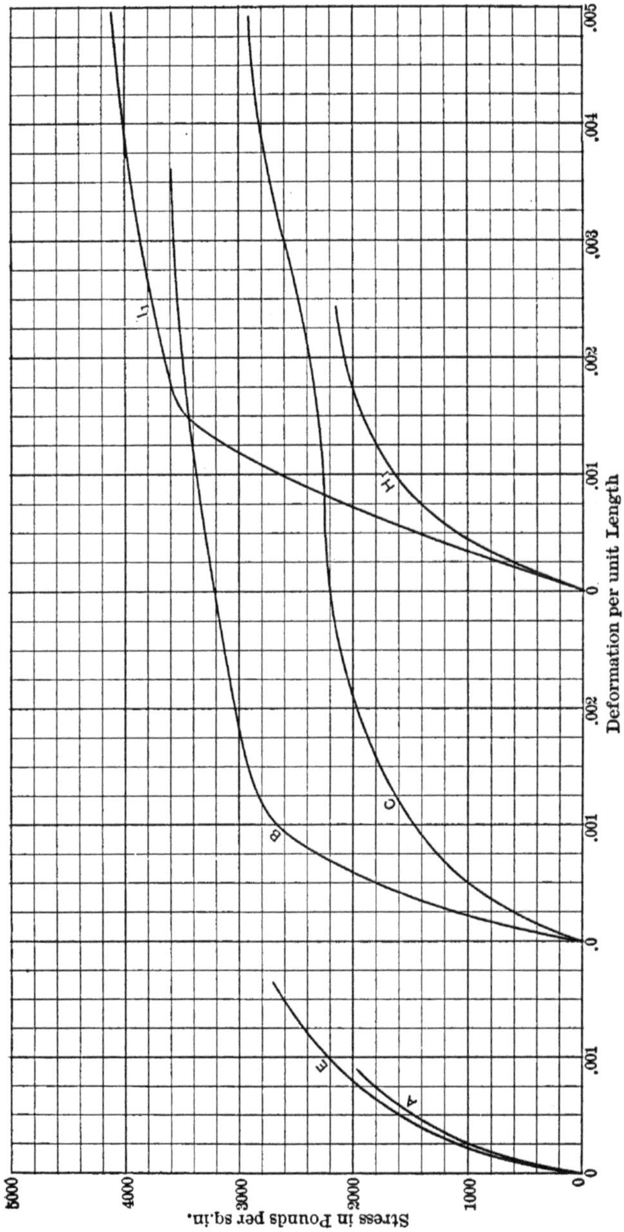

FIG. 56a.—Tests of Reinforced Columns (Withey).

TABLE No. 18A.

TESTS OF REINFORCED COLUMNS.

UNIVERSITY OF WISCONSIN.*

Length of columns *A* . . . *C* = 120″, *H* . . . *U* = 100″. Age 2 months.

No.	Reinforcement. Kind.	Reinforcement. Per Cent Vertical.	Per Cent Lateral.	Concrete. Mix.	Concrete. Compressive St'nth of Cyl., lbs/in²	ross-Csection, sq.in.	Average Strength of Columns, lbs/in². At Yield-point.	At Max. Load.	No. of Tests
A	None	0	0	1:2:4	2360	142	. . . .	2070	4
E	9 ⅝″ rods with ¼″ ties, 1 ft c. to c.	2.35	0.11	1:2:4	2300	118	. . . .	2470	3
B	4 2″×2″×3/16″ latticed angles	4.5	. . . .	1:2:4	2200	64	2860	3740	2
C	¼″ spiral, 1″ pitch	0	2.0	1:2:4	2180	78.5	1950	3660	1
H	No 7 wire spiral, 2″ pitch	0	0.46	1:2:3½	1750	78 5	1850	2230	2
G	8 ½″ rods and No 7. wire spiral, 2″ pitch	2.0	0.46	1:2:3½	1760	78.5	2710	3300	2
I	8 11/16″ rods and No. 7 wire spiral, 2″ pitch	3.78	0.46	1:2:3½	2180	78.5	3470	4160	2
J	8 ⅞″ rods and No. 7 wire spiral, 2″ pitch	6.11	0.46	1:2:3½	2050	78.5	4240	5120	2
L	No. 7 wire spiral, 1″ pitch	0	0.92	1:2:3½	1770	78.5	1370	2640	2
K	8 ½″ rods and No. 7 wire spiral, 1″ pitch	2.0	0.92	1:2:3½	2000	78.5	2610	3900	2
N	8 11/16″ rods and No. 7 wire spiral, 1″ pitch . . .	3.78	0.92	1:2:3½	1800	78.5	3370	4190	2
M	8 ⅞″ rods and No. 7 wire spiral, 1″ pitch	6.11	0.92	1:2:3½	1680	78.5	3760	4680	2
P	8 1″ rods and No. 7 wire spiral, 1″ pitch	8.0	0.92	1:2:4	2360	78.5	5660	6920	2
O	8 ⅞″ rods and ¼″ spiral 1″ pitch	6 11	1.96	1:2:4	2480	78.5	4480	6580	2
R	8 1″prods and ¼″ spiral 1″ pitch	8.0	1.96	1:2:4	2380	78.5	5190	6960	2
Q	8 1⅛″ rods and ¼″ spiral, 1″ pitch.	10 12	1.96	1:2:4	2300	78.5	5760	7090	2
S	No 7 wire spiral, 1″ pitch .	0	0.92	1:1:2	4070	78.5	4050	5850	2
T	8 ⅞″ rods and No. 7 wire spiral, 1″ pitch	6.11	0.92	1:1:2	4400	78.5	5760	7290	2
V	No. 7 wire spiral, 1″ pitch .	0	0.92	1:1⅓	4870	78.5	3570	5340	2
U	8 ⅞″ rods and No. 7 wire spiral, 1″ pitch	6.11	0.92	1:1⅓	4550	78.5	5950	8150	2

* *Engineering Record*, July 10, 1909.

but does not affect the "yield-point" greatly. Fig. 56*b* gives all results from columns *H* to *U*. The yield-point is taken to be the point on the stress-strain diagram where the rate of deformation increases rapidly, as illustrated by Curve I, Fig. 56*a*. It is very nearly the deformation at which plain concrete would fail. It was found that this "yield-point" is closely given by the formula $P_1 = A[(1-p)f_c + pf_s]$, in which f_c is the ultimate strength of the concrete, f_s is the yield-point of the longitudinal steel, and p is the steel ratio for the longitudinal steel. The spiral steel is not included as its effect is very small. The ultimate strength is given approximately by the formula

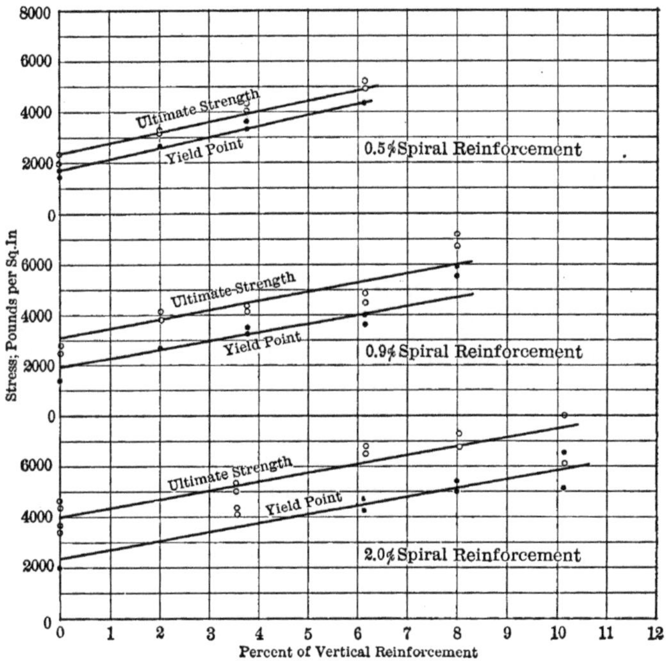

FIG. 56*b*.—Tests of Reinforced Columns (Withey).

$P = A[(1-p)f_c + pf_s + 0.12f_s'\sqrt{p'}]$, in which p' is the steel ratio for the spiral steel and f_s' is the yield-point of this steel.

116. Conclusions as to Strength of Columns.—From the results of tests quoted we may draw the following conclusions: that the strength of plain concrete columns of 1:2:4 mixture at 60 days may be taken at from 1600 to 1800 lbs/in^2; that very great gain in strength is shown for both plain and reinforced concrete by the use of richer mixtures; that the strength of columns reinforced with longitudinal rods only (or when fastened together at wide intervals) may be estimated in accordance with theory, but that the density and rigidity of the concrete itself is apt to be less in the reinforced than in the plain column, so that for small percentages of longitudinal reinforcement the gain in strength is small; that hooped columns, with or without longitudinal steel, show greatly increased deformation before rupture and a much higher ultimate strength than columns without such hooping, but the yield-point is not materially affected; that more than one per cent of hooping does not appear to be necessary; that the addition of longitudinal steel to hooped columns increases greatly the elastic limit and the ultimate strength of the columns.

116a. Effect of Length of Column on Compressive Strength.—Comparing the results on plain concrete columns, p. 186, with the tests on cubes, pp. 11–14, it is evident that the strength of the column is materially less. While there is thus a very considerable reduction of strength as compared to the cube, there appears to be little difference in the strength of columns of various lengths up to 15 to 20 diameters. A series of tests made at the Watertown Arsenal * for the Aberthaw Construction Co. on 12″×12″ columns gave practically the same results for all lengths from 2 ft. to 14 ft., the average of all being 957 lbs/in^2 for hand-mixed and 1099 lbs/in^2 for machine-mixed concrete. The temperatures were, however, low, and the results are not a fair criterion as to absolute strength.

In the tests of Table No. 15 the difference in average results upon the 8″×8″ columns and those on the 10″×10″ size is marked. But comparing results for each size among themselves there is little or no effect noticeable up to 25 diameters. Numbers 2 and 3 are reported as having failed by buckling, but these average practically the same as Nos. 1 and 4. From

* Tests of Metals, 1897.

these tests it would appear therefore that no account need be taken of length of column below about 20 diameters, although caution should be used in accepting these results as conclusive. In the case of hooped columns the effect of buckling is evident for shorter lengths, inasmuch as this kind of column has a sufficient toughness to permit of considerable deformation before failure.

Considering the comparatively brittle nature of the column with longitudinal reinforcement only, its use for columns of slender proportions should be discouraged. The banded or hooped columns is much more reliable for such work.

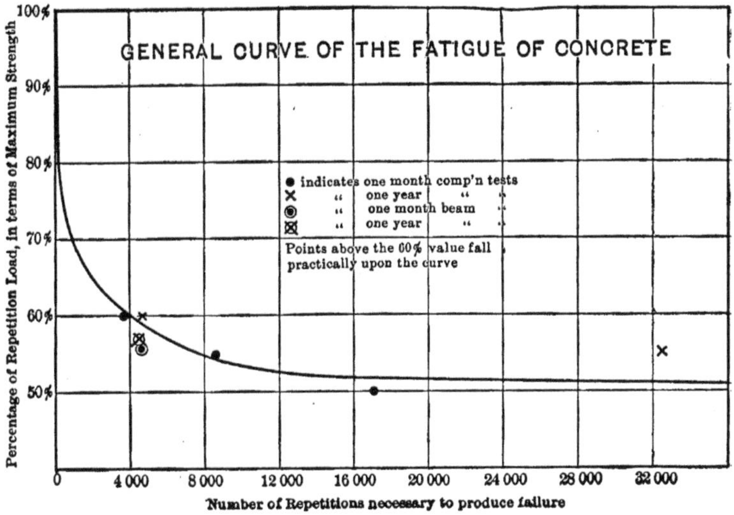

Fig. 57

117. Fatigue Tests of Reinforced Concrete.—Important experiments conducted by Professor J. L. Van Ornum * on reinforced beams indicate an effect under repeated application of loads similar to that which he found for mortar and concrete

* Trans. Am. Soc. C.E., 1907, LVIII, p. 294.

in compression as mentioned on p. 25. In the case of beams the failure under repeated loads appeared to be largely a gradual fracture in diagonal tension, ending with a compression failure. The number of repetitions required to produce failure varied with the load applied, rupture being ultimately produced after several thousand repetitions for loads as low as 55 and 60% of the usual ultimate strength. The most important of his results are indicated in Fig. 57, taken from his paper, showing the number of repetitions required to produce failure at various values of maximum load in percentage of the usual ultimate load.

The change in the modulus of elasticity was also investigated, and it was found that under repeated loads not ultimately causing rupture the concrete soon became perfectly elastic, with a value of the modulus of about two-thirds of its initial value. At loads ultimately causing rupture the modulus became for a time nearly constant, but rapidly decreased as rupture was approached.

These tests indicate that concrete when repeatedly loaded beyond about 50% of its ordinary ultimate strength will not remain indefinitely elastic and will fail. This limit may be called the permanent elastic or fatigue limit of concrete. It is of much importance in relation to working stresses.

In some repeated load tests at the University of Pennsylvania,* loads were applied many thousands of times, producing stresses in the concrete of from 25 to 40% of its ultimate strength (2460 lbs/in^2 in cube form). Some set in the compressive concrete was observed in all cases after the first few applications of the load. At a stress of 25% of the ultimate the set remained practically constant for 360,000 repetitions; at a stress of 32% of the ultimate the set continued to increase, reaching a value, in the case of one beam, of about twice as much at 500,000 repetitions as at 50,000; and at a stress of

* Eng. Record, Vol. 58, 1908; p. 90.

40% of the ultimate the set increased somewhat more rapidly. It was not found, however, that at these stresses the ultimate strength of the beam was affected. In these tests hair cracks formed in the tension side at loads producing stresses of 10,000 to 18,000 lbs/in^2 in the steel. These gradually extended well towards the neutral axis.

CHAPTER V.

WORKING STRESSES AND GENERAL CONSTRUCTIVE DETAILS.

118. Working Stresses and Factors of Safety.—In the design of steel structures it has come to be the practice to make use of definite working stresses rather than factors of safety. These working stresses are based, for the most part, on the permanent elastic-limit strength of the material, although the margin of safety between the elastic-limit and the ultimate strength (indicated by strength and ductility) receives consideration. The working stresses are made sufficiently below the elastic limit to provide for:

(*a*) Variations and imperfections in material and workmanship.

(*b*) Uncalculated stresses, such as secondary stresses, stresses due to unequal settlement, and, usually, those due to temperature changes.

(*c*) Dynamic effect of live load if not provided for by an allowance for impact.

(*d*) Possible increase in live load over that assumed, or rare applications of excessive loads.

(*e*) Deterioration of the structure.

The more accurately the various elements are determined in any case the closer may the working stress approach the elastic limit. Where the dynamic effect of the live load does not enter, or is otherwise fully provided for, and where items (*d*) and (*e*) are of small moment, working stresses for steel structures will vary from about one-half to two-thirds the elastic-limit strength of the material. Were it absolutely certain that the elastic limit of the material would never be exceeded in any emergency, then the margin of strength between the elastic limit and the ultimate strength would be of no importance. This is, however, not the case, and under actual conditions of service there is a very considerable element of safety in the fact that the ultimate strength is in most materials much higher than the elastic limit. Stated in another way, a designer would never use a working stress of one-half or two-thirds the elastic limit in a material where the ultimate strength did not considerably exceed this limit. While therefore the working stresses are selected chiefly with reference to the elastic limit, the ultimate strength also receives consideration.

In recent years most designers base their calculations on certain working stresses selected as above indicated. Formerly, and to some extent now, calculations are based on specified "factors of safety" referred to ultimate strengths. In either case both the elastic limit and the ultimate strength must be considered in the design, and experienced designers will arrive at about the same results by either method. In reinforced-concrete design the problem is complicated by the use of two unlike materials whose elastic limits and ultimate strengths are not similarly related. Furthermore, as the materials are stressed beyond their elastic limits the stresses do not necessarily increase in proportion to the load, so that if working stresses of one-fourth the ultimate are selected, for example, the corresponding load may be considerably greater or less than one-fourth the ultimate load. This condition makes it especially desirable to consider ultimate strength, and is an argument for the use of the "factor-of-safety" method.

119. Relative Effect of Dead and Live Loads.—The tendency of practice in the treatment of live-load stresses is to reduce them to equivalent dead-load stresses by the application of some sort of impact formula or by other means of estimation. The resulting stresses are then considered on the same basis as the usual dead-load stresses and a single set of working stresses applied. This method is simple, logical, and tends to facilitate a proper adjustment of the design to the conditions. Separate working stresses will give equally satisfactory results when properly selected, but the system is not as flexible or convenient as the method of the single working stress with impact coefficients.

The question of impact coefficients, or the relation between live- and dead-load working stresses, requires little special attention in connection with reinforced concrete structures. It is essentially the same as it is in the case of steel structures, excepting as the amount of impact may be modified by the structure itself. In steel railroad structures of short span, for example, the impact, or dynamic effect of live load, is usually assumed to be about 100% of the live load stresses. Experiments show that this is probably not too high and that the actual stresses from live load may be 100% greater than the static stresses, due largely to the effect of unbalanced locomotive wheels. Where a large amount of ballast intervenes between the load and the structure the impact is doubtless much less. In the case of concrete structures the great mass of the concrete undoubtedly tends to reduce the effect of impact and vibration, or to localize such effect more than in a steel structure. The conditions involved in concrete designing, therefore, are likely to be favorable as regards impact and may permit the use of lower coefficients than are used for steel structures. The proper coefficient to use, or the relation between live- and dead-load working stresses, varies much under different conditions and must be left to the judgment of the designer, or to formulas or rules prepared especially for the purpose. Further discussion of this question will not be undertaken here.

In buildings it is the practice in steel construction to use a single working stress, no account being taken directly of any special effect of the live load. Allowance is made in the design of large girders and columns which receive their load from large areas for the fact that such large areas, especially if on two or more floors, are seldom or never loaded to the extent assumed for smaller areas. This allowance varies with different conditions, but relates solely to the selection of the amount of live load rather than to its effect. In a building, when heavily loaded with its live load, the portion of the load which is in motion and capable of producing a dynamic effect is generally but a very small percentage of the total live load. In most cases, therefore, in building construction it is not necessary to treat the live-load stresses differently from the dead-load stresses, and the design is based on a single set of working stresses. Special cases will arise, however, where the dynamic effect of the live load requires consideration, as, for example, in the case of floors supporting moving machinery.

Whatever the effect of live load may be it can more readily be taken account of by adding to the resulting live-load stresses a percentage which, in the judgment of the engineer, will reduce them to their dead-load equivalent, and then apply a single set of working stresses, or factor of safety, to the sum of the stresses. The discussion of working stresses in the following articles will relate to the proper basal working stress for dead load, or for live load suitably increased for impact.

BEAMS.

120. Working Formulas.—From the analysis and results of experiments discussed in preceding chapters there would appear to be no good reason why the rational formulas as developed in Chapter III should not be used in designing. No empirical formula is needed. Furthermore, in the judgment of the authors, the simple formulas based on the straight-line stress variation should be used for purposes of design, safe working stresses being employed. These formulas are practically correct for such working stresses, and there seems to be no more reason to use formulas designed only to express ultimate strength than there is in the case of wooden or cast-iron beams where the conditions are similar. It is, however, desirable that the working stresses be selected with some reference to ultimate strength, although with principal reference to elastic strength.

121. Working Stresses in Concrete and Steel.—The strength of a beam is limited usually by:

(*a*) The compressive strength of the concrete,

(*b*) The elastic-limit strength of the steel, or

(*c*) The strength of the beam in diagonal tension.

In this article the first two elements only will be considered.

From tests relative to elastic limit, such as those of Bach and Van Ornum (see Chapters II and IV), it would appear that the permanent elastic limit of concrete is from 50% to 60% of its ultimate strength as determined in the usual manner. If a factor of safety of two be applied to the elastic-limit strength to provide for items (*a*), (*b*), and (*c*) of Art. 118, we will have a dead-load basal working stress of 25% to 30% of the ultimate strength as determined by tests on cubes. Taking this ratio at 30%, the data of Chapter II show that the working stresses for concrete of the usual proportions (1:2:4 to 1:2½:5) should range from about 550 to 650 lbs/in^2. A value of 600 lbs/in^2 is commonly used and should imply a strength of about 2000 lbs/in^2 in cube form in 60 days. As explained further on, the stress in the concrete does not increase in proportion to the load on the beam, so that a working stress of 30% of the compressive strength will give a factor of safety against failure of about 4. It is to be noted also that the strength of the concrete increases with age. On the whole, therefore, a stress of 30% of the strength at 60 days may be considered as conservative practice.

With respect to the steel it is to be observed that its elastic limit, or more correctly speaking, its yield-point, determines not only the elastic limit strength of the beam, but also, approximately, its ultimate strength, and the working stress should be selected with this in view. If, for example, the working stress is taken at one-half the elastic limit strength of the steel, the factor of safety will be two as to elastic strength and slightly more than this as to ultimate strength, whereas with respect to the concrete the factor of safety will be more than four.

It is desirable to study somewhat more closely the variation in the stress in steel and outer concrete fibre in a beam subjected to an increasing load. Assume a concrete having an ultimate compressive strength of 2000 lbs/in², and let working stresses be assumed of 500 lbs/in² in the concrete and 15,000 lbs/in² in the sıeel. Suppose the beam loaded so as to cause these respective stresses. Represent this load on the axis *OX*, Fig. 57*a*, by *Oa* and erect ordinates *ab* and *ac* representing to some scale the stresses of 500 and 15,000 lbs/in², respectively. Let the load now be doubled and represented by the abscissa *Oa'*, equal to twice *Oa*. The stress on extreme fibre must now be calculated with reference to the true stress-strain diagram of the concrete, which will be assumed a parabola. Following the method of Art. 65 we find the stress in the concrete to be 950 lbs/in²; the stress in the steel will be 30,000 lbs/in². Proceeding in this way with increased loads, we finally reach the ultimate strength of the concrete of 2000 lbs/in². The corresponding stress in the steel will be 89,400 lbs/in². The ordinates of the curve *OAB* represent the progressive increase in outer fibre stress and those of the curve *OC* represents that in the steel, assuming for this purpose a steel whose elastic limie is not less than the maximum stress. This diagram shows clearly the relative change in stress in concrete and steel under increasing loads. It shows that the ultimate load, as fixed by the ultimate strength of the concrete, is about 5.75 times the load which produces the working stress of 500 lbs/in². The stress in the steel increases nearly in proportion to the load and the stress corresponding to the ultimate strength of concrete is about 89,400 lbs/in², or about six times the working stress of 15,000 lbs/in².

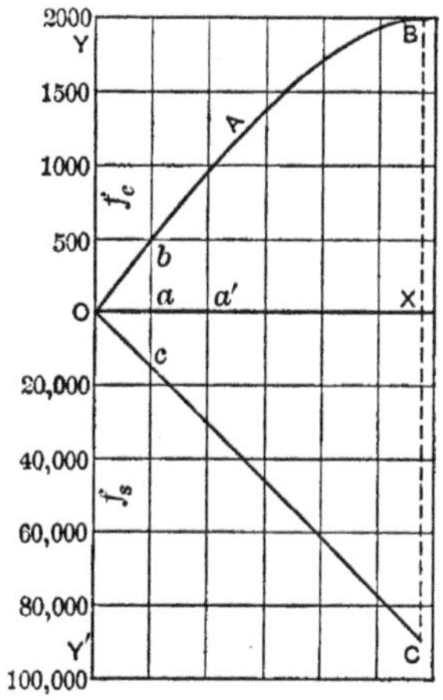

FIG. 57*a*.

Returning now to the question of working stresses in the steel, it is seen that if a stress is to be used so as to utilize fully the ultimate strength of the concrete, such stress cannot exceed one-sixth of the elastic limit of the steel. This would give a factor of safety of six as to ultimate strength of the beam, which is a much larger factor than necessary, especially as regards such a material as steel. On the other hand, suppose the working stress in the steel be selected at one-half its elastic limit, assumed in this case to be 30,000 lbs/in². Under increasing loads the beam will reach its elastic limit as to both concrete and steel at about double its working load, but as to ultimate strength there is still a large margin (about 60%) with respect to the concrete, but only a very small margin with respect to the steel. If it is desirable to utilize, for emergency purposes, a larger part of the ultimate compressive strength of the concrete, the working stress in the steel must therefore be selected so as to give the desired margin of strength without much exceeding *its elastic limit*. Considering the fact that in well-designed beams the steel stress at failure will considerably exceed its elastic limit, a working stress of one-half the elastic limit will give a factor of safety against ultimate failure of about 2½, and a working stress of one-third will give a factor of 3½ to 4. Under ordinary conditions a working stress of about four-tenths of the elastic limit, considered as the yield-point of the material, would appear to be desirable. With a working stress in the concrete of one-half its elastic limit the beam will then have a factor of safety as regards elastic limit of about two (determined by the concrete), and as regards ultimate strength its factor of safety will be at least five relative to the concrete and about three relative to the steel. Its elastic limit is thus determined by the concrete and its ultimate strength by the steel, which may be considered as satisfactory conditions. The greater uniformity and reliability of the steel, as compared to the concrete, should be noted in ths connection.

In determining the relative working stresses in steel and concrete some consideration should be given to the question of repeated loads. Where a large percentage of the load is live load and subject to frequent repetitions, a relatively low working stress in the concrete may well be employed in order that elastic conditions may be maintained. As regards the steel, more perfect elasticity exists up to a definite point, and hence repetition of load need not be considered in the selection of its working stresses.

The working stresses in the steel should also be considered with reference to its distortion. High working stresses involve large distortions, and hence a greater degree of incipient rupture in the concrete. This condition is probably of little moment in most cases so fas as it concerns the appearance of the concrete, but experiments, such as noted in Art. 117, show that these cracks may be of some consequence, and their influence on the possible corrosion of the steel is not yet well determined. In this connection it may be noted that under working conditions the actual stress in the steel is generally less than the calculated, owing to the tensile resistance of the concrete. The deformations will therefore also be proportionally less. The deformations of the concrete are also of importance with reference to their effect on diagonal tensile stresses, as explained in Art. 109. Low unit stresses in the steel are greatly to be preferred on this account. It will also be shown in Art. 133 that very little is to be gained in economy by using high stresses. Considering this fact and the objections above mentioned, it would seem that a stress of 16,000 lbs/in^2 should be considered about the maximum desirable value, irrespective of the quality of the steel used. A lower value is to be recommended. Finally, as the result of this analysis, we may conclude that the basal working stress in the steel should not exceed about 40% its elastic limit nor exceed 16,000 lbs/in^2.

122. Quality of Steel.—As stated in Art. 34, there exists considerable difference of opinion as to the quality of steel to be desired, especially with reference to the use of soft or hard material, or steel with low or high elastic limits. Certainly a material as hard as that formerly denominated "hard bridge steel" is entirely suitable for reinforced construction. Such material has an elastic limit of about 40,000 lbs/in^2. Much material has been used of an elastic limit of 45,000 to 50,000 lbs/in^2 and even higher, but a value beyond this is not to be desired. Practice tends to the use of one of two grades of material; the ordinary medium or mild steel having a yield-point, in the sizes suually employed, of from 35,000 to 40,000 lbs/in^2, and various special bars having a high elastic limit of 50,000 lbs/in^2 or more. The medium steel will permit, under ordinary conditions, a working stress of 14,000 to 16,000 lbs/in^2, and a steel of higher elastic limit is of doubtful wisdom unless a high factor of safety is desired. The ductility of the high elastic limit material of the usual quality is often not as great as desirable. As regards ductility and composition, a material of the quality used in buildings is satisfactory for most purposes. The requirements need not generally be as severe as for bridge steel, although the wide use of the standard bridge steel, as specified by the Maintenance of Way Association and by most of the railroad companies, tends to facilitate its adoption for all structural purposes.

123. Bond Stress.—The factor of safety with reference to the slipping of the rods should be at least 3, since the strength of a beam should not be limited by the strength of bond. From the data of Chapter IV, we may take the bond strength of plain steel at from 200 to 250 lbs/in^2. A working stress of from 60 to 80 lbs/in^2 is therefore suitable. Increase in age will increase the factor of safety in this respect very considerably. With a working bond stress of 60, say, and a tensile unit stress of 15,000 a round bar will need to be embedded a length of $15{,}000/4 \times 60 = 62.5$ diameters to develop its full strength. In the case of large bars of 1″ to 1½″ in diameter this length is very considerable and for short beams may be difficult to secure. The deformed bar, or the anchored bar, is of especial value under these conditions.

For deformed bars having a positive grip, a working stress of 150 lbs/in^2 gives an ample margin of safety. A larger value is undesirable, as it is preferable to keep the stress below the ultimate bond strength for plain bars so as to avoid initial slip under working conditions. At a value of 150 lbs/in^2 the required length of embedment, with a tensile unit stress of 15,000 lbs/in^2, would be 25 diameters.

In calculating bond stress the method of Art 92 should be used. For continuous beams the shear and the bond stress is a maximum near the support and it also changes sign at the support. This condition gives rise to a sudden change in the *direction* of bond stress. Thus in Fig. 57*b*, on the left of the support, the concrete pulls towards the left on the upper rod, and on the right it pulls towards the right, as shown by the small arrows. Any slipping increases the deformation of the concrete at once, and hence increases the tension in the concrete at the center. Likewise, at the bottom, any slip tends to increase the compressive stress in the concrete. It follows, therefore, that where rods continue over the support in continuous beams, the bond stress should be fully taken care of on each side of the center of support, otherwise the deformations and the concrete stresses will be increased. For compressive reinforcement the formulas of Art. 92*a* may be used, but generally it will be sufficient to consider simply the maximum compressive stress in the steel and provide a sufficient length from this point to the end of the bar to transmit this stress. The chance for an end bearing against the concrete in the case of compressive reinforcement reduces the danger from ultimate failure by failure of bond.

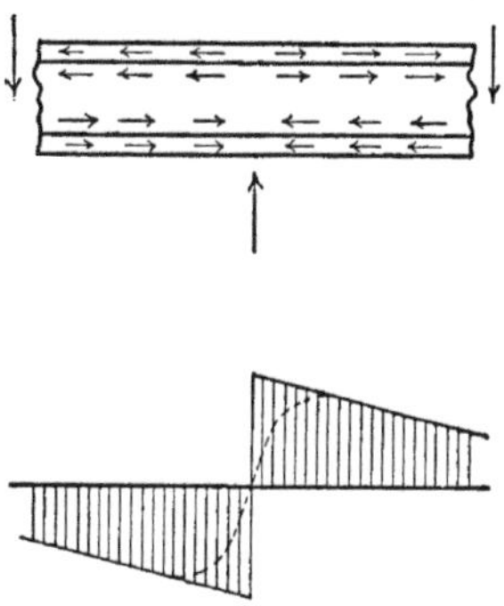

FIG. 57*b*.

Use of Anchored Bars.—Large bars are frequently anchored at their ends by nuts and washers, or partially anchored by means of sharp bends. A positive anchorage secures a reliable bond, but such an arrangement, if actually brought into action, results in a different distribution of stress than where adhesion is depended upon. So far as the anchorage is effective the resistance in the concrete is furnished by so-called "arch action." In the ideal case of anchorage, where the bond stress is zero and the full load comes upon the anchorage, there will be complete arch action. There will be no horizontal shear and cracks will tend to form nearly vertically. The total elongation of the rods will be greater than in the true beam and the cracks at the center will tend to open up wider, throwing the center of compression higher up. The compressive stresses in the concrete will therefore be somewhat higher than in the beam. In practice such complete arch action is not secured, and the effect of anchorage upon the concrete stresses, either tensile or compressive, need hardly be considered. Positive anchorage may well be employed for large rods where the length is insufficient to develop the necessary bond strength. The use of short square hooks upon the ends of bars is not of great value. The tests of Bach, quoted in Art. 40, show that the square bend is not very effective in preventing initial slip but affords a considerable factor of safety against ultimate failure. If hooks are used they should be in a bend of 180°, as explained in Art. 40.

124. **Shearing Stresses.**—From the results discussed in Chapter IV the ultimate shearing strength of a beam having no web reinforcement may be taken at about 100 lbs/in^2, calculated as average shearing stress on the cross-section. Inasmuch as a failure due to high shearing stresses is apt to be sudden, the factor of safety should be at least three. This gives a working stress of about 30 lbs/in^2. For beams in which the web is well reinforced the working stresses may be made 3 or 4 times as great, or about 100 lbs/in^2. The results of tests noted in Art. 110 show that in spite of ample web reinforcement visible cracks will form in the webs of beams at maximum shearing stresses about equal to the ultimate tensile strength of the concrete, or about 180 lbs/in^2 in the beams

tested. The working stresses should therefore keep well within this limit, and as the maximum shearing stress is about 15% more than the average, the value of 100 lbs/in² would appear to be about the maximum permissible.

The stresses here considered relate to shearing stresses involving large diagonal tensile stresses. Where such tensile stresses are not developed to any extent, as in "punching" shear, a higher value may be employed; but as it is almost impossible in practice to avoid altogether such tensile stresses it is not advisable to greatly increase the working stresses above the maximum value of 100 lbs/in² above suggested. A value of 150 lbs/in² should not be exceeded. This gives a factor of safety of about 6 relative to the shearing strength, as shown in Art. 22.

125. Calculation of Web Reinforcement.—*General Conditions.*—Before considering in detail the calculation of web reinforcement, or reinforcement against inclined tensile stresses, the reader is referred again to the discussion in Art. 46 and Fig. 12, showing the lines of maximum tension in a homogeneous beam. In the reinforced beam the intensity of the shearing stress is nearly uniform from the neutral axis down to the horizontal steel, so that the direction of maximum tension in the concrete is considerably inclined immediately above the steel. This inclination is greater the greater the shear and the less the horizontal tension. It will therefore increase from the center towards the end, being 45° where the horizontal tension becomes zero. From these considerations the ideal web reinforcement would be a system of rods arranged somewhat as shown in Fig. 57*c*, attached at their lower ends to the horizontal rods, or consisting of numerous horizontal rods bent up as indicated. The figure also indicates roughly the manner in which the inclination of diagonal cracks near the bottom tends to vary from nearly vertical at the center to a large inclination at the end. The exact conditions depend upon the nature of the loading, concentrated loads tending to extend the region of large shear to greater distances from the support. Generally speaking, then, the ideal web reinforcement should have greater inclination near the support than near the center. It is also evident that to be effective it is necessary that it be spaced at relatively close intervals and that it is chiefly effective in the region below the neutral axis. Attention is also called to the discussion of Art. 90 and the tests of Art. 109, showing the effect of low unit stresses in the horizontal steel in reducing the deformations and the tendency to the formation of inclined cracks. This condition makes it desirable to extend a considerable part of the horizontal reinforcement to the end of the beam: and if some of the rods are

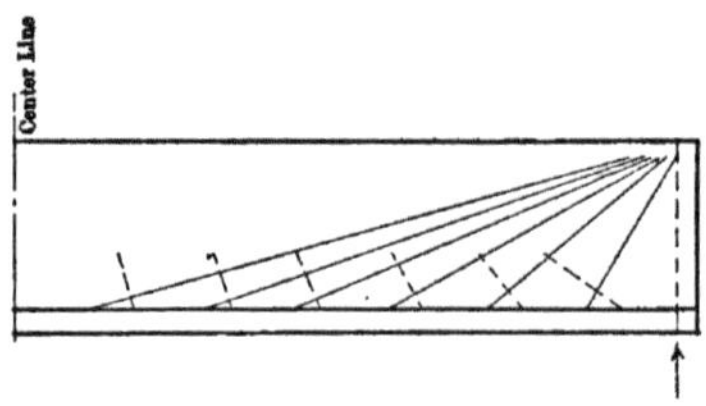

Fig. 57*c*.

bent up, the bends should be made somewhat beyond the theoretical points required for bending moment, so that the actual working stresses in the horizontal steel near the end of the beam will be low.

In practice, the method of reinforcement indicated in Fig. 57*c* cannot well be used. The number of horizontal rods is generally much too small and it is not convenient to handle the rods when bent up at a greater number of points and at various inclinations. Instead of this arrangement, various methods, as illustrated in Art. 107, are employed, the common practice being to use a few bent rods combined with vertical stirrups. From the considerations of the preceding paragraph it would appear that rods bent at a moderate angle would be well suited for sections near the center and vertical stirrups for sections near the end where the reinforcing members must be spaced closer together and at greater inclinations. This accords generally with the best practice.

Length of Horizontal Bars.—In determining the length of the various horizontal bars necessary to resist the bending moment, the same method may be used as in the design of plate girder flanges. If the bending moment is due to uniform load the parabolic formula may be used, as explained in Johnson's "Modern Framed Structures," Chapter XIX. It is

$$x_n=\frac{l}{\sqrt{A}}\sqrt{a_1+a_2+\ldots+a_n}, \quad \ldots \quad (1)$$

in which x_n=length of the nth rod in the order of length, counting the shortest as number one;

l=length of span;

A=total steel area at center; and

a_1, a_2, etc. =area of each rod up to the nth rod.

For unsymmetrical loading the maximum moments at various sections will need to be determined and the lengths obtained therefrom.

Web Reinforcement.—Sufficient experimental work has not been done to enable the proportioning of web reinforcement to be done with any degree of exactness. However, a rough estimate of the requirements can be determined on rational grounds. The tests already quoted in Chapter IV indicate that beams with horizontal bars only cannot be stressed safely beyond about 30 lbs/in^2 average shearing stress, the strength depending on the quality of concrete and the unit stresses adopted for the horizontal steel. In practice it will rarely happen that a beam need carry more than 100 lbs/in^2 average shearing stress, and tests of the best work indicate that this should be about the maximum limit, even with an effective system of web reinforcement.

Where it becomes necessary to provide web reinforcement, and the shearing stresses exceed a safe limit of say 30 lbs/in^2 on the concrete, some estimate must be made of the stresses in the steel, and an important question arises as to the mutual action of concrete and steel, and whether the concrete can still be counted upon for its safe stress or whether the steel must be proportioned to carry the entire load. In this connection certain tests made by Mr. Withey are instructive.* In the case of two rectangular beams and two T-beams reinforced by horizontal bars and vertical stirrups, failure was caused by the overstressing of the stirrups, in two cases the stirrups breaking. The results were as follows:

No. of Beam.	Cross-section of Stirrup. Sq. in	Spacing of Stirrups. Ins	Net Depth of Beam. Ins.	Width of Beam. Ins.	Av. Vert. Shearing Stress (v'). Lbs/in².	Calculated Stress in Stirrup. Lbs/in².
G_1	.049	$5\frac{1}{2}$	$13\frac{1}{2}$	8	222	100,000
G_2	.049	$5\frac{1}{2}$	$13\frac{1}{2}$	8	223	100,000
M_1	.049	6	16	8	272	133,000
M_2	.049	6	16	8	235	115,000

The stirrups were single loops of $\frac{1}{4}$-inch round steel having a yield-point of 47,000 lbs/in^2 and an ultimate strength of 62,000 lbs/in^2. The arrangement of rods in beams M is shown in Fig. 50*a*. The "calculated stresses" were determined on the assumption that the stirrups carried the vertical shear in a length equal to the depth of the beam, thus causing a stress in each stirrup equal to $v'bs$, where s=spacing of stirrups. As these stresses are much beyond the ultimate strength of the stirrups it is evident that a large amount of shear (about 40%) was carried by the concrete and by the bending resistance of the horizontal rods. Tests on beams without stirrups show the average shearing strength of concrete to be about 100 lbs/in^2, indicating that approximately correct results would be reached if the concrete be assumed to carry its full value and the stirrups the remainder. Similar results have been reached by other experimenters. If this is true at ultimate loads, it would be even more certain at working loads where the concrete is only slightly cracked at most and the distribution of stress

* Bull. No. 2, Vol. 4, 1908, Univ. of Wis.

more normal. From these considerations we may conclude that in calculating stresses in web reinforcement the concrete may be assumed to carry its safe load and the steel proportioned to carry the remainder. These stresses may be estimated in the following manner:

In Fig. 57*d* are represented two types of web reinforcement, vertical rods, and rods inclined at 45°. Let s represent the horizontal spacing in both cases and assume the line of failure at 45°. Let V represent the shear not carried by the concrete. Assume for simplicity, that the intensity of shear is uniform over the section and is equal to $V/bd = v'$. This will also be taken as the intensity of the diagonal tensile stress at 45°. (The maximum will be only 12% to 15% more than this, see

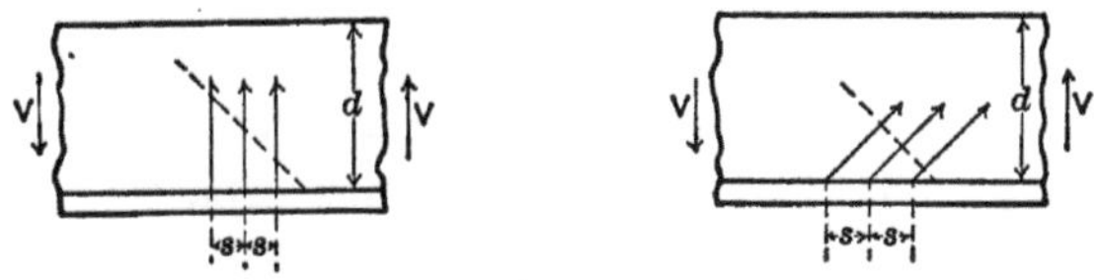

FIG. 57*d*.

Art. 89.) In the case of vertical stirrups they will be called upon to carry the vertical component only of this diagonal tension, the horizontal component being carried by the horizontal bars. This vertical component per unit of horizontal area will also be v'. Assuming equal stresses in each stirrup, represented by P, we have finally

$$P = v'bs = \frac{V}{d} \cdot s. \quad . \quad . \quad . \quad . \quad . \quad . \quad . \quad . \quad (2)$$

For inclined stirrups the most unfavorable assumption is that they carry the full diagonal tension. The spacing at right angles to the line of rupture is $s \cos 45°$, and the stress will therefore be

$$P = v'bs \cos 45° = 0.7 \frac{V}{d} \cdot s \quad . \quad . \quad . \quad . \quad . \quad . \quad (3)$$

The foregoing calculations must be considered as only roughly approximate, but they are doubtless on the safe side and are on a rational basis. In preventing initial rupture and distortions under low loads the inclined reinforcement is more effective than vertical, as it receives stress at an earlier stage, but in resisting larger distortions the vertical type appears to be equally good. Their relative efficiency, however, depends largely upon other elements, such as available bond strength, closeness of spacing and other practical considerations.

Spacing and other Details.—To be reasonably effective the web reinforcement should be so spaced that at least one rod will intersect any 45° line of rupture below the center of the beam.

As shown by the sketch (Fig. 57*e*) this requires a spacing of vertical reinforcement not greater than $d/2$, and for diagonal rods, a horizontal spacing not greater than d. Some advantage is gained by rods spaced somewhat further apart, but tests by Talbot show little or no value in vertical rods spaced a distance apart about equal to d. The working stress to be used in these calculations should be no higher than permitted elsewhere, and preferably lower, as it is desired to prevent large distortions so far as practicable. In bending up horizontal rods those remaining straight should be ample in number to take the moment stresses, and, preferably, at reduced intensities towards the ends of the beams. A small angle of bend near the center and larger angles near the end, as in Fig. 57*c*, should be observed so far as practicable. It will often be impracticable to provide as much reinforcement as desired by means of bent-up rods, and some vertical stirrups will be needed, especially near the end where the stresses are high. A combination of bent rods and vertical stirrups is common practice and readily lends itself to adequate and convenient treatment. For large beams, under heavy shearing stresses, both should be used.

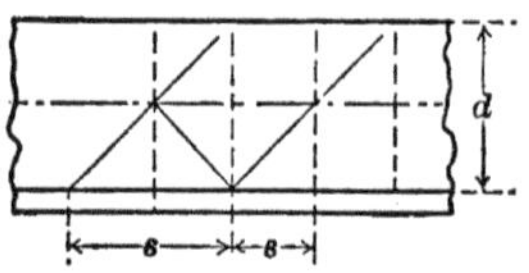

FIG. 57*e*.

Inclined stirrups are quite as effective as bent rods or vertical stirrups, but to prevent slipping on the horizontal rods they should be securely fastened thereto. For example see Art. 135.

In detail, stirrups may be made in various forms, as indicated in Fig. 50. Woven wire, bent around the rods, is a satisfactory and very effective reinforcement. Where compressive bars are used in the upper part of a beam the stirrups should hook around these bars also. In continuous beams the upper face becomes the tension face near the supports and this is also where the shear is large. Stirrups in this vicinity should loop about the upper bars.

The bond strength of web reinforcement must be carefully guarded, especially in the case of large bent-up rods. This strength should be provided in the upper portion of the beam. Plain rods bent up often lack sufficient bond strength to render them fully effective. Where bent up at a considerable angle they should be turned again horizontally and extend some distance along the upper part of the beam as in Fig. (*b*), p. 159. In non-continuous beams the ends of the bars should be bent into a hook.

The following simple graphical method may be used in important cases (large beams) for determining the stress or spacing of bent rods. It also serves to make clear the principles involved. Suppose *OB*, Fig. 57*f*, represent half of a simple beam uniformly loaded. Calculate the shearing stress v' at the end. Project the axis *OB* upon an axis *OC* at 45° inclination and lay off *CC'* equal to v', and draw the line *OC'*. Then the ordinates between *OC* and *OC'* will represent the shearing stresses v' along the beam; and the area between any two ordinates *DD'* and *EE'*, multiplied by the width b of the beam, will equal the product of the total average shear over the length s, times the projection of this length on the inclined axis *OC*. It will represent therefore the stress in a rod bent up at 45° at point *G*, in line with the center of gravity of the area *DEE'D'*. For, by eq. (3), this stress is equal to $0.7Vs/d=0.7\ v'bs$. But $0.7v's$ is the area *DEE'D'*, v' being the value of the average ordinate. Hence P_s = area $DEE'D' \times b$. If rods are bent at other angles, then the axis OC may be drawn at right angles thereto. If the concrete be assumed to carry a portion of the shear v' and stirrups another constant portion, these amounts may be subtracted graphically from the total shear as indicated in Fig. 57g, and the remainder taken by bent rods. If s=spacing of stirrups and P= safe stress, then the amount taken by stirrups will be found from eq. (2) equal to $v' = P_s s/b$. In no case should the area of the horizontal rods be taken into account. These offer resistance only through their bending strength.

Fig. 57f.

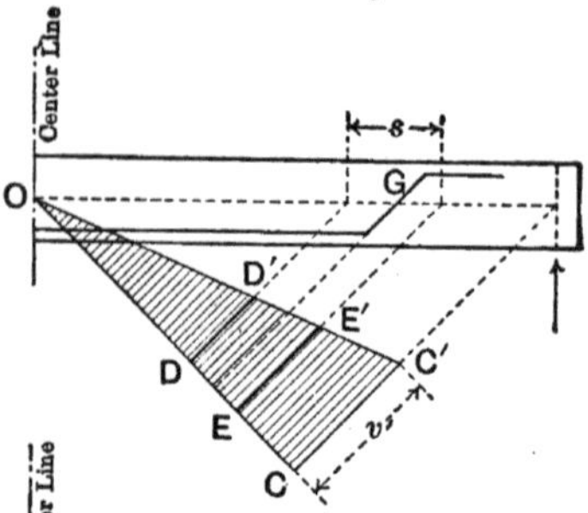

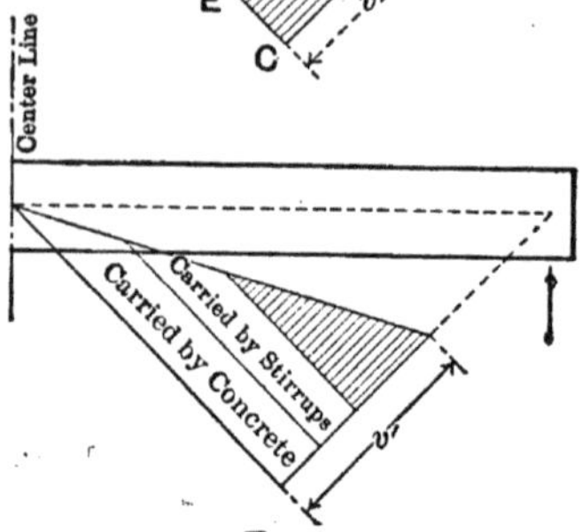

Fig. 57g.

126. Spacing of Bars.—In rectangular or T-beams the spacing of bars is important; in T-beams this consideration will largely control the width of the beam. The requirement in general as to spacing is that the amount of concrete left between the bars must be sufficient to transmit to the upper part of the beam the stress which the bars give over to the concrete below them. If the bars are circular it may be assumed that one-half of the stress in them is given over to the concrete below, hence the strength of the concrete on a longitudinal section through the center plane of the bars must equal one-half of the stress in the bars. If the shearing stress be taken as equal only to the bond stress then the clear space between bars must be one-half the circumference of a bar, or 1.57 diameters. In the sense here employed the shearing strength is at least twice the bond strength for smooth rods, so a clear spacing of less than one diameter is sufficient from this standpoint. In the case of square bars, on the same basis, the clear spacing would need to be 1½ diameters if the bars are placed with sides vertical, or one diameter if placed with sides diagonal.

But in addition to the shearing stresses there is likely to be developed more or less tension in the concrete surrounding the rods, so thatt here should be left ample areas of concrete between them, especially towards the end where the bond stresses are large. The space should also be sufficient to permit satisfactory manipulation of the concrete. A minimum clear spacing of at least $1\frac{1}{2}$ diameters should be provided, with an equal distance between the outside rod and the surface of the beam. Where some of the rods are bent up the spacing can readily be made more liberal towards the end of the beam. Between two horizontal layers of rods the spacing may be less but should be sufficient to insure good bond.

Liberal spacing, or large net section of concrete, favors large rods and few in number; good bond strength without waste of material favors small rods. If bent rods are to be used for web reinforcement, then numerous small rods are also advantageous. If the bond strength is not in question, or can easily be taken care of, then large rods are desirable, but more stirrups or other secondary reinforcement may be needed than where small rods are used.

127. Economical Proportions and Working Stresses.—For given unit prices, the cost of concrete beams per unit of resisting moment will vary with the proportions adopted for breadth and depth, and with the working stresses employed. Because of the mutual relations between the concrete and steel it may happen that the maximum economy of construction may be obtained by using less than the allowable working stresses in one or the other of the two materials. It will therefore be useful to investigate the effect on cost of variations in proportions and in the working stresses.

Consider a portion of a rectangular beam one unit in length.

Let c = cost of concrete per unit volume; r = ratio of cost of steel to cost of concrete per unit volume; p = ratio of steel area to concrete area; C = cost of beam per unit length.

Then $$C = c(bd + rpbd) = cbd(1 + rp). \quad . \quad . \quad . \quad . \quad (1)$$

From Art. 59 we have $bd^2 = M/R$, in which M = bending

moment and R=coefficient of strength of the beam, depending in value only upon f_s, f_c, and n. From this we may write $bd=M/Rd$, $bd=\sqrt{Mb/R}$, and $bd=\sqrt[3]{(b/d)(M^2/R^2)}$, whence we derive the three expressions for cost:

$$C=c(1+rp)\frac{M}{Rd}, \quad \ldots \ldots \quad (2)$$

$$C=c(1+rp)\sqrt{\frac{Mb}{R}}, \quad \ldots \ldots \quad (3)$$

and

$$C=c(1+rp)\sqrt[3]{\frac{b}{d}\cdot\frac{M^2}{R^2}}. \quad \ldots \ldots \quad (4)$$

128. *General Effect of Varying Proportions.*—Since the values of R and p depend only on f_s, f_c and n we note from (2) that the cost of a rectangular beam to support a given moment, M, varies inversely with the depth; and from (3) that the cost varies directly with $\sqrt{\text{breadth}}$; and finally, from (4) that it varies with the cube root of the ratio of breadth to depth. In all cases it is assumed that the two dimensions are made to correspond with each other as calculated from the selected values of f_s and f_c. It follows from (2) that with given values of f_s and f_c the deeper the beam the less the cost, so long as b can be reduced accordingly. The depth will, however, be limited in various ways. It may be limited by the requirement of shearing stress fixing the value of bd, or it may be limited by the head room required, or it may practically be limited by the fact that a certain breadth is necessary to give a convenient and proper covering of the steel reinforcement or to give a beam of satisfactory proportions. In the construction of continuous surfaces, such as floor slabs, the case is one of fixed width, since the width of beam to carry the load coming upon a strip one foot wide is also one foot. We may then consider four cases according to the particular feature of the design which is the controlling element. These cases are:

(a) When the area of cross-section is determined by the shear;

(b) When the depth of the beam is fixed:

(*c*) When the width of the beam is fixed;

(*d*) When the ratio of width to depth is fixed.

129. (*a*) *The Area of Cross-section is Determined by the Shear.*—A given value for shearing stress requires a fixed value of bd, but the requirement for bending moment is that $bd=M/Rd$; hence if a beam is designed for moment alone the area bd will be less the deeper the beam. Theoretically, therefore, for a given value of R the maximum depth permissible is that for which the resulting area bd is just large enough to carry the shear. If V is the total shear and v' is the permissible shearing stress, then $bd=V/v'$. Also $bd=M/Rd$. Hence for equal strength $M/Rd=V/v'$ and therefore

$$d=Mv'/RV \quad . \quad . \quad . \quad . \quad . \quad . \quad . \quad (5)$$

and

$$b=V/v'd. \quad . \quad . \quad . \quad . \quad . \quad . \quad . \quad . \quad (6)$$

These equations give the dimensions of a beam which will be of just the required strength in moment and shear. It remains to be determined, however, whether a still greater depth will result in greater economy.

If a greater depth be used, bd must remain constant; hence bd^2 will be increased and the concrete stress, f_c, decreased. Reference to Plate III, p. 215, shows that with constant f_s, a decrease in the value of f_c permits the use of a smaller percentage of steel. Hence with increasing depth and constant bd (or volume of concrete), the amount of steel will be reduced, and therefore the cost. The proportions of the beam will therefore not be determined by the shear excepting as to minimum cross-section.

130. (*b*) *The Depth of the Beam is Fixed.*—From eq. (2) it is seen that for given values of M and d the cost varies with $(1+rp)/R$. Now p and R depend only upon the working stresses f_s and f_c (n being constant), hence it will be convenient to determine the variation in cost due to variation in f_s and f_c, assuming certain values for r. Results of this analysis are shown in Fig. 58 for values of r of 60 and 80 and for various values of f_s and f_c. The results are very instructive and show

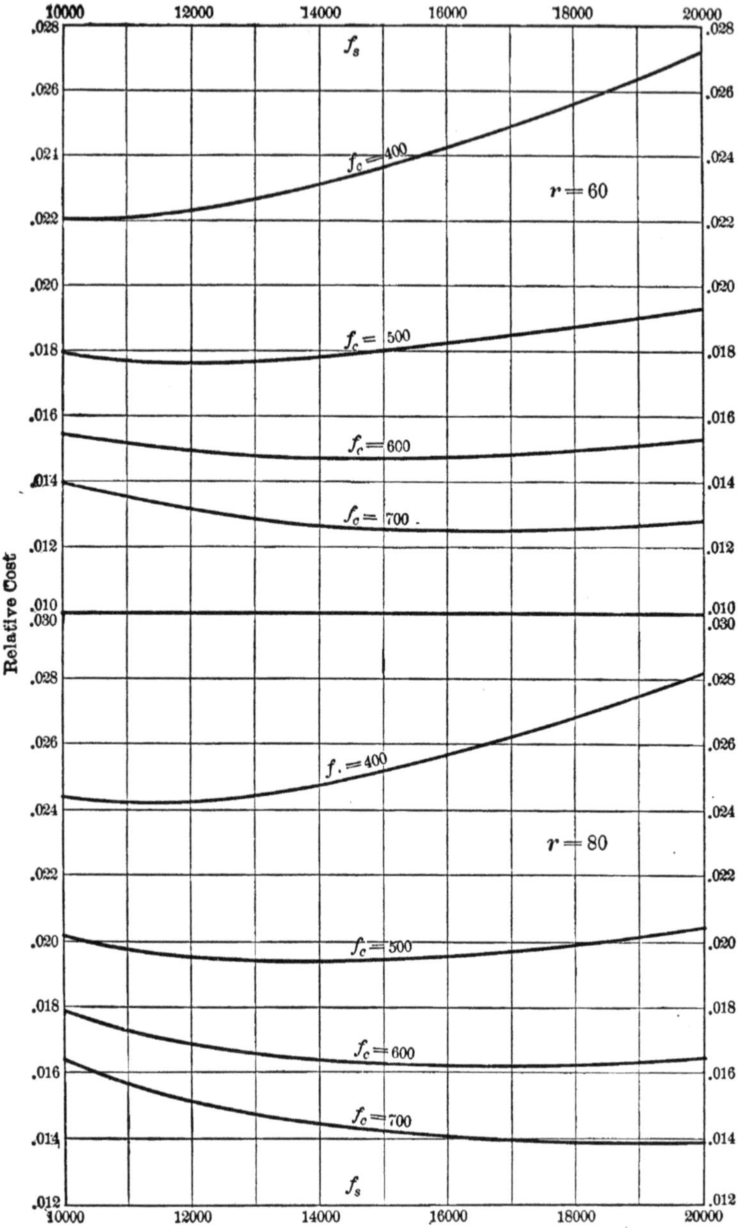

Fig. 58.—Relative Cost for Fixed Depth.

that for values of f_c of 500 or 600 lbs/in² no economy is secured by using values of f_s greater than 12–14,000 lbs/in². For larger values of r or of f_c, higher values can economically be used for f_s, but a value of 80 for r is not likely to be greatly exceeded. If the cost of concrete be as low as 20 c. per cu. ft. the corresponding cost of steel would be $16.00 per cu. ft., or 3.2 c. per pound. This is a low cost of concrete and a high cost of steel. The diagram shows that the cost is decreased by increased values of f_c.

131. (*c*) *The Width of the Beam is Fixed.*—From eq. (3) the cost for given values of M and b varies with $(1+rp)/\sqrt{R}$. Fig. 59 represents this quantity plotted for various values of f_s and f_c. Comparing this with Fig. 58 it is seen that somewhat higher values of f_s are warranted, but it is evident that the gain in economy is very small for values above 16,000 lbs/in², except where the steel is very expensive and the concrete cheap.

132. (*d*) *The Ratio of Width to Depth is Fixed.*—It is often desired to secure approximately a certain given ratio of breadth to depth. In this case we find from eq. (4) that the cost will vary with $(1+rp)/R^{\frac{2}{3}}$. Fig. 60 represents this quantity for various values of f_s and f_c. It is seen that the most economical values will lie between those of cases (*b*) and (*c*).

133. *Floor Slabs with Weight of Concrete Eliminated.*—In all the foregoing discussion the moment to be resisted has included that due to the weight of the beam itself. For large beams and girders this is unimportant in this connection, but with floor slabs, where the external load is small, the weight of the material itself modifies the results to a large extent. General results cannot be presented for all cases, but the analysis will be given for a single case representing ordinary conditions. A span length of 10 ft. has been taken and a net floor load of 150 lbs/ft². Then from Table No. 21, Chap. VI, the required cross-section and amount of steel has been determined for various values of f_s and f_c. The relative cost per unit floor area has been calculated for values of r of 40, 60, 80, and 100 and the results plotted in Fig. 61. Comparing these results with those of Fig. 59, where the weight of the beam has not been deducted,

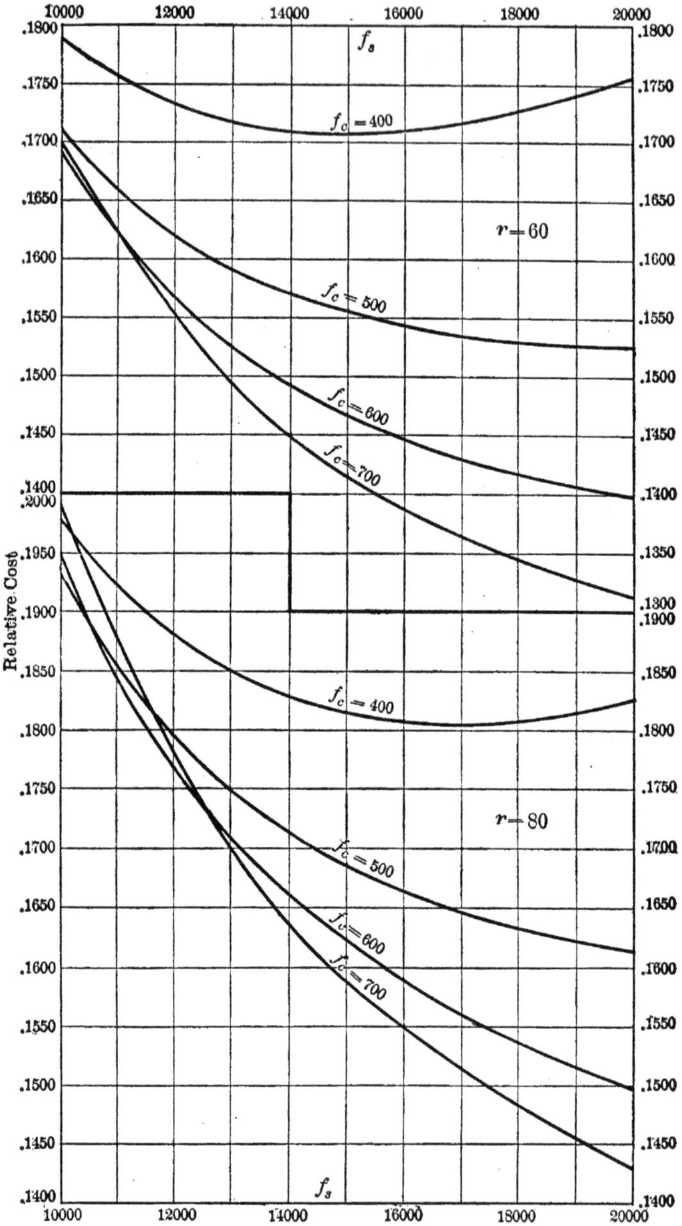

FIG. 59.—Relative Cost for Fixed Width.

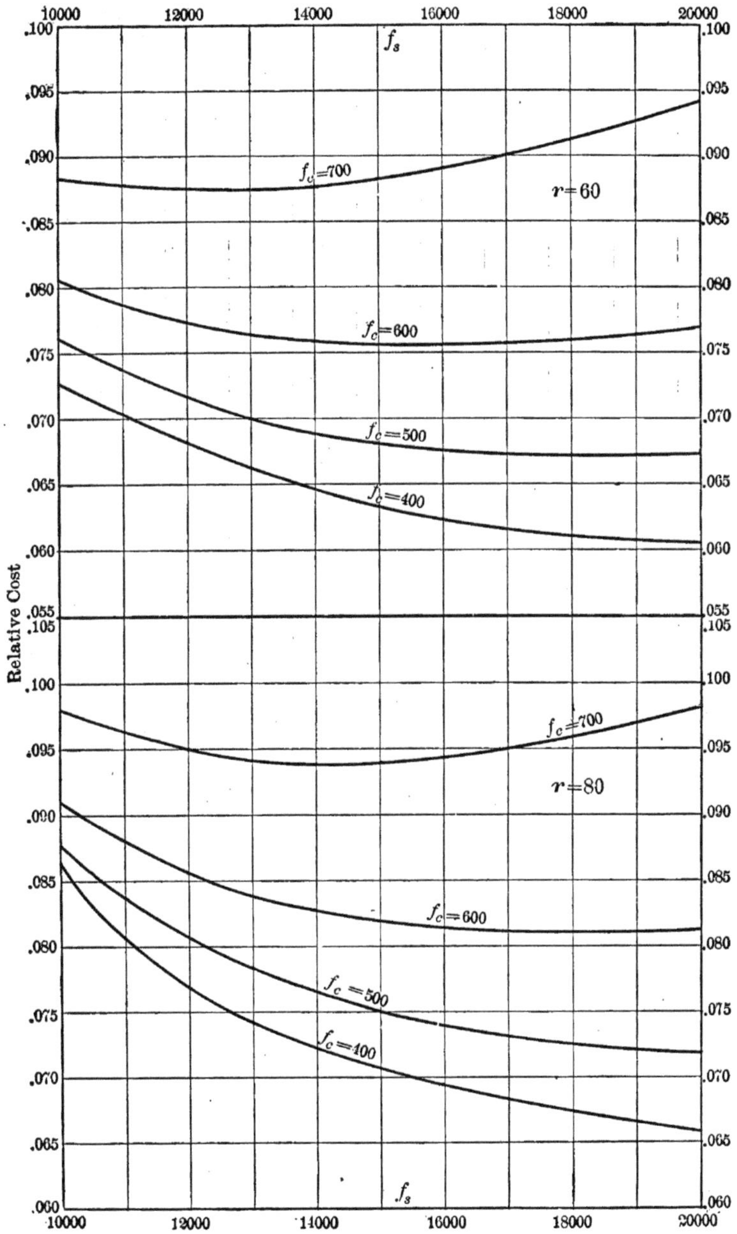

FIG. 60.—Relative Cost for Fixed Ratio, Breadth to Depth

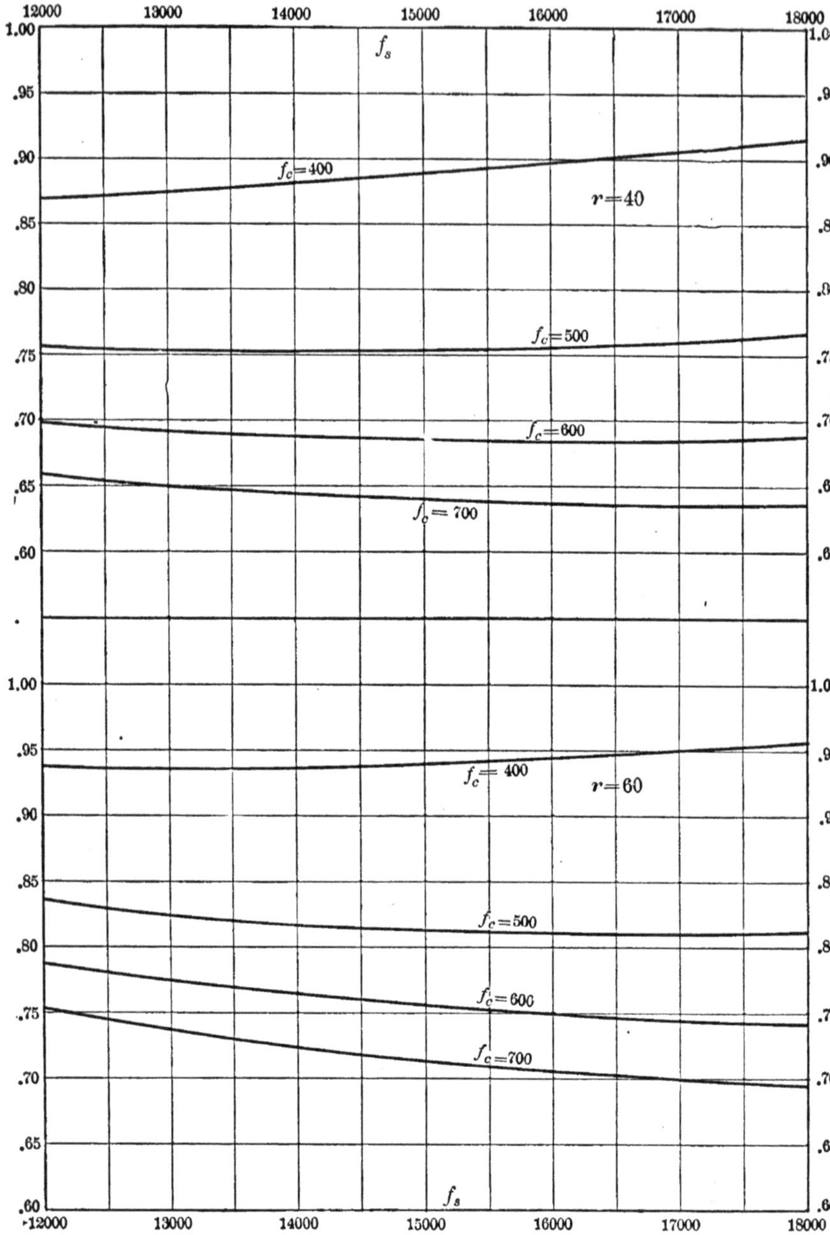

FIG. 61*a*.—Relative Cost for Fixed Breadth, Weight of Beam Deducted.

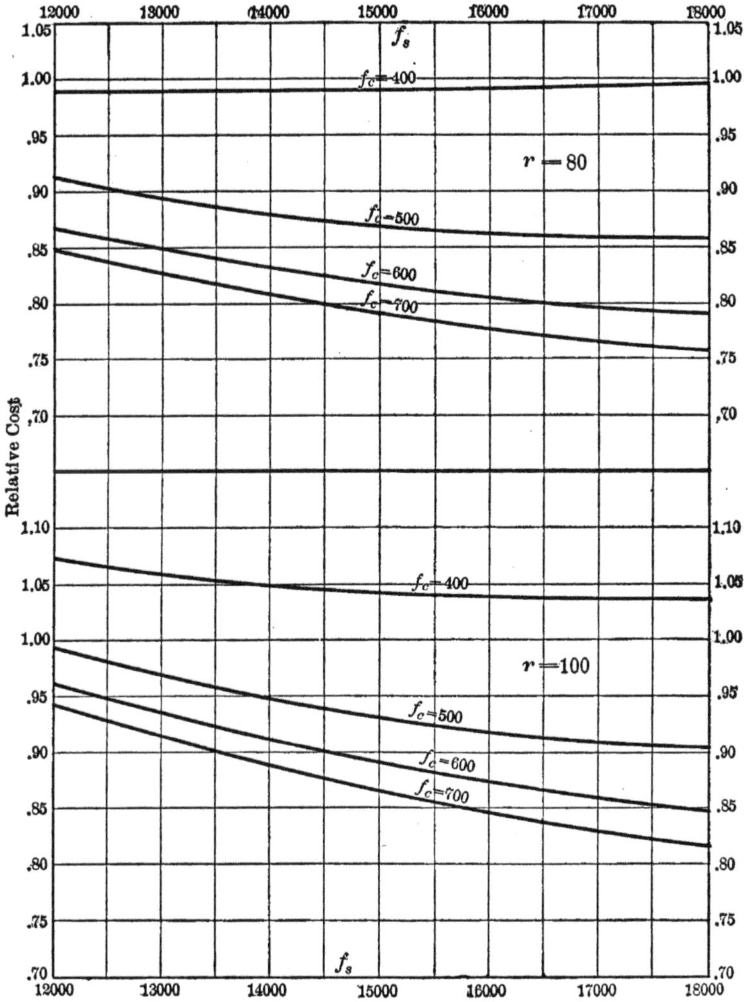

FIG. 61*b*.—Relative Cost for Fixed Breadth, Weight of Beam Deducted.

it is seen that the economical values of f_s are considerably less. For values of r not exceeding 60 and for f_c not exceeding 500 there is no reason for using a value for f_s higher than 14,000 lbs/in^2. For other spans and floor loads the results will be somewhat different, but the variation will not be great. Larger floor loads and shorter spans will give results more nearly approaching those of Fig. 59; smaller loads and longer spans will tend in the opposite direction.

Percentages of steel corresponding to any particular values of f_c and f_s are given by reference to Plate III, p. 215.

134. *Effect of Overlapping Bars.*—In most cases the reinforcing bars of slabs are made to overlap more or less; where negative moment over the beams is taken care of this overlapping may be 25 to 30 per cent. To take account of this in using the equations or diagrams of the preceding articles, the most convenient method is to increase the unit cost of steel, or the ratio r, by the same percentage that measures the overlap of the steel.

135. T-Beams.—*General Design.*—T-beams occur in practice generally where a floor slab and beam are built as a monolithic structure, as in floor construction. Occasionally, also, where heavy girders are required it is expedient to design the beam in the form of a T. Inasmuch as the only purpose of the concrete below the neutral axis is to bind together the tension and compression flanges, its section is determined by the shearing stresses involved, and a considerable saving can thus often be effected over the rectangular form. Where the flange is a part of the floor slab its thickness is already determined by other considerations, but the width of slab which can be taken as effective flange width must be estimated. A common rule of practice is to count a width of slab not greater than one third of the span length, but this should in fact depend also upon thickness of slab and of the stem of the T.

If made too wide and thin the shearing stresses along the line aa' and cc', Fig. 62, will be excessive and greater than those along the line $a'c'$. On this account it is desirable to limit the value of x to about four times the thickness of the slab, or three times the width of beam b'. Experiments show that a total flange width of 3 or 4 times the width of the web generally gives ample flange area so that the design of such a T-beam consists mainly in the design of the web or stem, and the proper arrangement of the steel. Of course the width of flange cannot exceed the spacing, center to center, of beams, and it is common to limit the effective width to three-fourths of the spacing of beams.

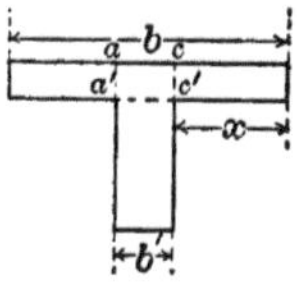

FIG. 62.

Where a T-beam is not connected with a floor system then the size of flange may be selected to meet the conditions at hand. In this case the stem of the beam should first be determined approximately, on the basis of the shearing stresses to be carried. A suitable flange can then be selected by a few trials, as explained in Art. 73. The deeper the beam the less the amount of steel required for constant cross-section. But T-beams should not be made too deep in proportion to width, as such forms are relatively weak at the junction of stem and flange. All re-entrant angles in rigid material such as concrete are points of weakness and such angles should therefore be modified by curved lines or be made obtuse by sloping the sides of the beam. A width of beam sufficient to carry the shear and to give plenty of space for the bars is usually ample. The maximum desirable ratio of depth to width may be taken at about two for small beams up to three or four for very large and massive work. Depths are often determined by available head room. Beams of excessive depths are objectionable as being more difficult and troublesome to reinforce properly; the cost of web reinforcement also becomes relatively greater. The flanges should be thorouglhy bonded to the web by means of web reinforcement running well up into the flange and, where the flange is wide, by additional cross-reinforcement in the plane of the flange. (For further details see Art. 165).

Economical proportions.—Where a floor-slab forms the flange of a T-beam, then the economical proportions of the stem may be considered. Here the slab forms practically all the compressive area, but does not enter into the cost of the beam. Using the approximate formula, eq. (7), of Art. 74 the area of the steel is equal to $M/f_s(d'+\frac{1}{2}t)$, in which d' is the depth of beam *below* the slab. The cost is then

$$C=c\left[b'd'+\frac{rM}{f_s(d'+\frac{1}{2}t)}\right]. \quad . \quad . \quad . \quad . \quad . \quad (1)$$

From this expression it is evident that the cost will decrease with increased values of f_s under all conditions, and that with a fixed value of $b'd'$ the cost decreases with increase in depth. If d' is fixed then the cost will be a minimum when b' is made as small as possible, and its value will then be determined by the shearing stress or by the space required for the bars. If the value of b' is assumed as fixed, then there is a definite value of d' which will give minimum cost. Considering d' as variable and b' as constant we find by differentiation that for minimum cost the value of d' is given by the equation

$$d'+t/2=\sqrt{rM/f_sb'}. \quad . \quad . \quad . \quad . \quad . \quad . \quad (2)$$

From this expression the best depth for various assumed widths can readily be determined and the desirable proportions finally selected.

Example of the Design of a T-beam.—To illustrate the principles discussed in the preceding article a design will be made of a large girder built in the form of a T-beam. Assume the following data:

Span length=40 ft.; dead load=1500 lbs/ft.; live load=2500 lbs/ft.; f_c=600 lbs/in.; f_s=15,000 lbs/in.; v'=30 lbs/in^2 for concrete alone, and 100 lbs/in^2 where the web is reinforced against tensile stresses. Bond stress=75 lbs/in^2, with an allowable increase of 50% for straight rods near the end, in accordance with Art. 39. The beam is to be simply supported at the ends and the flange is to be proportioned as well as the web; that is, the flange does not form a part of a floor system already determined.

Solution.—The total bending moment, $M, = \frac{40{,}000\times 40^2\times 12}{8} =$ 9,600,000 in-lbs. The maximum shear, $V, = 4000\times 20 = 80{,}000$ lbs. The required net web area $= b'd = 80{,}000/100 = 800$ in². This can be supplied by a web 16″×50″ or 18″×45″. To give better space for the steel the latter will be chosen for a preliminary value. A thickness of flange of 12 in. will be tried. For this thickness $t/d = 12/45 = .267$. Then, by means of Plate IX, p. 283, we find that $M/bd^2 = 93$, $bd^2 = 9{,}600{,}000/93 = 103{,}000$, and $b = 103{,}000/45^2 = 51$ in. From the diagram the value of $jd = .89\times 45 = 40$ in., and $A = 9{,}600{,}000/40\times 15{,}000 = 16$ in².

To illustrate the effect of varying proportions, calculations will also be made for a flange thickness of 8″, 10″, 14″, and 16″. The results are as follows:

t	b	jd	A	Overhanging Width of Flange.	Area of Flange Outside of Web.
8 in.	64 in.	41.5 in.	15.4 sq. in.	23 in.	368 sq. in.
10 "	56 "	40.5 "	15.8 "	19 "	380 "
12 "	51 "	40.0 "	16.0 "	16½"	396 "
14 "	50 "	39.5 "	16.2 "	16 "	448 "
16 "	49 "	39.5 "	16.2 "	15½"	496 "

It will be noted that the effect of variation of t upon the amount of steel is very small, but that the amount of concrete is less the thinner the slab. The saving in concrete is measured by the reduction in the areas of the flange exclusive of the width of web. The 10-in. flange gives 16 sq. in. less material than the 12-in. and the 8-in. 12 sq. in. less than the 10-in. At the same time considering the fact that the girder is not a part of a floor system and therefore that the flanges are unsupported at their outer edges, and also that some transverse steel will be required to bond the flange well together, it is evident that a compact beam having a relatively thick flange is desirable. The choice would probably lie between the 10-in. and the 12-in. flanges. The 12-in. flange will be adopted.

The steel area required is 16 sq. in. This will be made up of five rods 1⅜ in. diameter, and seven rods 1¼ in. diameter, giving a total area of 16.0 sq. in. To provide a spacing of 2½ diameters the rods will be placed in three rows; the five 1⅜ in. rods in the lower row, five 1¼ in. rods above, and two 1¼ in. rods in a third row. In bending up the rods the two uppermost rods will be bent up nearest the center. The arrangement of rods in cross-section is shown in Fig. 62*a*. Taking moments of areas about the center of the lowest row, the center of gravity of the group is found to be $\frac{5\times 1.23\times 2+21.23\times 4}{16}=1.4$ in. above this row. Hence the lower row should be placed about $46\frac{1}{2}$ in. below the top of the beam, thus giving a total depth, including the protective covering of 49 in.

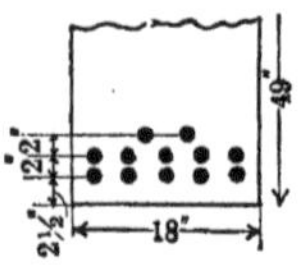

Fig. 62*a*.

To provide shearing reinforcement, as many of the rods will be bent up as practicable considering their necessary lengths for resisting the bending moment. This length may be found by a diagram of maximum moments, or, if the load is uniform so that the moment diagram is a parabola, by use of eq. (1), Art. 125. Applying this equation in the present example, we have $l/\sqrt{A}=10$. The necessary lengths of the rods are then as follows:

No. of Rod	$(a_1+\ldots a_n)$	$x=10\sqrt{(a_1+\ldots a_n)}$
1	1.23	11.1 ft.
2	2.46	15.7 "
3	3.69	19.5 "
4	4.92	22.2 "
5	6.15	24.8 "
6	7.38	27.2 "
7	8.61	29.3 "
8	10.09	31.8 "

The rods may be bent up at any point beyond the required lengths, as given above.

Shearing Stress and Reinforcement.—The maximum end shear is $4000\times 20=80{,}000$ lbs.; the maximum center shear is $2500\times 20\times\frac{1}{4}=12{,}500$ lbs. The average shearing stresses in the web at these two points are, respectively, 100 lbs/in² and 16 lbs/in². Under the specifications the concrete is good for 30 lbs/in² without reinforcement. Assuming the shear to vary uniformly from center towards end the value of 30 lbs/in² will be reached at a distance of $\frac{14}{84}\times 20=3.3$ ft. from the center. Beyond this point reinforcement will be required. It will be designed on the assumption that the concrete may be considered as carrying 30 lbs/in² and that the remainder must be carried by bent rods and stirrups. It will be desirable to use relatively low working stresses in the steel in order to avoid all danger of cracks.

It will be desirable to decide first upon a convenient arrangement of bent-up rods. Fig. 62*b* illustrates such an arrangement, which provides lengths somewhat in excess of those required for bending moment. It spaces bends at distances apart not greater than the depth of the beam and closer near the end. Two rods are bent up at each place (to bend the rods singly complicates the handling materially). For the same reason the bends are all made at 45° and the bent ends are extended far enough to give ample strength of bond. The resulting

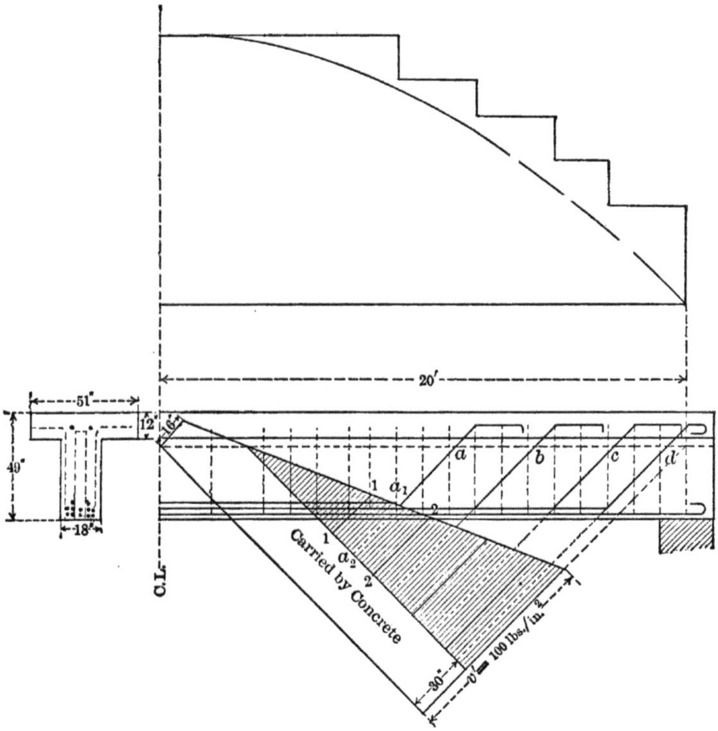

Fig. 62*b*.

lengths of straight portions are as follows: Two rods, 18 ft. long; two rods, 24 ft. long; two rods, 30 ft. long; two rods, 34 ft. long; and four straight rods running the entire length of the girder. These lengths all exceed somewhat the requirement for moment. This is shown graphically in the moment diagram of Fig. 62*b* where the resisting moment of the beam is shown by the stepped line, and the bending moment by the curve. The necessary length of rod to develop a bond strength equal to its full working stress at the specified values is $\frac{15,000}{4\times75}=50$ diameters. For the $1\frac{3}{8}$-in. rods this is equal to $50\times1\frac{3}{8}=69$ in. This is provided in all cases with the bent rods. For the four straight rods the maximum possible bond stress will be found from eq. (1), Art. 92. In this case $U=\frac{80,000}{40}=2000$ lbs. per lineal inch. The bond stress $=\frac{2000}{4\times4.32}=115$ lbs/in². The allowed value is $75\times1\frac{1}{2}$ or 112.5 lbs/in². The calculated stress is very slightly above this, but considering the prolongation of the beam beyond the center of support it will be allowed. Additional strength will be furnished by hooks at the ends of the rods.

The effectiveness of the bent rods in carrying shear will now be determined. A diagram of shearing stress is shown in Fig. 62*b* projected on a line at 45° to the axis. This will be convenient in representing the effect of the diagonal bars. These bars, together with other reinforcement, must be sufficient to carry the shear represented by the shaded area, the concrete carrying the remainder. The point *a* where the first rod intersects the neutral axis is about 11.5 ft. from the center and the unit shearing stress $=16+(100-16)\frac{11.5}{20}=64$ lbs/in². This is shown by the ordinate a_1a_2. The concrete will carry 30 lbs/in², leaving 34 lbs/in² to be carried by the steel. This amounts to $34\times18=610$ lbs. for each lineal inch of beam. Considering these rods effective over a distance of 3 ft. (the space between the first and second), the stress in each rod is, by eq. (3), Art. 125, equal to $\frac{1}{2}\times.7\times610\times36=7700$ lbs. This gives a unit stress of $7700/1.23=6270$ lbs/in², a low value. Graphically, the total amount carried may be represented by the shaded area between the ordinates 1–1 and 2–2 in Fig. 62*b*. In a similar manner it is found that at point *b*, 14.5 ft. from the center, the shearing stress $=16+84\times\frac{14.5}{20}=77$ lbs/in², the concrete carrying 30 lbs. and the rods 47 lbs., giving a stress in the rods of 8650 lbs/in². At *c* the shear is 90 lbs/in², and the stress in the rods 11,500 lbs/in², and at *d* the shear is about 98 lbs/in², and the stress about 11,500 lbs/in².

This analysis indicates that the bent rods are sufficient to carry all the shear except to the left of *a*. The maximum value $=58$ lbs/in², thus requiring 28 lbs/in² to be carried by steel. This will be supplied in the form of stirrups. If $\frac{3}{8}$-in. stirrups are used in a double loop, their strength, at say, 12,000 lbs/in², is $4\times12,000\times.11=5280$ lbs., and the spacing to carry 28 lbs/in² $=\frac{5280}{28\times 18}=10.5$ in. Inasmuch as stirrups assist in supporting the rods and in binding together web and flange, and as they add greatly to the security and reliability of the construction, they will be used throughout and spaced 12 in. apart, except along the center 8 ft. where they will be spaced 2 ft. apart.

If stirrups alone were used to carry the shear they would need to be spaced about 4 in. apart near the end.

In the design of web reinforcement it should be understood that the methods of calculation here used can be considered as only roughly approximate. They are based broadly on theoretical considerations and the results of experiment and lead to satisfactory and safe designs, but they cannot be considered as being in any sense precise methods or as representing the best that may be developed.

COLUMNS.

136. Working Stresses.—In determining the proper working stresses for columns it is necessary to consider the question mainly with reference to the stress in the concrete, for under ordinary working stresses in the concrete the stress in the steel will be relatively low. From the tests and discussion of preceding chapters it appears that with reference to the behavior of the concrete, columns may be divided into two classes: (1) columns reinforced with longitudinal reinforcement only, and (2) columns reinforced with hoops or bands and with or without longitudinal reinforcement. These types will be considered separately.

(1) *Columns Reinforced with Longitudinal Steel Only.*—In this form of column the concrete fails in a manner similar to the failure of an unreinforced column. When a load is reached which stresses the concrete to about the same value as in a plain concrete column, failure takes place suddenly and by shearing action. The ultimate strength of the entire column is, however, increased by the steel and approximately as theory would indicate. Considering the manner of failure and the lack of "toughness" in such a column the factor of safety should be relatively large and determined on about the same

basis as for a short column of plain concrete. The strength of a 1:2:4 concrete at 60 days in the form of a short column or cylinder ranges from about 1600 to 1800 lbs/in^2, and applying a factor of safety of four, gives a working stress of 400 to 450 lbs/in^2. This may be taken as a suitable value for the concrete in the type of column here considered. For richer and stronger mixtures the working stress may be increased accordingly. Where the concrete is depended upon to fire-proof the steel, a certain thickness should be deducted in calculating strength. As shown in Art. 141 the necessary thickness for fire-proofing is about two inches, but if 1½ inches be deducted all around in calculating strength this will amply provide for the weakening effect of fires.

The working stress in the steel is a function of the working stress in the concrete and the ratio, n, of the moduli of elasticity of the two materials. If this ratio is taken at 12, then with a stress of 400 in the concrete, the stress in the steel will be $12 \times 400 = 4800$ lbs/in^2. Under working loads the steel is therefore stressed only to a very low value.

Let us consider the variation in the stresses in a column subjected to increasing loads, following the same general method of analysis as for the beam in Art. 121. Assume a concrete having a compresive strength in the form of a short column of 1600 lbs/in^2, and assume, further, that the stress-strain diagram is parabolic, as shown in Fig. 63a, with a value of E at the origin of 3,000,000 lbs/in^2. The value of E at other stresses will be 2,625,000 at 400 lbs.; 2,250,000 at 800 lbs.; and 1,875,000 at 1200 lbs/in^2. Consider a column composed of this concrete and 2% of steel reinforcement. From Art. 65, Chapter III, the total load on a column is given by the formula $P' = Af_c[1 + (n-1)]p$, in which p=steel ratio, A= total area, and f_c=unit stress on the concrete. The average unit stress on the column will be $f = \frac{P'}{A} = f_c[1 + (n-1)]p$. Assume a working stress in the concrete of 400 lbs/in^2. For this stress the value of n is $30/2.625 = 11.4$, and the average unit stress on the column section will therefore be $400(1+10.4\times.02)$ $=483$ lbs/in². In Fig. 63*b*, let this be represented on the axis OX by the distance Oa, which is conveniently taken as a unit. Let the ordinate ab represent the corresponding unit stress in the concrete of 400 lbs/in², and ac the unit stress in the steel, $=400\times11.4=4560$ lbs/in². Now assume the load to be increased so as to cause a stress of 800 lbs/in² in the concrete.

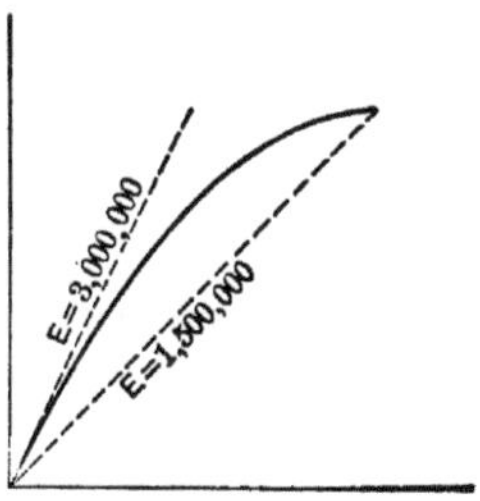

FIG. 63*a*.

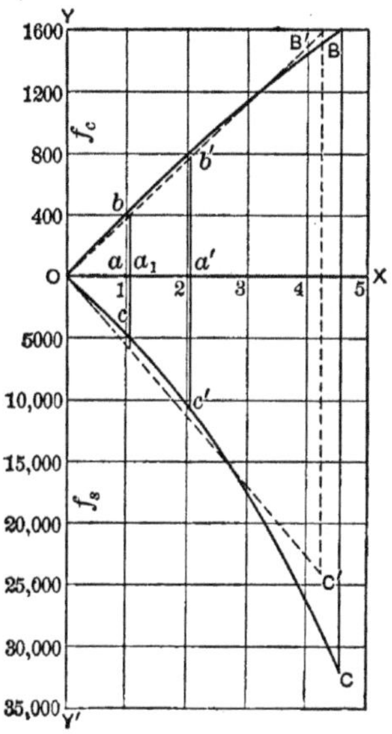

FIG. 63*b*.

The value of n at this stress$=30/2.250=13.33$. The value of $f_s=800\times13.33=10,670$ lbs/in². The average unit stress is $f=800(1+12.33\times.02)=997$ lbs/in². This will be represented by the abscissa Oa' of a value of $997/483=2.06$ units. The ordinates $a'b'$ and $a'c'$ (to the heavy lines)$=800$ and 10,670 lbs/in², respectively. In the same manner continue the calculations and plot the curves OB and OC, which will then represent the variation of concrete and steel stress throughout the entire range of load to the ultimate strength of the concrete. At this point the stress in the steel will be $20\times1600=32,000$

lbs/in², and the average unit stress will be $1600(1+19\times.02)=$ 2208 lbs/in². This is 4.57 times the load causing the stress of 400 lbs/in². From this it is plain that with increasing loads the steel receives a greater proportionate stress, the variation in the amount carried by the steel depending on the variation in the value of n. It is also evident from this diagram that the ultimate load on the column is greater than four times the load (4.57 times in the assumed case) which produces the stress of 400 lbs/in² in the concrete. Hence if the working stress in the concrete is based on a factor of safety of four relative to plain concrete, then the factor of safety of the reinforced column will be greater than four. The case is somewhat similar to that of the beam. Obviously the total load increases more rapidly than the value of the stress f_c, the exact rate depending on the relative amount of steel and the variation in n.

In order to secure a more uniform factor of safety, and to take some account of the fact that under increasing loads the steel receives an increasing proportion, it is desirable to use a value of n in the calculations somewhat larger than that which is obtained by taking a value of E_c corresponding to very low stresses. On this basis the actual stress in the concrete at working load will be slightly greater than assumed, and that in the steel somewhat less, but the calculated and actual stresses will coincide at about one-half of the ultimate load and the factor of safety will still be somewhat greater than the ratio of ultimate strength of the concrete to its assumed working strength. In Fig. 63*b* the dotted straight lines AB' and AC' represent the assumed variation of stress, using a constant value of $n=15$. The average unit stress on the column, at working loads, will be $f=400(1+14\times.02)=512$ lbs/in². This is 6% greater than the load represented by the abscissa Oa, and is represented by the distance Oa_1. The ordinates to the curves OB and OC show the actual stresses in the concrete and steel. The ultimate strength of the column being 2208 lbs/in², the real factor of safety $=2208/512=4.3$. A value of 15 for n

may well be used for all ordinary mixtures and for all types of columns.

(2) *Columns Reinforced with Hoops or Bands and with or without Longitudinal Steel.*—From the tests given in Chapter IV, it is seen that in general the effect of hooping is to increase the "toughness" and the ultimate strength of the column. The elastic limit and rigidity of the column appears to be decreased if anything. When used with longitudinal steel the hooping acts in much the same way, but is of greater importance in this case as it keeps the concrete intact up to a degree of deformation that enables the longitudinal steel to be stressed to its elastic limit. It thus renders such reinforcement very effective.

Concerning the proper working stress for hooped columns, it would seem that this should be selected mainly with reference to the elastic limit, as in the case of structural steel; but the greater toughness of the hooped column, as compared to the other type, insures a much larger and more certain margin of safety, and hence the working stress may be made a greater proportion of its elastic limit strength than in the other case. The two types of columns may be compared to mild steel and cast iron; a much higher relative working stress may be used in the former than in the latter, chiefly because of its larger margin of safety against deformation beyond the elastic limit. This is of great importance, especially with respect to effects of unequal settlement, eccentric loading and secondary stresses.

For hooped columns, without longitudinal steel, the elastic limit is about the same as for plain concrete and varies but little for various percentages of steel. Hence the same working stress may be used for all percentages of hooping; but for reasons already stated this value may be made greater than for plain or for longitudinally reinforced columns. For this type of column, therefore, the authors would suggest a working stress about 20% greater than for plain concrete, or from 500 to 550 lbs/in^2 for an amount of hooping of 1% or more. This value is to be applied to the concrete alone, and the hooping is not to be taken directly into account. For large amounts of hooping, somewhat higher stresses might be used, but increased strength and rigidity can be provided more effectively by adding longitudinal reinforcement.

For hooped columns containing longitudinal reinforcement, the elastic limit of the column tends to approach a point corresponding to the elastic limit of the longitudinal steel, the exact effect depending upon the effectiveness of the hooping and the amount of longitudinal steel. If this were fully accomplished, the working stress might be placed as high as 15,000 lbs/in^2 on the steel, corresponding to a stress on the concrete of about 1000 lbs/in^2. This is beyond the normal elastic limit strength of the material, and is not to be recommended. Considering all the factors involved, it would seem that the stress on the concrete could safely be taken at 50% more than for plain concrete. This would give a value of 600 to 675 lbs/in^2, with 9000 to 10,000 lbs/in^2 on the longitudinal steel. To render such stresses safe, an amount of hooping equal to 1% would appear from the results of tests to be sufficient. More hooping will increase the ultimate strength, but not materially the elastic limit, and hence it will not permit the use of higher stresses. For rich concrete still higher values may be used, but not to exceed about 800 lbs/in^2 on the concrete, corresponding to 12,000 lbs/in^2 on the steel. With our present knowledge, also, it would be unwise to depend upon the steel to carry its full share of stress as here calculated for very large percentages. With 5% of steel and 700 lbs/in^2 in the concrete, the average unit stress on the column would be $f=700(1+14\times .05)=1190$ lbs/in^2.

In the determination of the strength of hooped columns, only the section within the hooping should be considered. The shell outside is of the same character as plain concrete and it is found to crack and split off at deformations correspondnig to the ultimate strength of plain concrete. It is useful as fireproofing, but its limitations of deformation is another reason for not selecting too high values for the working stress on the core.

It should be said that the above treatment of the hooped column is quite different from that of Considère and of the French Commission on Reinforced Concrete. These authorities recommend that the hooping be counted upon directly to a much larger extent than the longitudinal reinforcement. The formula recommended by the French Commission is

$$f=f_c(1+15p+32p'),$$

in which f_c is the safe strength of plain concrete, taken at 28% of the ultimate strength in the form of cubes, p=ratio of longitudinal reinforcement, and p'=ratio of spiral reinforcement. It is also recommended that the maximum stress shall not exceed 0.6 of the ultimate strength of the concrete. These values are based chiefly on a consideration of ultimate strength.

(3) *Columns Reinforced by Structural Steel Column Units.*—Where a large amount of reinforcement is desired, certain advantages are gained by arranging it in the form of structural column units, such as four angles latticed together, which in themselves are capable of acting as columns.* The construction can be so arranged that the steel columns will carry the false work and dead load of two or more floors, thus enabling the placing of concrete to proceed simultaneously on several floors. In this way, also, some initial dead load stress can be applied to the steel of the column before the concrete of the column is placed, thus enabling higher steel stresses to be used. On the other hand, such steel is much more costly per pound than rods. Furthermore, the results of experiments show that the adhesion of concrete to steel where the latter presents broad flat surfaces is not good, and the presence of numerous lattice bars hinders the production of a dense homogeneous concrete.

* For a good example of such a design, see paper by Wm. H. Burr on "The Reinforced Concrete Work of the McGraw Building," Trans. Am. Soc. C. E., Vol. 60, 1908.

The resulting column is therefore likely to be less of a monolithic character than one in which the reinforcement consists of small rods. In order that the concrete may be counted upon in such a column, it should be well enclosed either by the structural form itself or by means of bands or hooping. All concrete not so enclosed can be considered only as fire-proofing.

Where designed in accordance with these principles, and the steel and concrete receive their load simultaneously, the working stresses may be taken about the same as for the second class of columns here discussed. If, however, a partial load is applied to the steel before the concrete is placed, such initial stress need not be counted, excepting that the total stress in the steel should not exceed the usual working stress for steel columns of about 16,000 lbs/in^2. Where the amount of steel becomes very large the relative value of the concrete becomes more uncertain and its consideration as an element of strength is of doubtful wisdom and unsupported by experimental evidence.

137. Long Columns.—The tests of Chap. IV indicate that for lengths of 20 to 25 diameters little or no difference in strength is shown for different lengths. Very long columns should, however, be avoided, and it is important to adopt a conservative practice in this regard. It is therefore advisable to apply the long column formula of Art. 96*b* for lengths exceeding 12 or 15 diameters. It would also seem that a ratio of l to r greater than 100 should be not used, or a ratio of length to least width greater than about 30. Banded columns are much to be preferred for slender proportions.

It is important to note that plain concrete is entirely suitable for short columns up to lengths of 6 to 10 diameters, and for such columns the addition of steel is not in general economical.

138. Column Details.—In the construction of columns great care should be exercised to place and hold the steel in its proper position and to secure sound work. In this respect poor workmanship is more serious perhaps than in any other structural form. Eccentricity of steel or uneven quality of concrete not only causes weakness at the section in question, but also results in eccentricity of load and lateral deflections. Reinforcing rods must be arranged concentrically and held securely in place until the concrete is set. This is usually accomplished by wiring or banding the rods together at intervals of a foot or so, but such banding cannot be considered as hooping in the sense usually employed. Where hooping is used as reinforcement it may consist of wire spirally wound or otherwise, or of separate bands of welded or riveted steel. To be effective such hooping should be spaced relatively close, so as to serve to confine the concrete within the cylinder formed by the hooping and to effect the "toughening" assumed in the previous discussion. What such spacing may be is not well determined, but until further evidence is available a clear spacing of about one-fifth to one-fourth the diameter of the hoops or bands may be considered the maximum. A total amount of hooping or banding at least equal to 1% of the enclosed volume should be used. The French Commission recommends a spacing of spirals of one-eighth to one-fifth the column diameter, but in this case the spiral reinforcement is counted upon directly in the calculation of strength.

In the case of hooped columns or columns in which lateral reinforcing members are used, such as lacing on structural units, special care should be taken to secure as dense concrete as possible, and to reduce the settlement of the material to a minimum. Any settlement tends to create vacant spaces or porous material underneath the reinforcement. In splicing columns large rods or structural shapes should be accurately fitted and well spliced; small rods may be spliced at floor levels by overlapping a sufficient distance to develop the requisite bond strength. At the base of a column large rods or shapes should rest upon suitable base plates in the foundation concrete.

139. Economy in the Use of Reinforced Columns.—From eq. (1), Art. 95, we see that with a value of $n=15$, the use of each 1% of longitudinal steel adds 14% to the strength of a column. If the ratio of cost of steel to cost of concrete per unit volume be 50, then the increased cost of a column with 1% of steel will be $50 \times 1\% = 50\%$. The gain in strength being only 14%, the relative economy of the reinforced column is only $^{114}/_{150} = 76\%$ that of the plain concrete. Again, take a very strong mixture, such as 1:1 mortar, whose working stress may possibly be taken as high as 800 lbs/in². Such a mortar will cost perhaps $12.00 per cu. yd. (not including forms, etc.) or 45 c. per cu. ft. Placing steel at the low value of 2 c. per lb., the cost ratio becomes 22.5. Such concrete will have a value of E_c of at least 3,000,000, giving $n=10$. Hence 1% reinforcement will add 9% to the strength and 22.5% to the cost. If a cheap concrete be taken with a low modulus the steel will add a larger percentage of strength, but at the same time a much greater percentage of cost. Another way of considering this question is from the standpoint of the relative working stresses in concrete and steel. Using a value of $n=15$ the load carried by the steel per square inch is fifteen times that taken by the concrete. If the cost ratio is 50 then the steel is $50/15 = 3\frac{1}{3}$ times as costly as the concrete for the same service.

The above analysis shows that from the standpoint of theoretical economy the use of steel in columns is undesirable, and were this the only consideration it would not be used, at least in the form discussed. While no economy can be figured for the use of steel in columns it is by no means valueless. In practice, columns are subjected to bending moments uncertain in amount, but for which something more than plain concrete is desired, especially where the column is of considerable length. It is in such columns that tensile stresses are most apt to occur and where steel is most needed. Furthermore, steel is a more reliable material than concrete, and in small sections where the danger of weak or imperfect spots in the concrete is greatest, steel reinforcement is of great value in producing a more reliable structure. Then, again, great strength may be desired from small sections in order to save space, in which case steel may be used. In very large (relatively short) columns little is to be feared from bending stresses, as in such a case no resultant tensile stress is likely to occur.

DURABILITY OF REINFORCED CONCRETE.

140. The Protection of Steel from Corrosion.—A continuous coating of Portland cement has been found by experience to be a practically perfect protection of steel against corrosion. The rusting of iron requires the presence of moisture and carbon dioxide. Portland cement not only forms a coating which excludes the moisture and CO_2, but in hardening it absorbs CO_2, tending to remove any of this gas which may be present. In practice the protective nature of Portland-cement concrete has long been known, and its use as a paint was adopted by the Boston Subway Engineers after careful investigation.

While an unbroken coating of cement offers what appears to be a perfect protection, the value of a concrete as actually deposited may be very much less. A series of experiments made by Professor Charles L. Norton gives valuable information on this subject. In one series, small specimens of steel 6″ long were embedded in blocks 3″×3″×8″ in size of various mixtures of cement, sand, and stone or cinders. The blocks were then exposed for three weeks to various corrosive atmospheres consisting of steam, air, and CO_2. The results were as follows: The neat cement furnished perfect protection. The specimens embedded in mortars and concretes showed spots of rust at voids or adjacent to a badly rusted cinder. He concludes that concrete to be an effective protection should be mixed quite wet so as to furnish a thin coating on the metal, and must be free from voids and cracks. He finds that dense cinder concrete mixed wet is as effective as stone concrete.

In a second series of experiments on steel already rusted,

from a slight stain to a deep scale, the following results were obtained: The concrete was 1:2½:5 (stone) and 1:3:6 (cinders). After one to three months in corroders and one to nine months in damp air no specimen showed any change except where the concrete was poorly applied. Some of the concrete was purposely made very dry and the rods were not well covered. These specimens were seriously corroded. Unprotected steel specimens subjected to the same treatment were almost entirely corroded. While the experiments of Professor Norton provided for a covering of 1½ inches, there is no reason to suppose that a much thinner covering, if intact, will not furnish as good protection.

Many cases have been cited of steel removed from concrete after the lapse of 20 years or more and found to be in perfect condition. A test by Mr. H. C. Turner,* in which steel bars embedded to a depth of 3 inches in blocks of 1:2:4 and 1:3:5 concrete and exposed to sea-water and air for nine months showed perfect preservation.

Perfect protection of the steel by concrete was demonstrated in the case of a building at New Brighton, N. Y., built in 1902 and partly torn down in 1908. All steel was found to be in perfect condition excepting in a few cases where column hoops came closer than ¾ inch to the surface. The footings were covered by the tide twice daily but the bars therein showed no corrosion.†

In view of such tests and observations as here noted it may be concluded that when well placed the concrete affords complete protection against corrosion.

141. Fireproofing Effect of Concrete.—Severe fire tests show that when concrete is subjected to red-hot temperatures (about 1700°) for three or four hours and then is quenched by hose streams, it is likely to show pitting but that it will still offer a sufficient protection to the steel.‡

* *Eng. News*, Aug. 1904, p. 153.

† *Eng. Record*, Vol. 57, 1908, p. 105.

‡ See tests by Professor Ira W. Winslow in *Eng. Record*, Nov. 26, 1904, p. 634, and by Professor F. P. McKibben in *Eng. News*, Nov. 21, 1901, p. 378.

A reinforced-concrete building at Bayonne, N. J., was subjected to a very hot fire in the burning up of its contents but with no injury to the building.*

In the Baltimore fire of 1904 the value of concrete as a fireproofing material, and of reinforced-concrete construction, was fully demonstrated. Professor C. L. Norton of the Insurance Engineering Experiment Station, after a careful study of the damage done by the fire, states as follows:†

"Where concrete floor arches and concrete-steel construction receive the full force of the fire it appears to have stood well, distinctly better than the terra-cotta." The reason for this he considers to be the fact that terra-cotta expands about twice as much as steel, but that concrete expands about the same. Little difference was observed between stone and cinder concrete. High temperatures long continued dehydrate and soften concrete, but this process in itself gives off water and absorbs the heat, thus protecting the interior. The layer of changed material is then a better non-conductor than before, so the process goes on very slowly. Captain J. S. Sewell, reporting to the Chief of Engineers ‡ on the Baltimore fire, states that, with reference to concrete construction subjected to very high heats: "Exposed corners of columns and girders were cracked and spalled, showing a tendency to round off to a curve of about 3 in. radius. Where the heat was most intense the concrete was calcined to a depth of $\frac{1}{4}''$–$\frac{3}{4}''$, but showed no tendency to spall, except at exposed corners. On wide, flat surfaces the calcined material was not more than $\frac{1}{4}$-in. thick and showed no disposition to come off. The terra-cotta fireproofing showed up much poorer." In his general conclusions he considers it at least as desirable as steel work protected by the best commercial hollow tiles, and preferable to tile for floor slabs and fire-proof covering.

---

* *Eng. Record*, April 12, 1902, p. 341.

† *Eng. News*, June 2, 1904, p. 524.

‡ *Eng. News*, March 24, 1904.

In a report of a committee of members of the American Society of Civil Engineers on the effects of fire in the San Francisco conflagration, similar conclusions were reached as to the value of concrete as a fireproofing material. It was also found far preferable to tile for floors. With respect to the injury to the concrete itself the committee was of the opinion that it was sufficient in many cases to require reconstruction.* Additional evidence of the value of concrete as a fireproofing material is contained in a report of a committee of the National Fire Protection Association.† This report also goes far to indicate the necessary thickness of the protective covering.

The necessary thickness of concrete to furnish adequate fire protection depends somewhat upon the character and importance of the member. Such members as main girders, where a failure would involve a considerable portion of the building and where the steel is concentrated in a few rods, should be more thoroughly protected than floor slabs of small span, where a few local failures would be of no importance, and where additional covering would add largely to the expense. Results of fire tests and experience in conflagrations indicate that $2''$–$2\frac{1}{2}''$ will offer practically complete protection, and that a minimum of $\frac{1}{2}''$–$\frac{3}{4}''$ for floor slabs will usually be sufficient. Large flat surfaces, such as floor slabs, are less exposed than the corners of projecting forms like beams and columns.

While satisfactory protection of the steel can thus be secured the effect of fire upon the concrete itself, and its usefulness after more or less calcination, is a question of much importance. Where a sufficient allowance has been made for such damaged material it would appear that the removal of the soft or loosened portions and replastering by cement mortar would generally secure effective repair.

---

* Trans. Am. Soc. C. E., Vol. 59, 1907.

† *Eng. News*, Vol. 59, 1908, p. 627.

142. Reinforcing Against Shrinkage and Temperature Stresses.—Where a reinforced structure is unrestrained by outside forces the only stresses arising from shrinkage and temperature changes are those due to the mutual action of steel and concrete. As the two materials have nearly equal rates of expansion temperature changes will cause very little stress. Shrinkage in hardening will cause more important stresses, as shown in Art. 43, but still not unduly large unless the steel ratio is very high.

When the structure is restrained by outside forces so that it is not free to contract or expand, as in the case of a long wall, then the resulting stresses are likely to be high. When not reinforced, concrete will, under such circumstances, crack at intervals, its maximum deformation under stress not being equal to its maximum temperature deformations. If it be assumed that concrete when reinforced will not stretch more than plain concrete, as seems probable (Art. 42), then no amount of reinforcement can entirely prevent contraction cracks. The reinforcement can, however, force such cracks to take place as they do in a beam—at such frequent intervals that the requisite deformation takes place without any one crack becoming large. Laboratory tests on beams would indicate that if steel is used in sufficient quantities the cracks may easily remain quite invisible and be of no consequence from any practical standpoint. Thus if the coefficient of expansion be .000006 a change of temperature of 50° causes a change of length (if free) of .0003 part. A deformation of this amount in a beam (corresponding to a steel stress of 9000 lbs/in^2) would not cause cracks easily detected. The prevention of large cracks by means of reinforcement is then a matter of using sufficient steel to force the concrete to crack at small intervals. No one crack will open up far until the steel is stressed beyond its elastic limit, hence we may say approximately that the amount of steel used must be such that the concrete will crack elsewhere before the steel is stressed beyond its elastic limit. A larger amount of steel will serve to keep the cracks smaller.

The size and distribution of the cracks will also depend upon the bend strength furnished by the rods. If we assume the cracks to develop successively the distance between cracks must be sufficient to develop a bond strength equal to the tensile strength of the concrete. Hence, in general, the size and spacing of the cracks will vary inversely with the bond strength of the reinforcing steel per unit of concrete section.

In calculating the requiste amount of steel the temperature stress in the steel itself must be considered. This will add to its skrinkage stress, so that its total stress will equal its temperature stress plus the stress necessary to crack the concrete. If, for example, the assumed drop in temperature be 50° the temperature stress in the steel $=50\times.0000065\times30{,}000{,}000=9750$ lbs/in². If the tensile strength of the concrete be 200 lbs/in² and the assumed allowed stress (elastic limit) in the steel be 40,000 lbs/in², then the stress available $=40{,}000-9750=30{,}250$ lbs/in², and the required percentage of steel $=p=\frac{200}{30{,}250}=$ .0066. If the elastic limit be 60,000 lbs/in² the steel ratio $=p=\frac{200}{60{,}000-9750}=.004$. For the purposes here considered obviously a high elastic-limit steel is desirable, and in order to distribute the deformation as much as possible a mechanical bond is advantageous.

CHAPTER VI.

FORMULAS, DIAGRAMS, AND TABLES.

143. Rectangular Beams; Linear Variation of Stress.

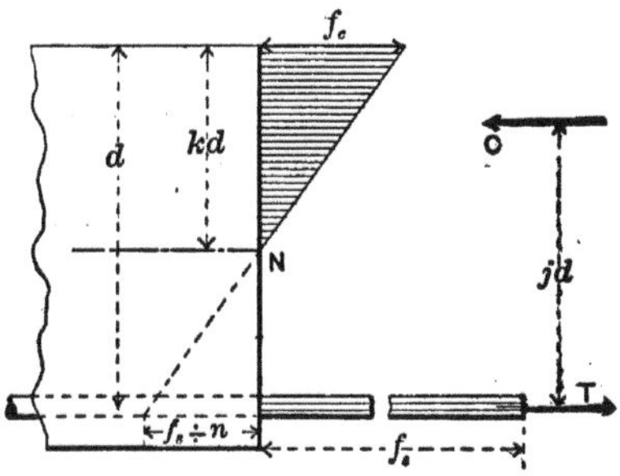

FIG. 64.

Notation.

f_s = unit stress in steel;
f_c = " " " concrete;
E_s = modulus of elasticity of steel;
E_c = " " " " concrete;
$n = E_s/E_c$;
T = total tension;
C = " compression;
M_s = moment of resistance relative to the steel;
M_c = " " " " " " concrete;
M = bending moment or moment of resistance in general;
A = steel area;
b = breadth of beam;
d = net depth of beam;
k = ratio of depth of neutral axis to depth d;
j = ratio of lever-arm of resisting couple to depth d;

p = steel ratio = A/bd;
$R_s = f_s pj$ = "coefficient of strength" relative to steel;
$R_c = \frac{1}{2} f_c kj$ = " " " " " concrete.

Formulas.

Position of neutral axis,

$$k = \sqrt{2pn + (pn)^2} - pn. \qquad (1)$$

Arm of resisting couple,

$$j = 1 - \tfrac{1}{3}k. \qquad (2)$$

Moment of resistance,

$$M_s = f_s pj \cdot bd^2 = R_s \cdot bd^2, \qquad (3)$$

$$M_c = \tfrac{1}{2} f_c kj \cdot bd^2 = R_c \cdot bd^2. \qquad (4)$$

Approximately,

$$M_s = f_s A \cdot \tfrac{7}{8}d, \qquad (3')$$

$$M_c = f_c \cdot \tfrac{1}{6}bd^2. \qquad (4')$$

Fibre stresses,

$$f_s = \frac{T}{A} = \frac{M \div jd}{A}, \qquad (5)$$

$$f_c = \frac{2C}{bkd} = \frac{2M \div jd}{bkd}. \qquad (6)$$

Steel ratio,

$$p = \frac{1}{2} \cdot \frac{1}{\frac{f_s}{f_c}\left(\frac{f_s}{nf_c} + 1\right)}. \qquad (7)$$

Cross-section of beam for given bending moment M,

$$bd^2 = \frac{M}{f_s pj} = \frac{M}{R_s}, \qquad (8)$$

$$bd^2 = \frac{M}{\frac{1}{2} f_c kj} = \frac{M}{R_c}. \qquad (9)$$

Diagrams.—Plates I–IV, pp. 274–277, are diagrams of values of k and j for various values of p; and values of R_s and R_c (called simply R) for various values of p and of f_s and f_c. The value of n is taken at 10, 12, 15, and 18 respectively.

The use of the diagrams in finding moments of resistance (Eqs. (3) and (4)) and in determining cross-sections (Eqs. (8) and (9)) is obvious. The proper steel ratio, p, to use for given values of f_s and f_c (Eq. (7)) is determined from the intersection of the curves for the given values of f_s and f_c. Finally, the actual fibre stress, f_s or f_c, resulting from a given M, p, and bd^2 will be found by first calculating M/bd^2 from the given values. Call this R. Then with this value of R and the given value of p enter the diagram and find the corresponding values of f_s and f_c.

ILLUSTRATIVE EXAMPLES. — 1. *Moment of Resistance.* — Given the following: $b=12''$, $d=20''$, $f_s=14{,}000$, $f_c=600$, and $p=0.8\%$; find M_s and M_c. Assume $n=15$. *Solution.* From Plate III, p. 276, we find for $p=0.8\%$ and $f_s=14{,}000$, $R_s=96$; and for $f_c=600$, $R_c=100$. Hence $M_s=96bd^2=460{,}800$ in-lbs., and $M_c=100bd^2=480{,}000$ in-lbs.

2. *Fibre Stresses.*—Given $b=12''$, $d=20''$, $p=0.8\%$, and $M=450{,}000$ in-lbs., to find f_s and f_c. *Solution.* Use Eqs. (5) and (6) directly; or, find M/bd^2 and use the diagrams. Thus $M/bd^2=450{,}000/4800=93.75$. Then from Plate III, with $R=93.75$ and $p=0.8\%$ we find f_s=about 13,500 and f_c=about 560 lbs/in^2.

3. *Cross-section of Beam and Steel Ratio.*—Given $M=500{,}000$ in-lbs., $f_s=12{,}000$, $f_c=500$, to find bd^2. *Solution.* From Plate III we find at the intersection of the curves for $f_s=12{,}000$ and $f_c=500$, a value of R of 84. Hence $bd^2=500{,}000/84=5950$. The required amount of steel is also found from the diagram to be 0.8%.

144. Rectangular Beams; Parabolic Variation of Stress; for Ultimate Loads.

Notation.—

As in Art. 143, but here $R_c=\frac{2}{3}f_ckj$.

Formulas.

Position of neutral axis,

$$k=\sqrt{3pn+(\tfrac{3}{2}pn)^2}-\tfrac{3}{2}pn. \quad . \quad . \quad . \quad . \quad . \quad (10)$$

Arm of resisting couple,

$$j=1-\tfrac{3}{8}k. \qquad (11)$$

Moment of resistance,

$$M_s=f_s pj\cdot bd^2=R_s\cdot bd^2, \qquad (12)$$

$$M_c=\tfrac{2}{3}f_c kj\cdot bd^2=R_c\cdot bd^2. \qquad (13)$$

Approximately,

$$M_s=f_s A\cdot 0.8d, \qquad (12')$$

$$M_c=f_c\cdot 0.28bd^2. \qquad (13')$$

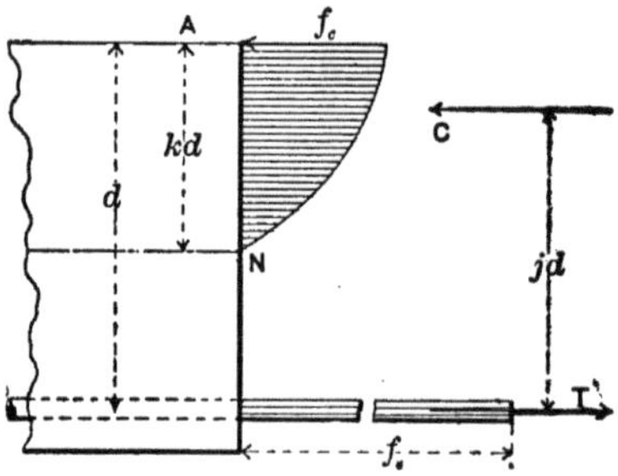

FIG. 65.

Fibre stresses,

$$f_s=\frac{T}{A}=\frac{M\div jd}{A}, \qquad (14)$$

$$f_c=\frac{\frac{3}{2}C}{bkd}=\frac{\frac{3}{2}M\div jd}{bkd}. \qquad (15)$$

Steel ratio,

$$p=\tfrac{2}{3}\cdot\frac{1}{\dfrac{f_s}{f_c}\left(\dfrac{f_s}{2nf_c}+1\right)}. \qquad (16)$$

Cross-section of beam for given bending moment M,

$$bd^2=\frac{M}{f_s pj}=\frac{M}{R_s}, \qquad (17)$$

$$bd^2=\frac{M}{\frac{2}{3}f_c kj}=\frac{M}{R_c}. \qquad (18)$$

Diagrams.—Plate V, p. 279, is a diagram of values of k and j for various values of p; and values of R_s and R_c for various values of p, f_s, and f_c. The full lines are drawn for $n=15$; the dotted lines for $n=12$. The fibre stresses are here assumed as representing ultimate strengths, and the diagram is supposed to give results pertaining to ultimate strength. To use it for purposes of designing, the given loads or moments should be multiplied by the selected factor of safety, or the value of R obtained from the diagram divided by such factor of safety.

145. T-Beams; Linear Variation of Stress.

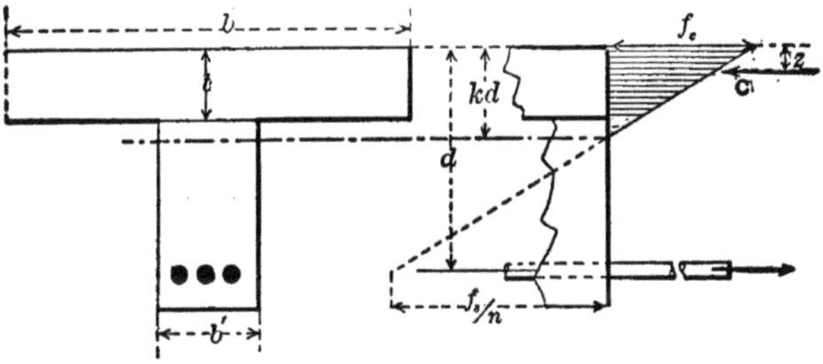

FIG. 66.

Notation. (In addition to that of Art. 143.)

b = width of flange;

b' = width of web;

t = thickness of flange;

z = depth of resultant of compressive stress;

$d-z$ = arm of resisting couple.

Formulas.

Case I. Neutral axis in the flange.

Use formulas (1)–(9) as for rectangular beams; formula (1) for k will determine whether the case is I or II.

Approximately,

$$M_s = f_s A(d - \tfrac{1}{3}t), \qquad (19)$$

$$A = \frac{M}{f_s(d - \frac{1}{3}t)}. \qquad (20)$$

Case II. Neutral axis in the web; compression in web neglected.

Position of neutral axis,

$$k=\frac{nA+\frac{1}{2}bt\cdot\frac{t}{d}}{nA+bt}=\frac{pn+\frac{1}{2}\left(\frac{t}{d}\right)^2}{pn+\frac{t}{d}}. \quad . \quad . \quad . \quad . \quad (21)$$

Position of resultant of compressive stress,

$$z=\frac{3k-2\frac{t}{d}}{2k-\frac{t}{d}}\cdot\frac{t}{3}. \quad . \quad . \quad . \quad . \quad . \quad . \quad . \quad (22)$$

Moment of resistance,

$$M_s=f_sA(d-z), \quad . \quad . \quad . \quad . \quad . \quad . \quad . \quad (23)$$

$$M_c=f_c\left(1-\frac{t}{2kd}\right)bt(d-z). \quad . \quad . \quad . \quad . \quad (24)$$

Steel area,

$$A=\frac{M}{f_s(d-x)}. \quad . \quad . \quad . \quad . \quad . \quad . \quad . \quad (25)$$

Approximately,

Assume $(d-z)=\frac{7}{8}d.$

Diagrams.—Values of k and j, for various values of p and t/d, are given in Plate VI. Plates VII to XI give values of M/bd^2 from eq. (24) for various values of f_s, f_c, and t/d.

146. Beams Reinforced for Compression.

Notation. (In addition to that of Art. 143.)

A' = area of compressive steel;

p' = steel ratio of compressive steel;

f_s' = unit stress in " "

C' = total stress in the compressive steel;

d' = distance from compressive face to the plane of the compressive steel;

x = depth to resultant compression, $C+C'$.

Formulas.

Position of neutral axis,

$$k=\sqrt{2n\left(p+p'\frac{d'}{d}\right)+[n(p+p')]^2}-n(p+p'). \quad . \quad . \quad (26)$$

Position of resultant of compressive stress, $C+C'$,

$$x=\frac{\frac{k}{3}+\frac{d'C'}{dC}}{1+\frac{C'}{C}}.d; \text{ in which } \frac{C'}{C}=\frac{2p'n\left(k-\frac{d'}{d}\right)}{k^2}. \quad . \quad . \quad (27)$$

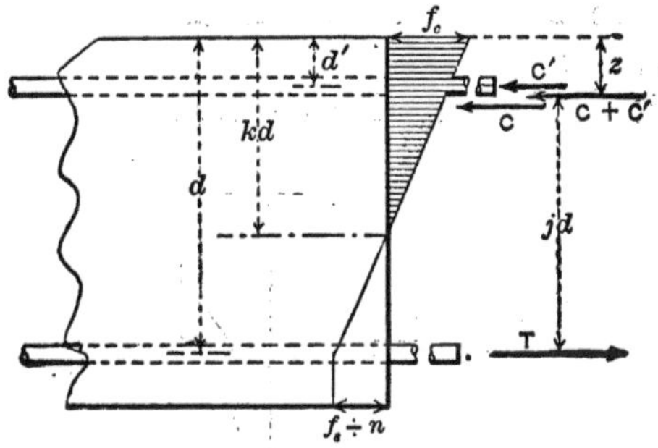

FIG. 67.

Arm of resisting couple,

$$j=\left(1-\frac{z}{d}\right). \quad . \quad . \quad . \quad . \quad . \quad . \quad (28)$$

Moment of resistance,

$$M_s=f_spj\cdot bd^2, \quad . \quad . \quad . \quad . \quad . \quad . \quad . \quad . \quad (29)$$

$$M_c=\tfrac{1}{2}f_ck(1-\tfrac{1}{3}k)bd^2+f_s'p'bd(d-d'). \quad . \quad . \quad . \quad (30)$$

Fibre stresses,

$$f_s=\frac{M\div jd}{A}, \quad . \quad . \quad . \quad . \quad . \quad . \quad . \quad . \quad . \quad . \quad (31)$$

$$f_c=\frac{k}{n(1-k)}\cdot f_s, \quad . \quad . \quad . \quad . \quad . \quad . \quad . \quad . \quad . \quad (32)$$

$$f_s'=\frac{n\left(k-\frac{d'}{d}\right)}{k}\cdot f_c=\frac{kd-d'}{d-kd}\cdot f_s. \quad . \quad . \quad . \quad . \quad (33)$$

Diagrams.—Values of k and j are given in Fig. 29, p. 94, for various values of p and of p'. It is assumed that $d'/d = 1/10$ and $n = 15$. Plate XII, p. 285, gives the amount of compressive steel (values of p') necessary to use in order to reduce the compressive fibre stress, f_c, any given percentage below the value it would have with no compressive reinforcement. The effect of this compressive steel upon the value of the tensile stress in the steel is also given in the diagram for various values of p and p'.

147. Flexure and Direct Stress.—There are two cases:

I. Where there is compression on the entire cross-section (Figs. 68 and 69);

II. Where there is some tension on the cross-section (Fig. 70).

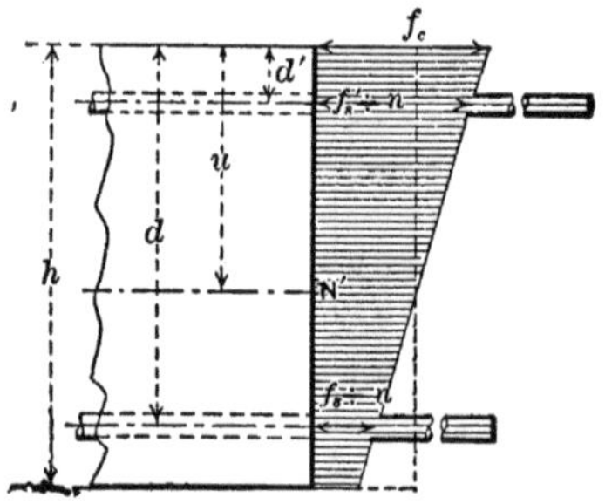

FIG. 68.

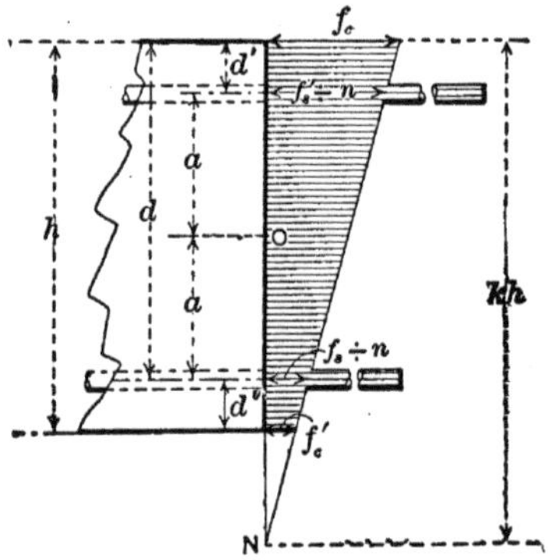

FIG. 69.

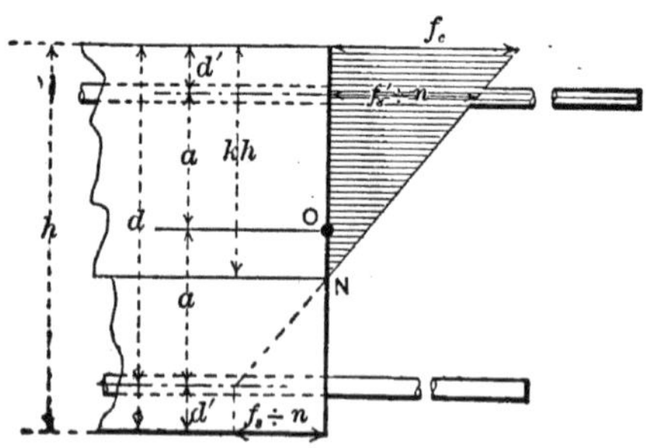

FIG. 70.

Notation.—The lower side of the beam in the figures on the preceding page is called the "tension face".

R = resultant force acting on the section;
N = component of R normal to section;
e = eccentric distance of R, e/h = eccentricity;
M = bending moment $= Ne$;
A' = area of steel near compressive face;
$p' = A'/bh$;
A = area of steel near tension face;
$p = A/bh$;
d' = distance of compressive steel from face;
u = distance from compressive face to centroid of transformed section;
a = distance from steel to center of section for symmetrical reinforcement;
A_t = area of transformed section;
I_c = moment of inertia of concrete about central axis of transformed section;
I_s = moment of inertia of steel about central axis of transformed section;
I_t = moment of inertia of transformed section;
f_c = maximum compressive fibre stress in concrete;
f_c' = maximum tensile fibre stress in concrete;
f_s' = stress in steel near compressive face;
f_s = stress in steel near tension face;

Formulas.

General.

$$A_t = bh + n(A + A'), \quad \dots \dots \quad (34)$$

$$I_t = I_c + nI_s, \quad \dots \dots \quad (35)$$

$$u = \frac{\frac{1}{2}h + npd + np'd'}{1 + np + np'}. \quad \dots \dots \quad (36)$$

Case I. Compression on the entire cross-section.

Fibre stresses:

$$f_c=\frac{N}{A_t}+\frac{Mu}{I_t}, \quad \ldots \quad (37)$$

$$f_c'=\frac{N}{A_t}-\frac{M(h-u)}{I_t}, \quad \ldots \quad (38)$$

$$f_s'=n\left(\frac{N}{A_t}+\frac{M(u-d')}{I_t}\right), \quad \ldots \quad (39)$$

$$f_s=n\left(\frac{N}{A_t}-\frac{M(d-u)}{I_t}\right). \quad \ldots \quad (40)$$

If f_c' is negative, then the case is Case II.

For symmetrical reinforcement and for $d'/d=1/10$,

$$\frac{M}{bh^2f_c}=\frac{1}{12k}\left(1+24np\frac{a^2}{h^2}\right), \quad \ldots \quad (41)$$

$$f_s'=nf_c\left(1-\frac{d'}{kh}\right). \quad \ldots \quad (42)$$

Case II. Tension on part of the cross-section.

If the tension in the concrete is considered, use the formulas of Case I.

For symmetrical reinforcement and for $d'/d=1/10$,

$$f_c'=f_c\frac{1-k}{k}, \quad \ldots \quad (43)$$

$$f_s=nf_c'\left(1-\frac{d'}{(1-k)h}\right). \quad \ldots \quad (44)$$

If the tension in the concrete is neglected.

For symmetrical reinforcement and for $d'/d=1/10$,

$$\frac{M}{bh^2f_c}=\frac{1}{12}k(3-2k)+\frac{2pn}{k}\cdot\frac{a^2}{h^2}, \quad \ldots \quad (45)$$

$$f_s'=nf_c\left(1-\frac{d'}{kh}\right), \quad \ldots \quad (46)$$

$$f_s=f_s'\frac{(1-k)h-d'}{kh-d'}=nf_c\frac{(1-k)h-d'}{kh}. \quad \ldots \quad (47)$$

Diagrams.—Values of $1/k$ for Case I, Eqs. (41) and (42), and Case II, Eqs. (43) and (44), are given in Fig. 33, p. 103; and values of k for Case II, Eqs. (45) and (46), are given in Fig. 35, p. 106. Plate XIII, p. 287, is a diagram for values of M/bh^2f_c for Case I, Eq. (41); and Plate XIV for the same quantity for Case II, Eq. (45), given in terms of the eccentricity e/h and the steel ratio p. The diagrams are constructed for $n=15$.

ILLUSTRATIVE EXAMPLES.—I. An arch ring is 24 in. deep and is symmetrically reinforced. For each side $p=0.9\%$. On a width of 12 in. $N=75{,}000$ lbs.; $e=3$ in.; what is the maximum stress f_c? *Solution.* The eccentricity $=3/24=.125$. The diagram of Plate VII will be used, and the case is Case I. This diagram gives at once $M/bh^2f_c=.097$. We also have $M=75{,}000\times3=225{,}000$ in-lbs. Hence $f_c=225{,}000/(12\times24^2\times.097)=336$ lbs/in².

2. If, in Ex. 1, the eccentricity be 6 in., find the maximum compressive stress f_c and the maximum tensile stress f_c', the concrete being considered as carrying tension if necessary. *Solution.* Use Plate XIII. The eccentricity is $6/24=.25$. From the diagram we find $M/bh^2f_c=.141$, whence $f_c=572$ lbs/in². From Eq. (10), p. 102, the value of $k=.9$. This being less than unity there will be tension on the section. From Eq. (43) the tensile concrete stress $=f_c'=64$ lbs/in².

3. If in Ex. 2 the tension in the concrete be neglected, find f_c, f_s', and f_s. *Solution.* Use Plate XIV. $e/h=.25$. The value of $M/bh^2f_c=.14$, whence $f_c=576$ lbs/in². The compressive stress in the steel, f_s', is always less than nf_c; in this case it is, from Eq. (46), equal to $nf_c\times\left(1-\frac{1}{11k}\right)=nf_c\times.88$, k being found from Fig. 35. The tensile steel stress, f_s, is less than the compressive. From Eq. (47) it is found to be 276 lbs/in².

148. Shearing and Bond Stress.

Notation.

V = total vertical shear at any section;
v = maximum horizontal or vertical shearing stress per unit area;
v' = average shearing stress per unit area;
U = bond stress per unit length of beam;
b and d = dimensions of a rectangular beam;
b' = width of web of T-beam;

d = net depth of T-beam;
t = flange thickness of T-beam;
jd = arm of resisting couple for any beam.

Formulas.

Rectangular beams:

$$v'=\frac{V}{bd}, \qquad (48)$$

$$v=\frac{V}{bjd}, \qquad (49)$$

$$U=\frac{V}{jd}. \qquad (50)$$

Approximately,

$$v=\frac{V}{b\cdot\frac{7}{8}d}=\frac{8}{7}v'. \qquad (49')$$

$$U=\frac{8}{7}\frac{V}{d} \qquad (50')$$

T-beams:

$$v'=\frac{V}{b'd}, \qquad (51)$$

$$v=\frac{V}{b'jd}, \qquad (52)$$

$$U=\frac{V}{jd}. \qquad (53)$$

Approximately,

$$v=\frac{V}{b'(d-\frac{1}{2}t)}, \qquad (52')$$

$$U=\frac{V}{d-\frac{1}{2}t}. \qquad (53')$$

149. Columns.

Notation.

A = total cross-section;
A_c = cross-section of concrete;
A_s = " " longitudinal steel;

$p = A_s/A$;
P = strength of plain concrete column;
P' = " " reinforced column;
f_c = unit stress in concrete;
f_s = " " " steel (not exceeding its elastic limit);
f_{el} = elastic-limit strength of steel;
f = average unit stress for entire cross-section;
p' = steel ratio of the hoops of hooped columns.

Formulas.

For short columns; ratio of length to least width not exceeding 20:

$$f_s = nf_c, \quad \ldots \ldots \ldots \quad (54)$$

$$P' = f_cA_c + f_sA_s, \quad \ldots \ldots \ldots \quad (55)$$

$$P' = f_cA[1 + (n-1)p], \quad \ldots \ldots \quad (56)$$

$$\frac{P'}{P} = 1 + (n-1)p. \quad \ldots \ldots \ldots \quad (57)$$

If nf_c is greater than the elastic-limit strength of the steel, then

$$P' = f_cA_c + f_{el}A_s. \quad \ldots \ldots \ldots \quad (58)$$

French Commission's formula for hooped columns:

$$P' = f_cA(1 + 15p + 32p'). \quad \ldots \ldots \quad (59)$$

For long columns:

$$f = \frac{f_c[1 + (n-1)p]}{1 + \frac{1}{20{,}000}\left(\frac{l}{r}\right)^2}. \quad \ldots \ldots \quad (60)$$

Diagrams. — Plate XV is a diagram of the function $1 + (n-1)p$ $(= f/f_c)$ of Eqs. (56) and (57) for various values of p and values of n equal to 10, 12, 15, 20, and 25. The average working stress, f, for any column is then found by multiplying the corresponding ordinate from this diagram by the selected working stress f_c.

150. Stresses in Circular Plates.—The exact determination of stresses in floor systems, such as the "mushroom" system described in Art. 168, and in the ordinary foundation-plate supporting a single column, involves very complex analytical processes. As an aid in estimating the stresses in such cases, Plates XVI and XVII have been prepared. They give the bending moments in circular plates supported rigidly over any given area at the center. Plate XVI gives the moments for the case of a uniformly distributed load on the entire area, and Plate XVII the moments for a load uniformly distributed along the periphery. In each case the full lines give the coefficients for the radial bending moments, and the dotted lines those for the circumferential bending moments. The curves are drawn for five different ratios of r_1 to r_0, or radius of plate to radius of fixed support. For other ratios interpolations may be made.

The calculations for the diagrams are based upon the analysis presented by Prof. H. T. Eddy* for homogeneous plates. The value of Poission's ratio assumed in the numerical substitutions has been 0.1, as approximately determined in recent experiments by Prof. A. N. Talbot.

Example.—A circular plate 10 ft. in diameter is rigidly supported by a column 24 in. in diameter. It supports a load of 150 lbs/ft² over the area and a load of 500 lbs/ft along its outer circumference. Required, the radial and circumferential bending moments.

Solution. The ratio of $r_1 : r_0 = 120 : 24 = 5$. (The upper diagram of Plate XVII may be used in finding this ratio.) In Plate XVI we then obtain the coefficients Q_1 and Q_2 for any desired point in the plate, using the curves corresponding to $r_1 \div r_0 = 5$. The value of Q_1 (ordinate to dotted curve) is seen to be a maximum at a distance from the center equal to about $1.7r$; its value is about 4.7. Hence the maximum circumferential moment due to the load of 150 lbs/ft is $4.7 \times 150 \times 1^2 = 705$ ft-lbs per foot width of section. The value of Q_2 (ordinate to full curve) is a maximum at the edge of the support and has a value of 16. The radial bending moment is therefore equal to $16 \times 150 \times 1^2 = 2400$ ft-lbs per foot

* Year Book, Engrs. Soc., Univ. of Minn., 1899.

width of section. The radial moment rapidly falls off with increased distance from the support.

The moments due to the peripheral load of 500 lbs/ft are found from Plate XVII to be respectively $M_1=3.1\times500\times1=1550$ ft.-lbs., and $M_2=9.6\times500\times1=4800$ ft.-lbs.

151. Coefficients and Working Stresses.—The following is a résumé of the coefficients and working stresses suggested in the discussion of Chapter V. They may be considered as applicable to ordinary conditions on the basis of equivalent dead-load stresses and with concrete of 1:2:4 to 1:2½:5 composition.

Beams.

	Working Stress.	
Concrete in compression...........	550–650	lbs/in²
Concrete in shear, average stress:		
a. Without shear reinforcement.	30–40	
b. With shear reinforcement...	60–100	
Bond stress:		
a. Smooth rods...............	60–80	
b. Deformed rods.............	100–175	
Steel in tension...................	14,000–16,000	"
Value of $n=E_s/E_c$................	12–15	

Columns.

Concrete in compression...........	400–500	
Value of $n=E_s/E_c$................	15	

152. Tables.—*Areas of Steel Rods.*—Table No. 19 gives sectional areas and weights per foot of round and square rods of various sizes, and the total area per foot of width of slab when the rods are spaced various distances apart.

Materials Required for One Cubic Yard of Concrete.—Table No. 20 gives the quantities of material required for one cubic yard of concrete of various proportions. The table is based on *Thatcher's Tables.** As conditions vary greatly, these tables should be used only for approximate values.

Safe Loads for Floors.—Table No. 21 gives span lengths for floor-slabs for various live loads per square foot, and for various values of working stresses f_s and f_c. The tables have been calculated for bending moments equal to $\frac{1}{8}wl^2$ and also $\frac{1}{12}wl^2$. The value of n has been taken at 15. For continuous slabs $\frac{1}{12}wl^2$ may generally be taken as the bending moment. The table also gives the amount of steel required per foot of slab, so that by reference to Table No. 19 a suitable size and spacing can readily be determined. The moment of resistance of a beam one foot wide is also given for general use.

* Johnson's *Materials of Construction*, p. 610*a*.

n = 10

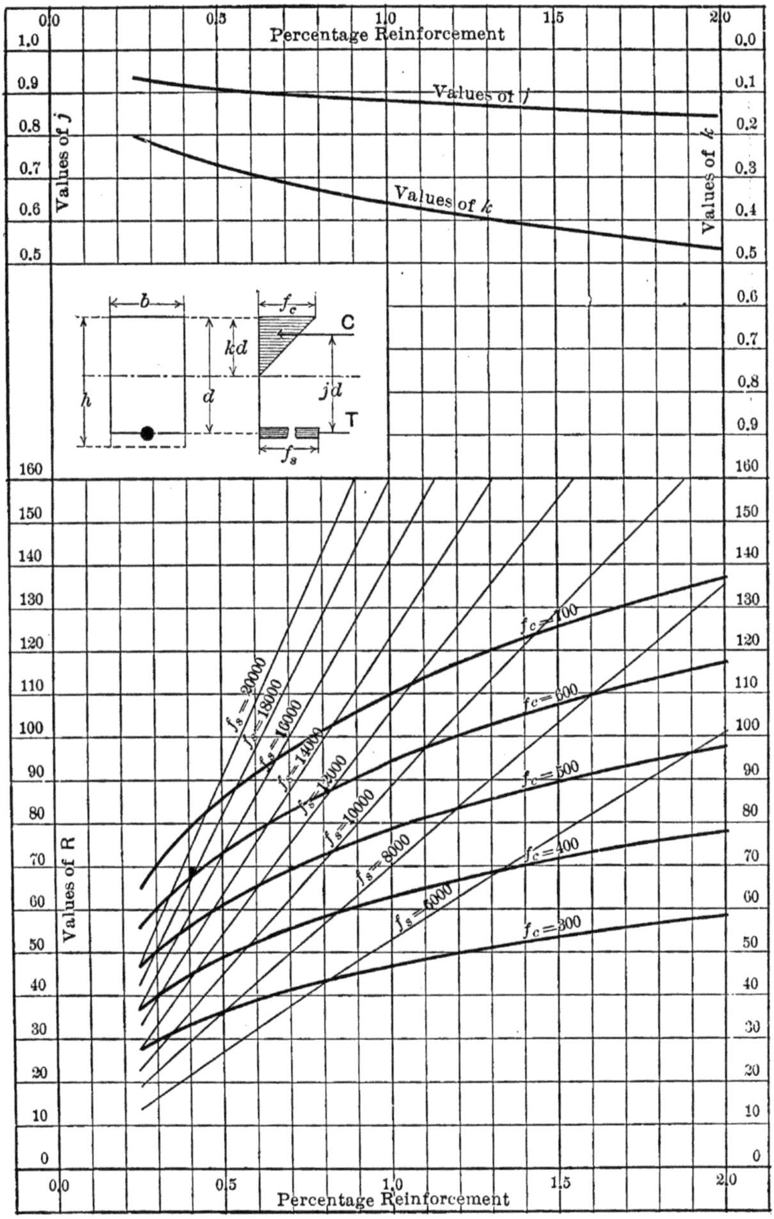

PLATE I.—Coefficients of Resistance of Beams.

n = 12

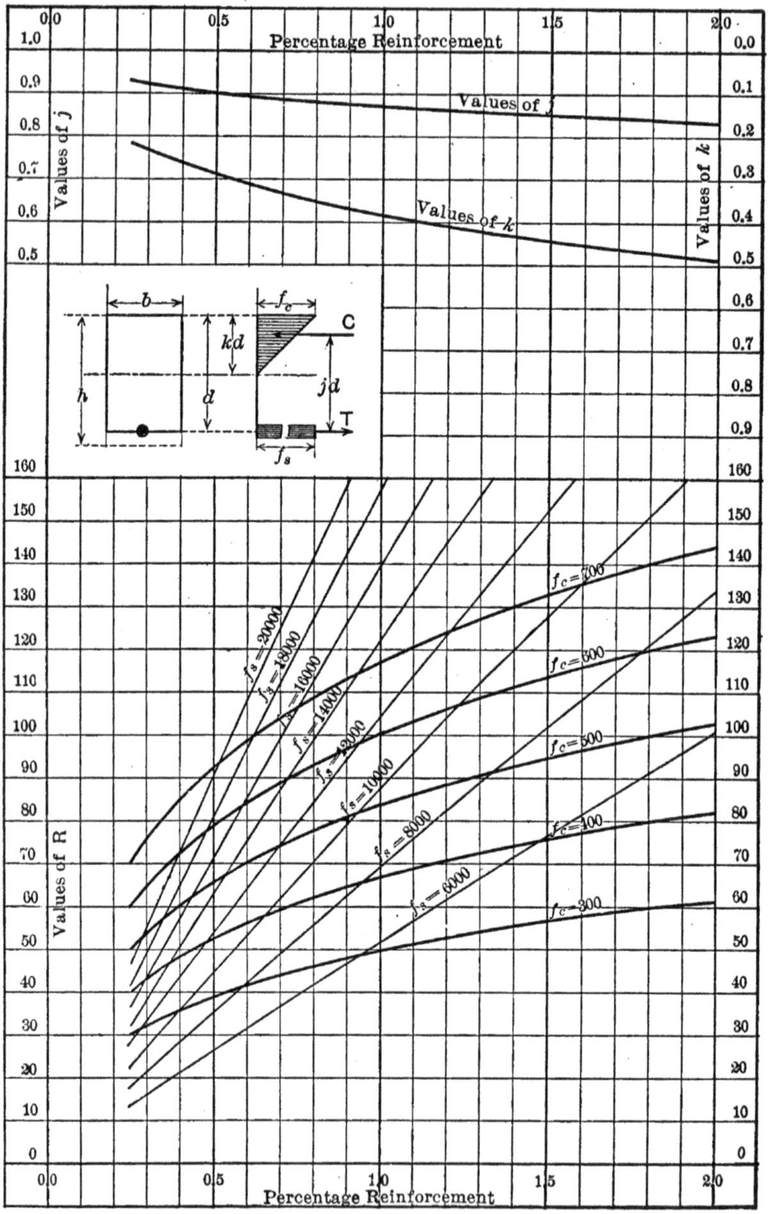

PLATE II.—Coefficients of Resistance of Beams.

n = 15

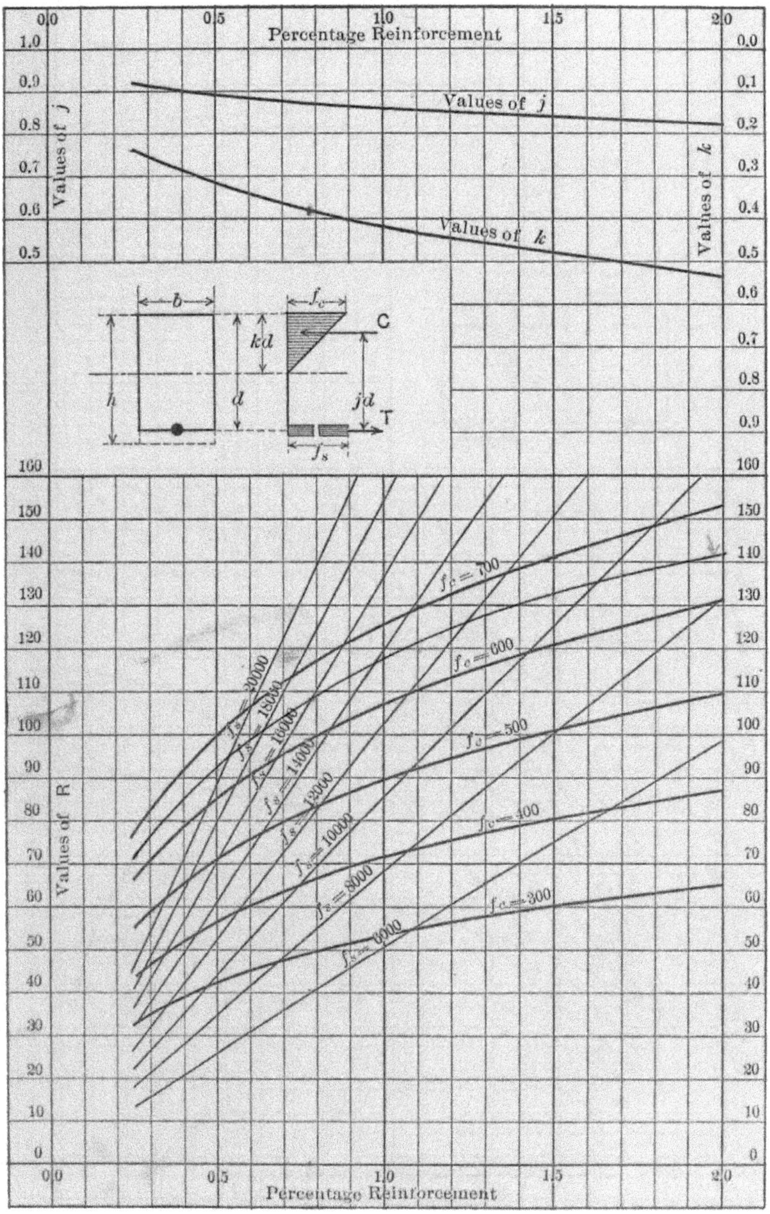

PLATE III.—Coefficients of Resistance of Beams.

n = 18

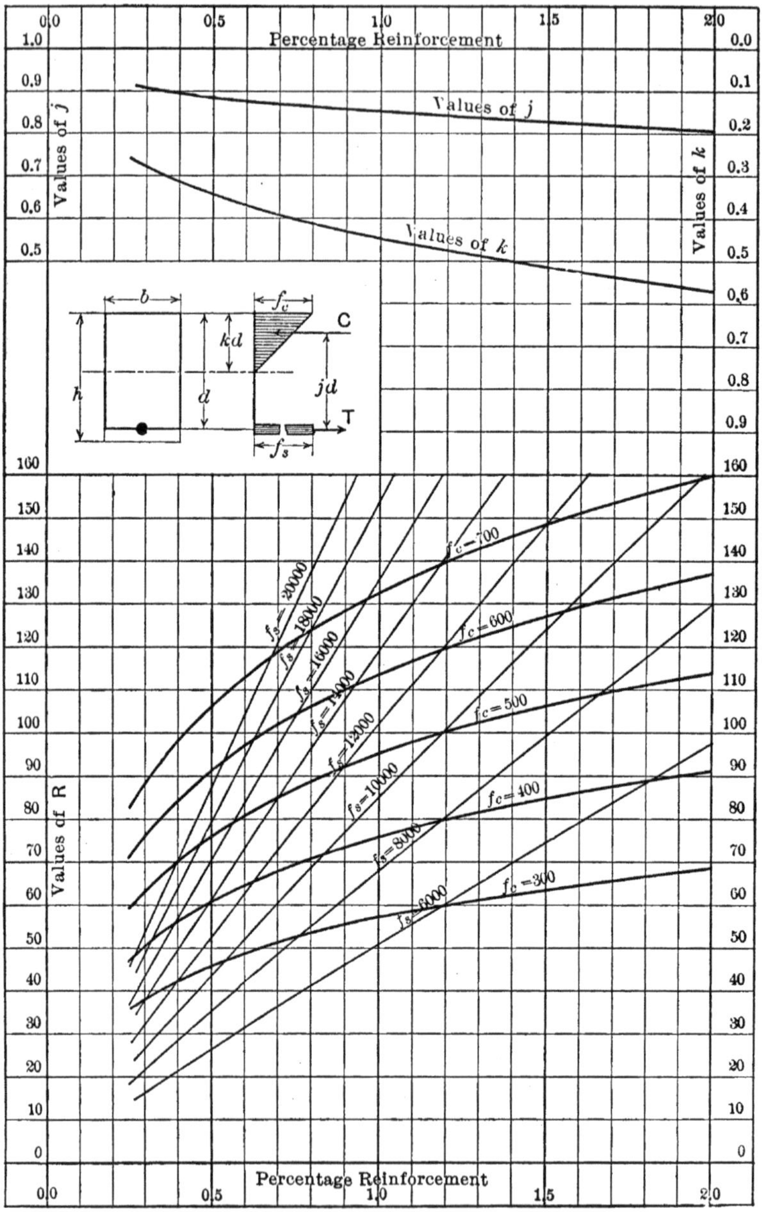

PLATE IV.—Coefficients of Resistance of Beams.

Full lines for n = 15; dotted lines for n = 12.

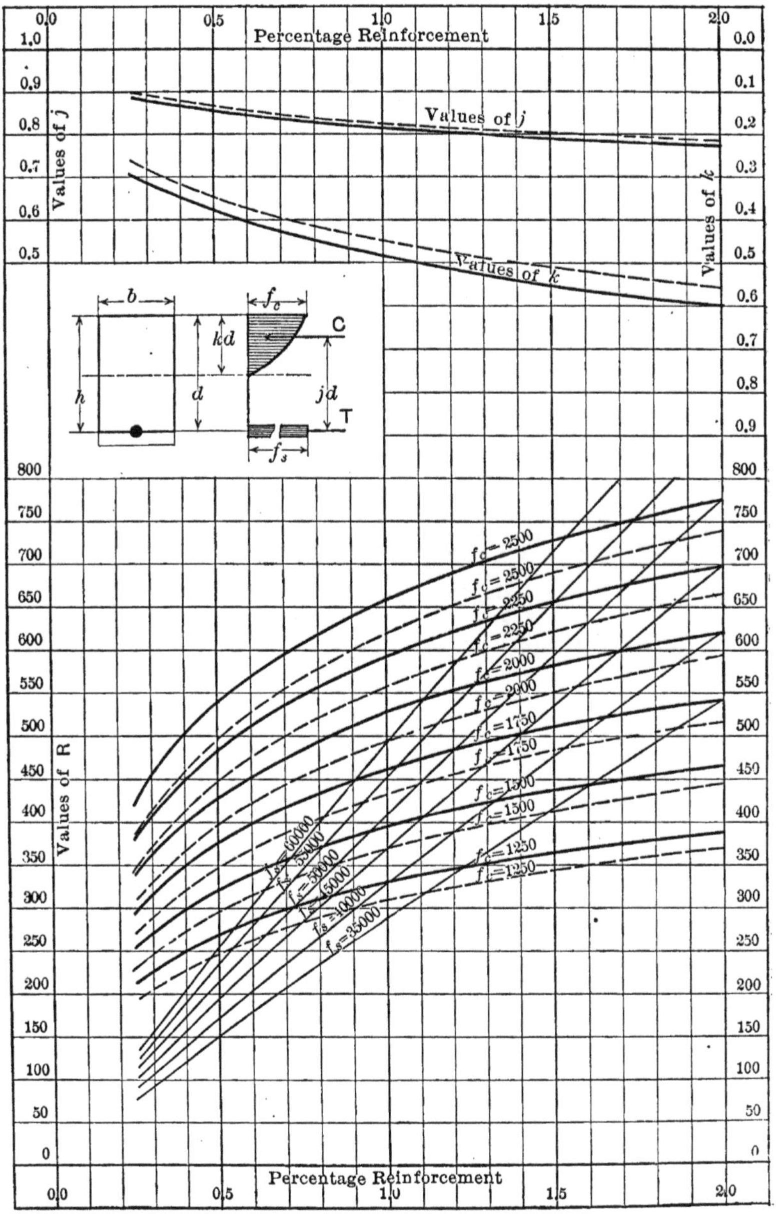

Plate V.—Coefficients of Resistance of Beams.

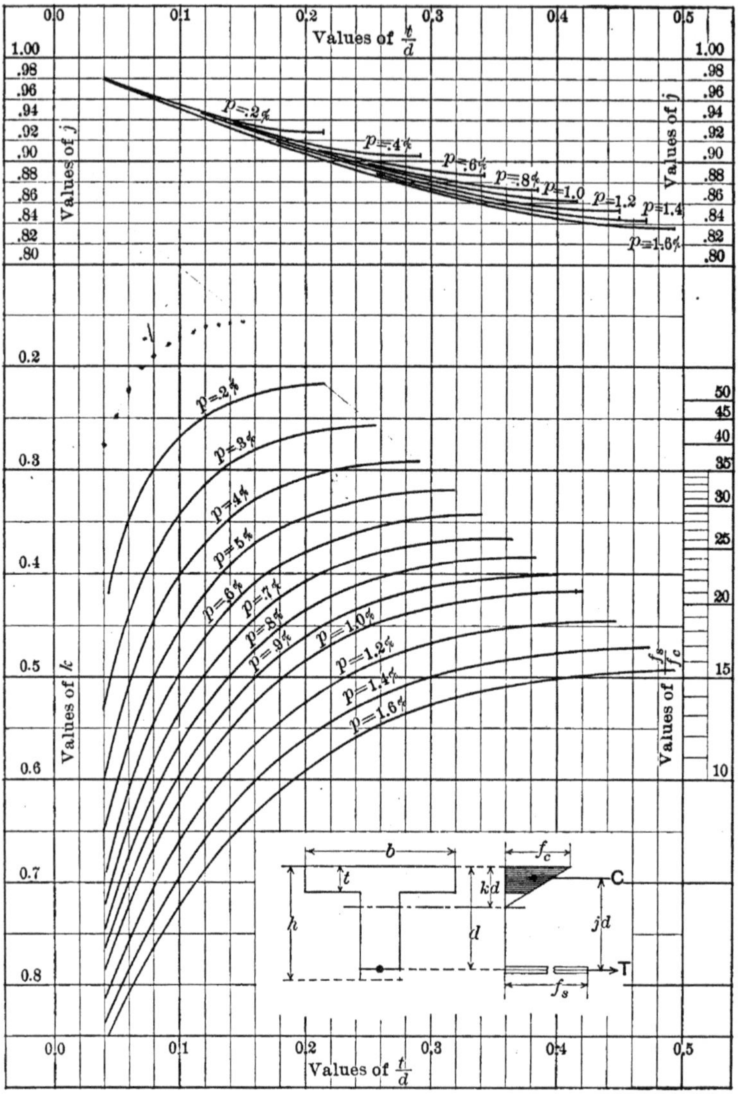

PLATE VI.—Values of k and j for T-beams.

$f_s = 12{,}000$

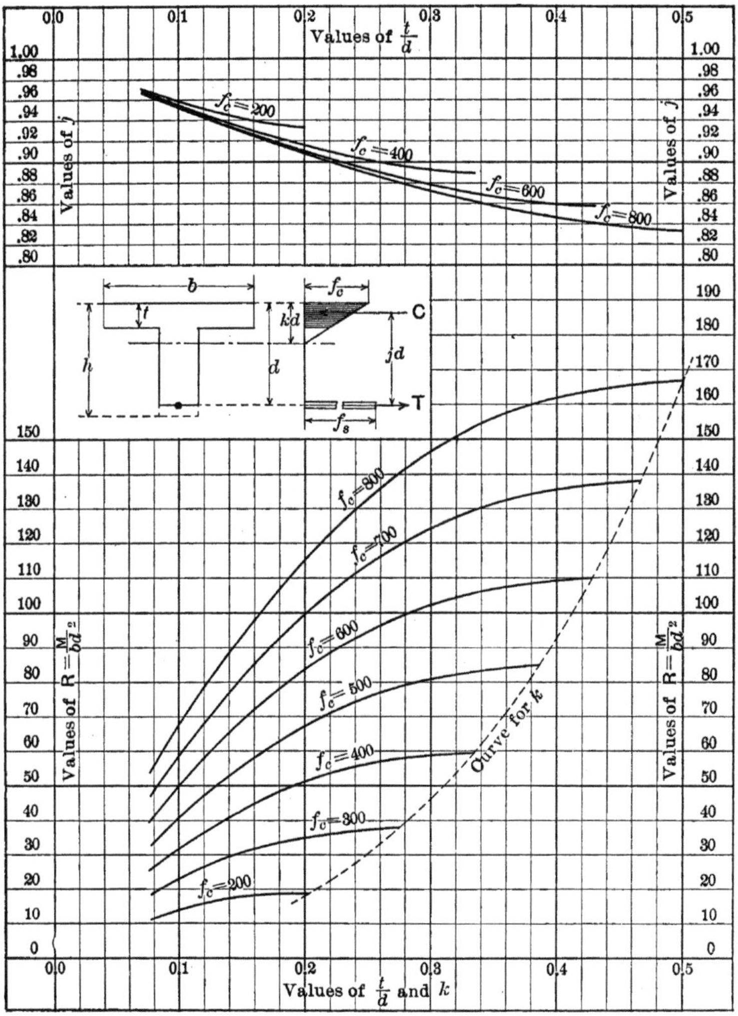

PLATE VII.—Coefficients of Resistance of T-beams.

$f_s = 14,000$

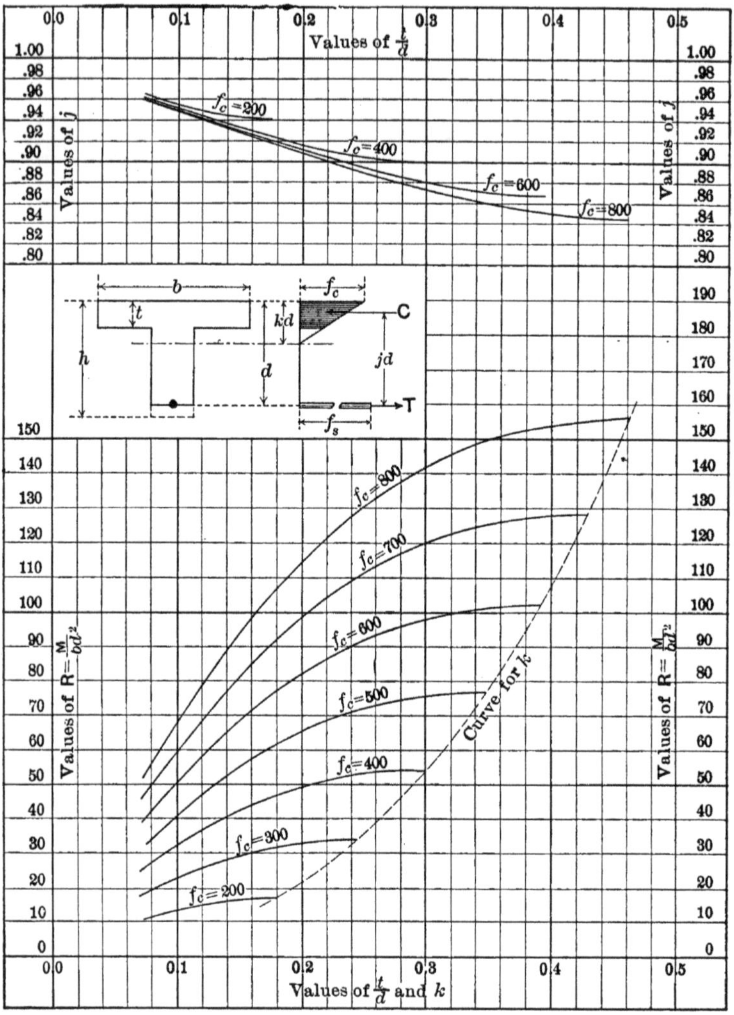

PLATE VIII.—Coefficients of Resistance of T-beams.

f = 15,000

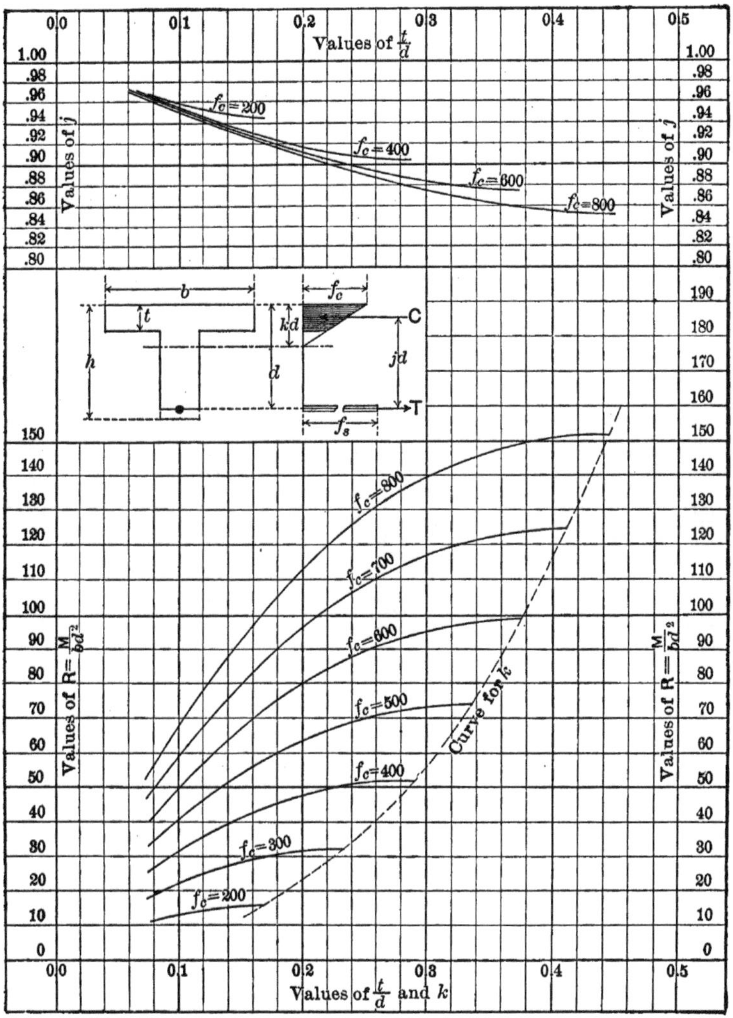

Plate IX.—Coefficients of Resistance of T-beams.

$f_s = 16,000$ n = 15

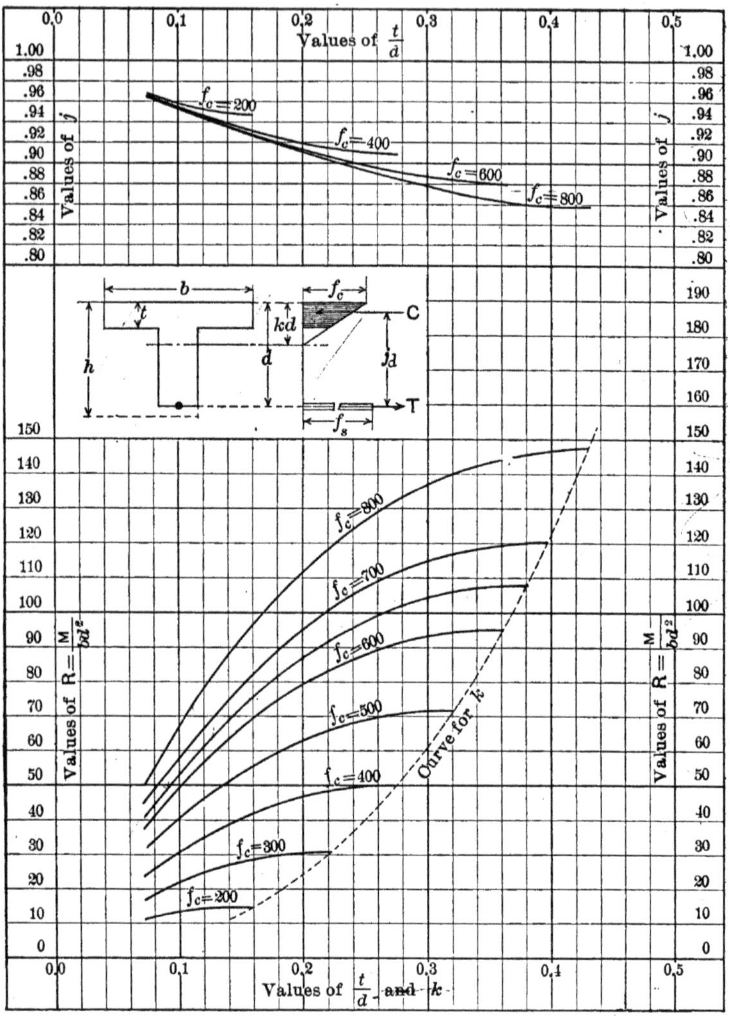

PLATE X.—Coefficients of Resistance of T-beams.

$f_s = 18,000$

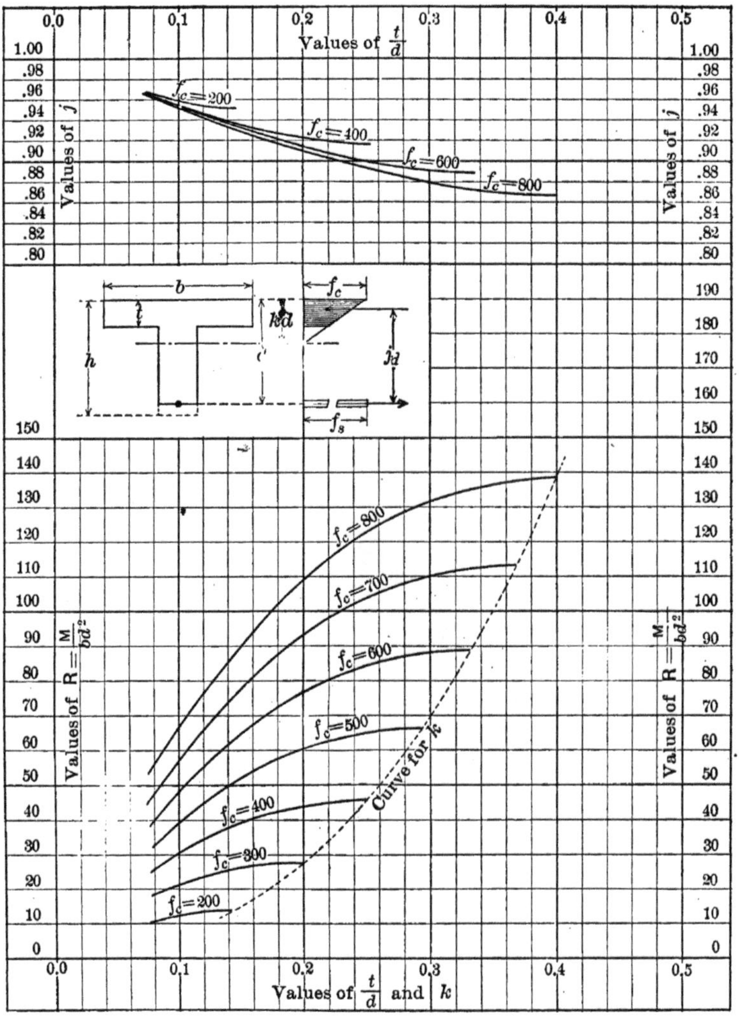

PLATE XI.—Coefficients of Resistance of T-beams.

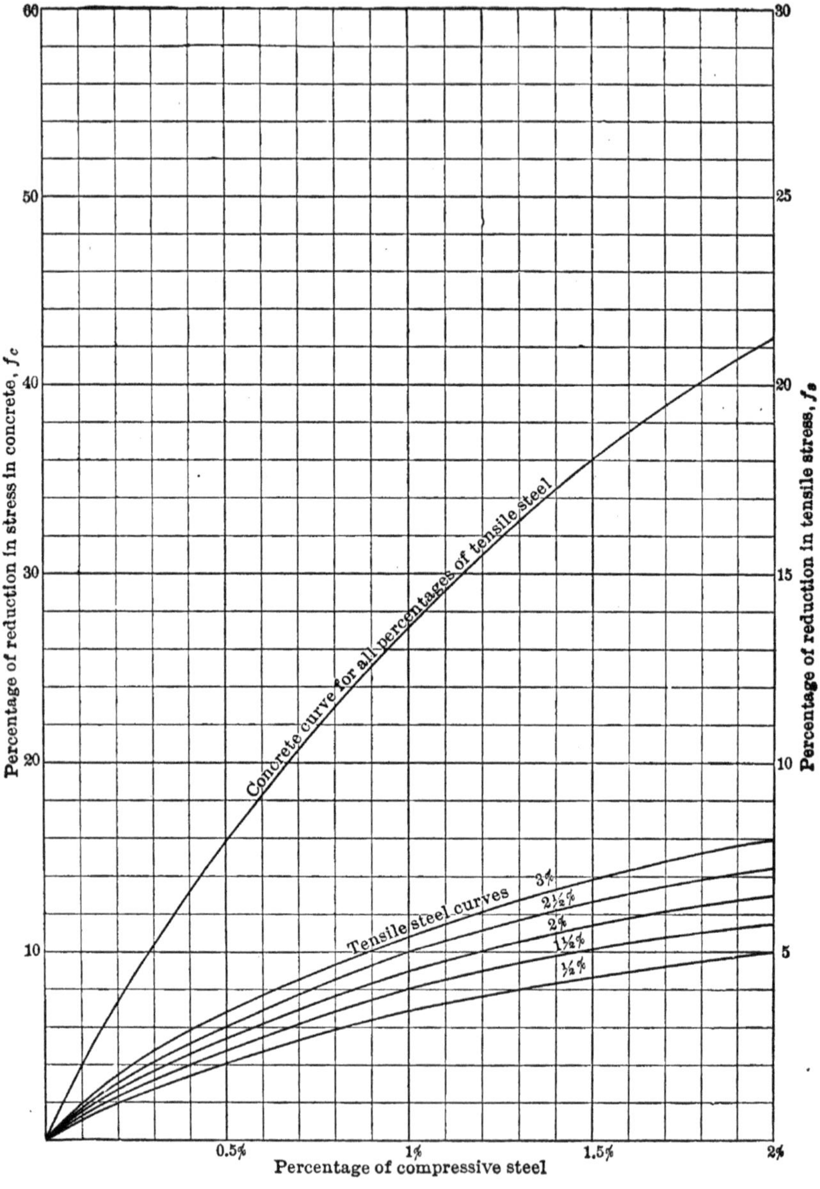

PLATE XII.—Compressive Reinforcement of Beams.

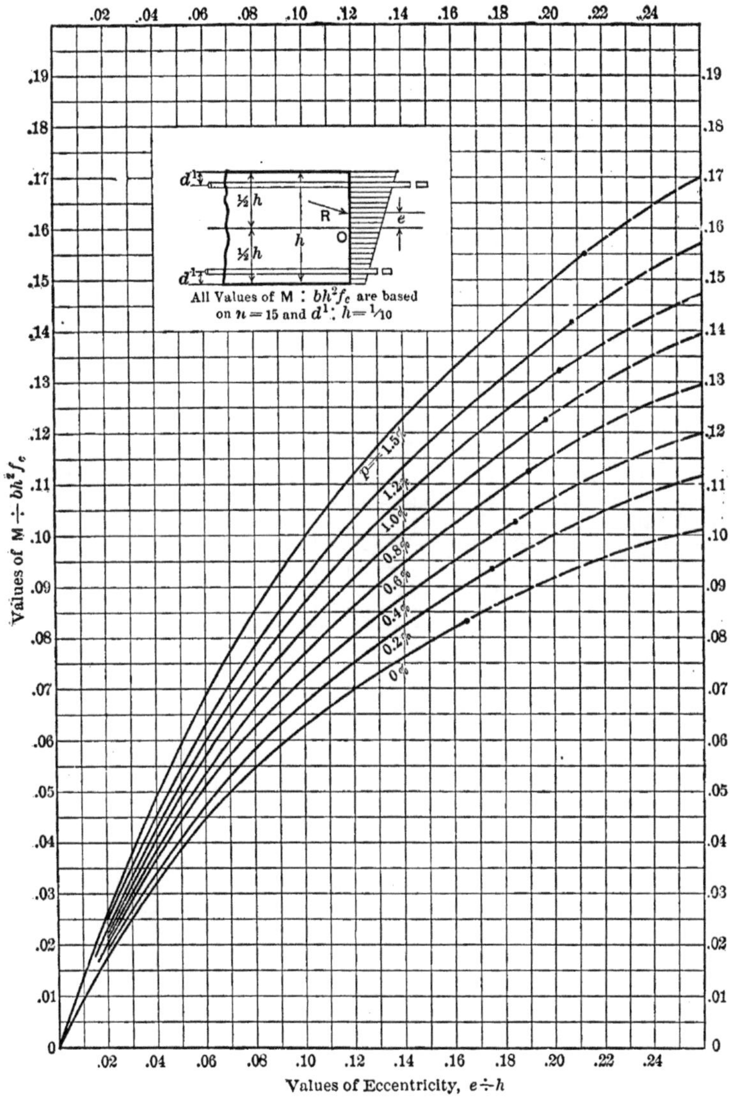

PLATE XIII.—Flexure and Direct Stress.

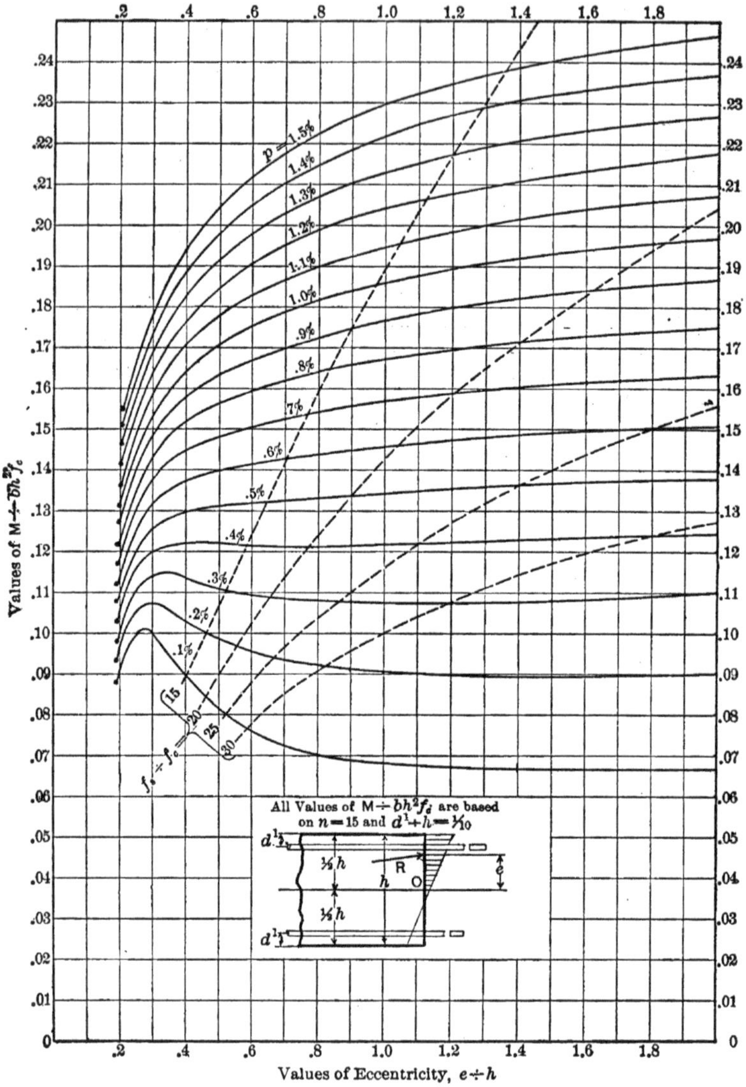

Plate XIV.—Flexure and Direct Stress.

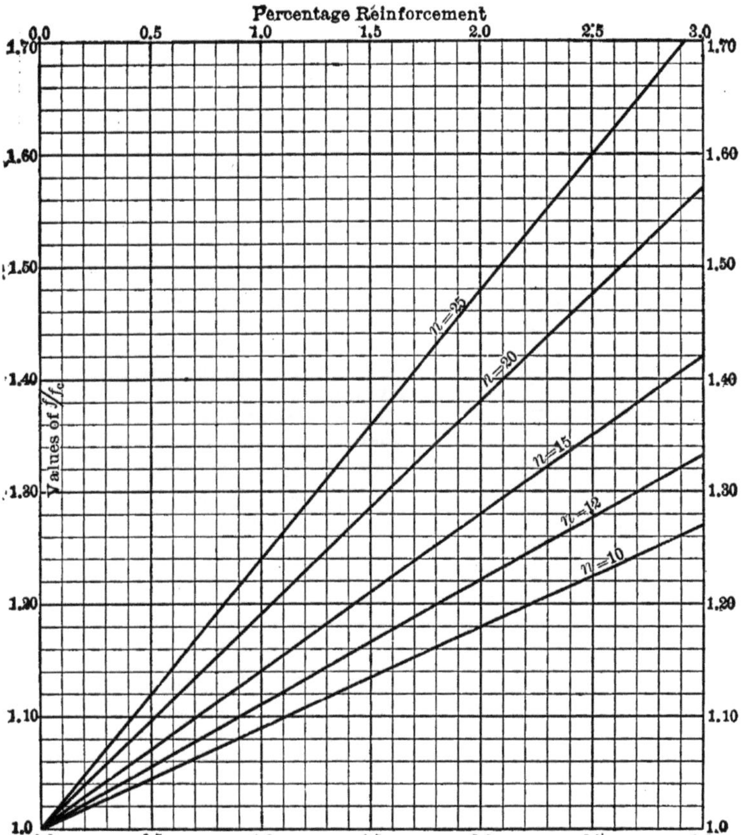

PLATE XV.—Working Stresses in Columns.

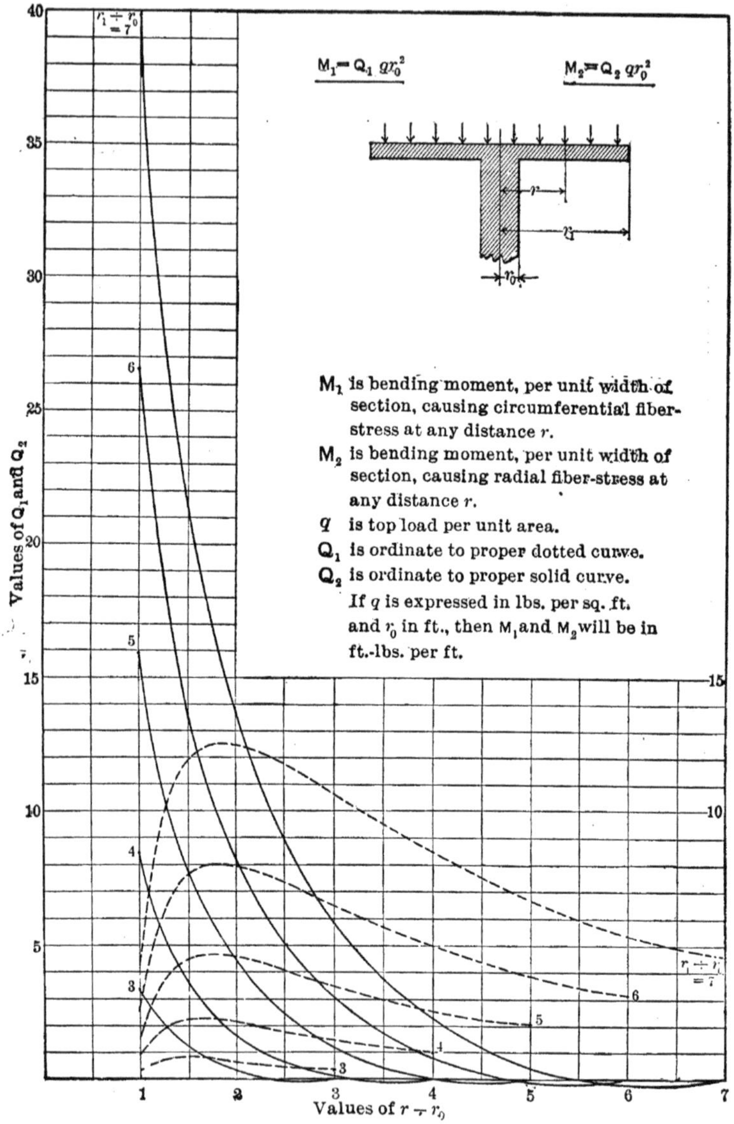

PLATE XVI.—**Stresses in Circular Slabs.**

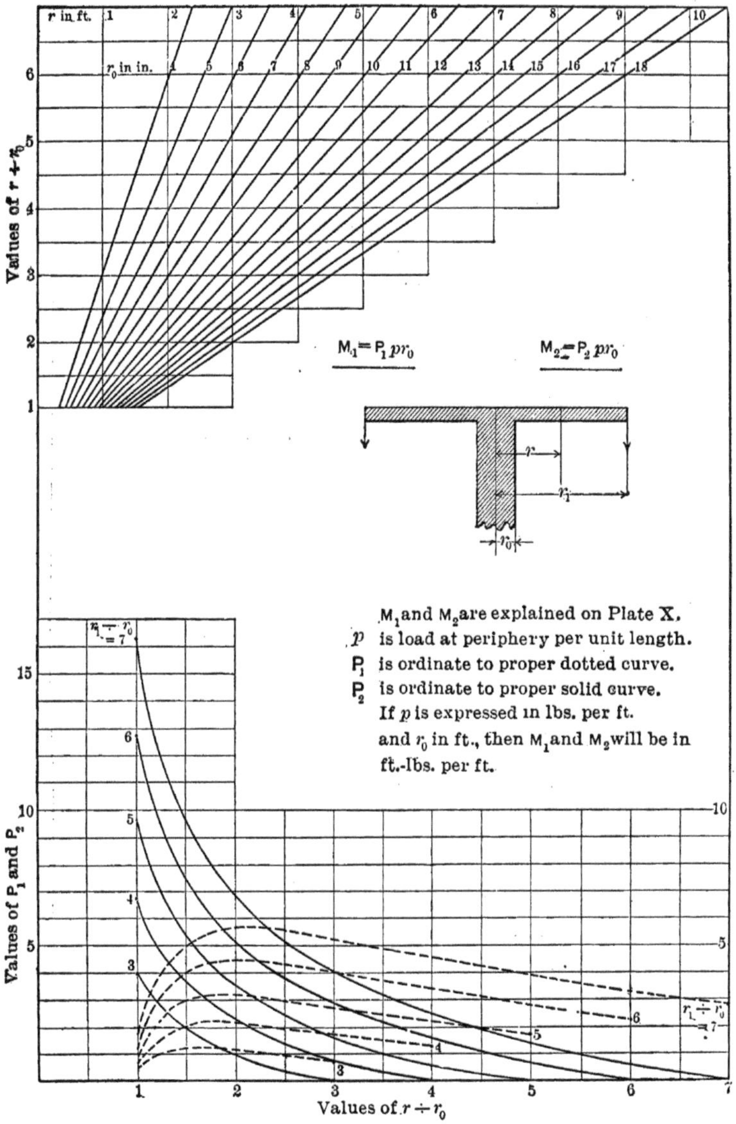

PLATE XVII.—Stresses in Circular Slabs.

Table No. 19.

AREAS, WEIGHTS, AND SPACING OF RODS.

ROUND RODS.

Diameter, Inches.	Area, Square Inches.	Circumference, Inches.	Weight per Foot, Pounds.	Sectional Area of Steel per Foot of Slab when Spaced as follows:													
				2″	2½″	3″	3½″	4″	4½″	5″	5½″	6″	7″	8″	9″	10″	12″
¼	.0491	.7854	.167	.29	.23	.20	.17	.15	.13	.12	.11	.10	.08	.07	.07	.06	.05
5/16	.0767	.9818	.261	.46	.36	.31	.26	.23	.20	.18	.17	.15	.13	.11	.10	.09	.08
⅜	.1104	1.1781	.376	.66	.53	.44	.38	.33	.29	.26	.24	.22	.19	.17	.15	.13	.11
7/16	.1503	1.3745	.511	.90	.72	.60	.51	.45	.40	.36	.33	.30	.26	.23	.20	.18	.15
½	.1963	1.5708	.668	1.18	.94	.78	.67	.59	.52	.47	.43	.39	.34	.29	.26	.24	.20
9/16	.2485	1.7672	.845	1.49	1.19	.99	.85	.75	.66	.60	.54	.50	.43	.37	.33	.30	.25
⅝	.3068	1.9635	1.043	1.84	1.47	1.23	1.05	.92	.82	.74	.67	.61	.53	.46	.41	.37	.31
11/16	.3712	2.1599	1.262	2.23	1.78	1.48	1.27	1.11	.99	.89	.81	.74	.64	.56	.49	.45	.37
¾	.4418	2.3562	1.502	2.65	2.12	1.77	1.51	1.32	1.18	1.06	.96	.88	.76	.66	.59	.53	.44
13/16	.5185	2.5526	1.763	3.11	2.48	2.07	1.78	1.56	1.38	1.24	1.13	1.04	.89	.78	.69	.62	.52
⅞	.6013	2.7489	2.044	3.61	2.88	2.40	2.06	1.80	1.60	1.44	1.31	1.20	1.03	.90	.80	.72	.60
15/16	.6903	2.9453	2.347	4.14	3.31	2.76	2.37	2.07	1.84	1.66	1.51	1.38	1.18	1.03	.92	.83	.69
1	.7854	3.1416	2.670	4.71	3.77	3.14	2.69	2.36	2.09	1.88	1.71	1.57	1.35	1.18	1.05	.94	.78
1⅛	.9940	3.5343	3.380	5.96	4.77	3.98	3.41	2.98	2.65	2.39	2.17	1.99	1.70	1.49	1.33	1.19	.99
1¼	1.2272	3.9270	4.172	7.36	5.89	4.91	4.21	3.68	3.27	2.95	2.68	2.45	2.10	1.84	1.64	1.47	1.23
1⅜	1.4849	4.3197	5.049	8.91	7.12	5.94	5.09	4.45	3.96	3.56	3.24	2.97	2.55	2.23	1.98	1.78	1.48
1½	1.7671	4.7124	6.008	10.60	8.48	7.07	6.06	5.30	4.71	4.24	3.86	3.53	3.03	2.65	2.36	2.12	1.77

Table No. 19—*Continued.*

AREAS, WEIGHTS, AND SPACING OF RODS.

SQUARE RODS.

Dimension, Inches.	Area, Square Inches.	Perimeter, Inches.	Weight per Foot, Pounds.	Sectional Area of Steel per Foot of Slab when Spaced as follows:													
				2″	2½″	3″	3½″	4″	4½″	5″	5½″	6″	7″	8″	9″	10″	12″
¼	.0625	1.00	.212	.37	.30	.25	.21	.19	.17	.15	.13	.12	.11	.10	.08	.07	.06
5/16	.0977	1.25	.332	.59	.47	.39	.33	.29	.26	.23	.21	.19	.17	.15	.13	.12	.10
⅜	.1406	1.50	.478	.84	.67	.56	.48	.42	.37	.34	.31	.28	.24	.21	.19	.17	.14
7/16	.1914	1.75	.651	1.15	.92	.77	.66	.57	.51	.46	.42	.38	.33	.29	.25	.23	.19
½	.2500	2.00	.850	1.50	1.20	1.00	.86	.75	.67	.60	.55	.50	.43	.37	.33	.30	.25
9/16	.3164	2.25	1.076	1.90	1.52	1.27	1.08	.95	.84	.76	.69	.63	.54	.47	.42	.38	.32
⅝	.3906	2.50	1.328	2.34	1.87	1.56	1.34	1.17	1.04	.94	.85	.78	.67	.59	.52	.47	.39
11/16	.4727	2.75	1.607	2.84	2.27	1.99	1.62	1.42	1.33	1.13	1.03	.94	.81	.71	.66	.57	.47
¾	.5625	3.00	1.913	3.37	2.70	2.25	1.93	1.69	1.50	1.35	1.23	1.12	.96	.84	.75	.67	.56
13/16	.6602	3.25	2.245	3.96	3.17	2.64	2.26	1.98	1.76	1.58	1.44	1.32	1.13	.99	.88	.79	.66
⅞	.7656	3.50	2.603	4.59	3.67	3.06	2.62	2.30	2.04	1.84	1.67	1.53	1.31	1.15	1.02	.92	.77
15/16	.8789	3.75	2.988	5.27	4.22	3.52	3.01	2.64	2.34	2.11	1.92	1.76	1.51	1.32	1.17	1.05	.88
1	1.0000	4.00	3.400	6.00	4.80	4.00	3.43	3.00	2.67	2.40	2.18	2.00	1.71	1.50	1.33	1.20	1.00
1⅛	1.2656	4.50	4.303	7.59	6.08	5.06	4.34	3.80	3.37	3.04	2.76	2.53	2.17	1.89	1.69	1.52	1.27
1¼	1.5625	5.00	5.313	9.37	7.50	6.25	5.36	4.69	4.17	3.75	3.41	3.12	2.68	2.34	2.08	1.87	1.56
1⅜	1.8906	5.50	6.428	11.34	9.08	7.56	6.48	5.67	5.04	4.54	4.12	3.78	3.24	2.84	2.52	2.27	1.89
1½	2.2500	6.00	7.650	13.50	10.80	9.00	7.71	6.75	6.00	5.40	4.91	4.50	3.86	3.37	3.00	2.70	2.25

TABLE No. 20.

MATERIALS REQUIRED FOR ONE CUBIC YARD OF CONCRETE.

Proportion of Mixture.				Required for One Cubic Yard.		
Cement.	Sand.	Stone.	Ratio: $\frac{\text{Mortar}}{\text{Stone}}$.	Cement, Barrels.	Sand, Cubic Yards.	Stone, Cubic Yards.
1	1	2.0	.70	2.57	0.39	0.78
1	1	2.5	.56	2.29	0.35	0.87
1	1	3.0	.47	2.06	0.31	0.94
1	1.5	2.5	.71	2.05	0.47	0.78
1	1.5	3.0	.60	1.85	0.42	0.84
1	1.5	3.5	.51	1.72	0.39	0.91
1	1.5	4.0	.44	1.57	0.36	0.96
1	2.0	3.0	.72	1.70	0.52	0.77
1	2.0	3.5	.62	1.57	0.48	0.83
1	2.0	4.0	.54	1.46	0.44	0.89
1	2.0	4.5	.48	1.36	0.42	0.93
1	2.5	4.0	.64	1.35	0.52	0.82
1	2.5	4.5	.57	1.27	0.48	0.87
1	2.5	5.0	.51	1.19	0.46	0.91
1	2.5	5.5	.46	1.13	0.43	0.94
1	3	4.5	.66	1.18	0.54	0.81
1	3	5.0	.60	1.11	0.51	0.85
1	3	5.5	.54	1.06	0.48	0.89
1	3	6.0	.50	1.00	0.46	0.92
1	3	6.5	.46	.96	0.44	0.95

Table No. 21.—STRENGTH OF FLOOR-SLABS.

Bold-faced type, $M=\frac{1}{8}wl^2$; light-faced type, $M=\frac{1}{12}wl^2$.

1. $f_c=500$ $f_s=14,000$ $R=77$ $p=.0062$

Total Thickness of Slab, Inches.	Thickness of Concrete below Steel, Inches.	Required Area of Steel per Foot of Slab, Sq. In.	Moment of Resistance per Foot of Slab, In.-lbs.	Weight of Slab per Square Foot, Lbs.	Span in Feet for Given Net Loads per Square Foot of Floor in Pounds. 50	75	100	150	200	250	300	400	500
2	¾	.094	1400	24.2	3.6 [illegible]	3.1 [illegible]	2.8 [illegible]	2.4 2.[illegible]	2.1 [illegible]	1.9 [illegible].3	1.7 2.1	1.5 1.8	1.4 1.[illegible]
2½	¾	.131	2800	30.3	. .		. .	3. 3.	. .	.6 .2	2.4 2.9	2.1 2.6	1. 2.
3	¾	.168	4700	36.4	. .		. .	4. 5.	. .	.3 .0	3.0 3.7	2.7 3.3	2. 2.
3½	¾	.200	7000	42.5	. .		. .	4. 6.	. .	.0 .9	3.7 4.5	3.3 4.0	2. 3.
4	1	.224	8300	48.5	. .		. .	5. 6.	. .	.3 .3	4.0 4.9	3.5 4.3	3. 3.
5	1	.299	14800	60.7	½ . .	1	. .	6. 8.	. .	.7 .0	5.2 6.4	4.6 5.6	4. 5.
6	1¼	.355	20900	72.9	. .	½	½ . .	7. 9.	. .	.5 .0	6.0 7.5	5.4 6.6	4. 5.
7	1¼	.430	30600	85.1	. .		. .	9. 11.	1 . .	.7 .4	7.2 8.8	6.5 8.0	5. 7.
8	1¼	.505	42100	97.4	. .		. .	10. 2.	. 1 .	1 .0 .0	8.3 10.2	7.5 9.2	6. 8.
9	1½	.561	52000	109.5	. .		. .	1. 4.	. .	.8 1 .0	9.2 11.2	8.3 10.2	7. 9.
10	1½	.636	66800	121.7	. .		. .	2. 5	. .	.0 .5	10.3 2.6	9 3 11 4	8. 10.
12	1½	.785	102000	145.9	. .		. .	5. 8.	. .	.1 .0	2.4 5.2	11 2 13 7	10. 12.

2. $f_c=500$ $f_s=15,000$ $R=74$ $p=.0056$

Total Thickness of Slab, Inches.	Thickness of Concrete below Steel, Inches.	Required Area of Steel per Foot of Slab, Sq. In.	Moment of Resistance per Foot of Slab, In.-lbs.	Weight of Slab per Square Foot, Lbs.	50	75	100	150	200	250	300	400	500
2	¾	.083	1400	24.2	3.5 4.3	3.1 [illegible]	2.7 [illegible]	2.3 [illegible].8	2.0 [illegible]	1.8 [illegible]	1.7 [illegible]	1.5 [illegible].8	[illegible].3 [illegible]
2½	¾	.117	2700	30.3	4.7 5.8	. .	. .	.2 .9	. .	. .	. .	.0 .4	. .
3	¾	.150	4500	36.4	5.9 7.2	. .	. .	.0 .9	. .	. .	. .	.6 .2	. .
3½	¾	.183	6700	42.5	7.0 8.6	. .	. .	.8 .9	. .	. .	. .	.2 .9	. .
4	1	.200	8000	48.5	7.4 9.0	. .	. .	.2 .4	. .	. .	. .	.4 .2	. 3.8
5	1	.267	14200	60.6	9.2 11.2	. 1 .	. .	.7 .2	. .	.[illegible] .	. .	.5 .5	4.1 [illegible]
6	1¼	.317	20100	72.8	½ 10.4 2.7	. ½ .	. 1 .	.7 .4	. .	. .	. .	.3 .5	. .
7	1¼	.383	29400	84.9	2.0 4.7	. .	. .	.1 1 .1	. 1 .	. .	. .	.4 .8	. .
8	1¼	.450	40500	97.2	3.5 16.5	. .	. .	.5 .8	. 1 .	. 1 .	. 1 .	.4 .0	. .
9	1½	.500	50000	109.	4.5 7.7	. .	. .	.3 .8	. .	. 1 .	. 1 .	.1 .9	. .
10	1½	.567	64200	121.5	5.8 9.3	. .	. .	.6 .4	. .	. .	. .	.1 1 .1	. 1 .
12	1½	.700	98000	145.7	8.2 2.2	. .	. .	.8 .1	. .	. 5 .	. .	.9 .3	. .

TABLE No. 21 (*Continued*).—STRENGTH OF FLOOR-SLABS.

Bold-faced type, $M = \frac{1}{8}wl^2$; light-faced type, $M = \frac{1}{12}wl^2$.

3. $f_c = 500$ $f_s = 16,000$ $R = 71$ $p = .0050$

Total [illegible]ess of Slab, Inches.	Thickness of [illegible]crete below Steel, Inches.	Required Area of Steel per Foot of Slab, Sq. In.	Moment of Resistance per Foot of Slab, In.-lbs.	Weight of Slab per Square Foot, Lbs.	Span in Feet for Given Net Loads per Square Foot of Floor in Pounds.								
					50	75	100	150	200	250	300	400	500
2	¾	.075	1300	24.2	**3.5** [illegible].3	**3.0** 3.[illegible]	2.7 3.3	**2.3** [illegible]	**2.0** 2.4	1.8 2.2	1.7 2.1	1.4 1.7	**1.3** 1.6
2½	¾	.105	2 00	30.3	**.7** .8	**4.** 5.	**3.7** 4.5	. .	**2.8** 3.4	**2.5** 3.1	**2.3** 2.8	**2.0** 2.4	**1.8** 2.2
3	¾	.135	4 00	36.4	**.8** .1	**5.** 6.	**4.6** 5.6	. .	**3.5** 4.3	**3.2** 3.9	**2.9** 3.6	**2.6** 3.2	**2.3** 2.8
3½	¾	.165	6 00	42.5	**.8** .3	**6.** 7.	**5.5** 6.7	. .	**4.2** 5.1	**3.8** 4.6	**3.6** 4.4	**3.1** 3.8	**2.8** 3.4
4	1	.180	7 00	48.5	**.2** .8	**6.** 7.	**5.9** 7.2	. .	**4.5** 5.5	**4.1** 5.1	**3.8** 4.6	**3.4** 4.2	**3.1** 3.8
5	1	.239	13 00	60.7	**.1** 1 .1	**8.** 10.	**7.5** 9.2	. .	**5.9** 7.2	**5.4** 6.6	**5.0** 6.1	**4.4** 5.4	**4.0** 4.9
6	1¼	.284	9 00	72.7	**.3** .6	**9.** 11.	**8.7** 10.6	. .	**6.9** 8.4	**6.3** 7.7	**5.9** 7.2	**5.2** 6.4	**4.7** 5.8
7	1¼	.344	8 00	84.8	**.9** .5	**10.** 3.	**10.1** 2.4	. 1 .	**8.2** 10.0	**7.5** 9.2	**7.0** 8.6	**6.3** 7.7	**5.7** 7.0
8	1¼	.404	9000	97.0	**1 .4** 1 .4	**2.** 5.	**1.5** 4.1	1 . .	**9.4** 11.5	**8.7** 10.6	**8.1** 9.9	**7.3** 8.9	**6.6** 8.1
9	1½	.449	48100	109.1	**1 .3** 1 .5	**3.** 6.	**2.4** 5.2	. .	**10.2** 2.5	**9.5** 11.6	**8.9** 10.9	**8.0** 9.8	**7.3** 8.9
10	1½	.509	61 00	21.3	**1 .6** .1	**3.** 7.	**3.7** 6.7	. .	**1.4** 3.9	**0.6** 2.9	**9.9** 12.1	**8.9** 10.9	**8.2** 10.0
12	1½	.629	94400	45.5	**.9** .8	**6.** 20.	**6.0** 9.5	. .	**3.5** 6.5	**2.6** 5.4	**11.9** 14.5	**10.7** 13.1	**9.9** 12.1

4. $f_c = 500$ $f_s = 18,000$ $R = 66$ $p = .0041$

Total Thickness of Slab, Inches.	Thickness of Concrete below Steel, Inches.	Required Area of Steel per Foot of Slab, Sq. In.	Moment of Resistance per Foot of Slab, In.-lbs.	Weight of Slab per Square Foot, Lbs.	50	75	100	150	200	250	300	400	500
2	¾	.061	1200	24.2	**3.3** 4.[illegible]	**2.9** [illegible].6	**2.6** 3.2	**2.2** 2.[illegible]	**1.9** [illegible].3	**1.7** [illegible].1	1.6 2.0	1.4 1.7	**1.3** 1[illegible]
2½	¾	.086	2400	30.3	**4.** 5.	**.9** .8	**3.**[illegible] 4.	**3.** 3.	**.6** .2	**.4** .9	**2.2** 2.7	**1.9** 2.3	**1** 2
3	¾	.110	4000	36.4	**5.** 6.	**.9** .0	**4.** 5.	**3.** 4.	**.4** .2	**.0** .7	**2.8** 3.4	**2.5** 3.1	**2** 2
3½	¾	.135	6000	42.4	**6.** 8.	[illegible]**.9** .2	**5.** 6.	**4.** 5.	**.0** .9	**.7** .5	**3.4** 4.2	**3.0** 3.7	**2** 3
4	1	.147	7200	48.4	**7.** 8.	**.2** .6	**5.** 7.	**4.** 6.	**.4** .4	**.0** .9	**3.7** 4.5	**3.3** 4.0	**2** 3
5	1	.196	12700	60.5	**8.** 10.	**.9** .7	**7.** 8.	**6.** 7.	**.7** .0	**.2** .4	**4.8** 5.9	**4.3** 5,3	**3** 4
6	1¼	.233	18000	72.6	**9.** 12.	**.0** 1 .0	**8.** 10.	**7.** 8.	**.6** .1	**.1** .5	**5.7** 7.0	**5.0** 6.1	**4** 5
7	1¼	.282	26300	84.7	**11.** 13.	**1 .4** .7	**9.** 11.	**8.** 10.	**.8** .6	**.2** .8	**6.7** 8.2	**6.0** 7.4	**5** 6
8	1¼	.331	36300	96.8	**12.** 15.	**.8** .4	**1.** 3.	**9.** 12.	**.0** 1 .0	**.3** 1 .2	**7.8** 9.6	**7.0** 8.6	**6** 7
9	1½	.368	44800	108.9	**13.** 16.	**.7** .5	**1.** 4.	**10.** 3.	**.8** 1 .0	**.1** 1 .1	**8.5** 10.4	**7.7** 9.4	**7** 8
10	1½	.417	57500	121.1	**14.** 18.	**.9** .0	**3.** 6.	**1.** 4.	**.9** .3	**.1** .4	**9.5** 11.6	**8.6** 10.5	**7** 9
12	1½	.515	87700	145.3	**17.** 20.	**.2** .8	**5.** 8.	**4.** 7.	**.0** .9	**.1** .8	**11.4** 13.9	**10.4** 12.7	**9** 11

TABLE No. 21 (*Continued*)—STRENGTH OF FLOOR-SLABS.

Bold-faced type, $M=\frac{1}{8}wl^2$; light-faced type, $M=\frac{1}{12}wl^2$.

5. $f_c=600$ $f_s=14,000$ $R=102$ $p=.0084$

Total Thickness of Slab, Inches.	Thickness of Concrete below Steel, Inches.	Required Area of Steel per Foot of Slab, Sq. In.	Moment of Resistance per Foot of Slab, In.-lbs.	Weight of Slab per Square Foot, Lbs.	Span in Feet for Given Net Loads per Square Foot of Floor in Pounds.								
					50	75	100	150	200	250	300	400	500
2	¾	.126	1900	24.3	4.1 5.0	3.6 4.4	3.2 [illegible]	2.7 [illegible]	2.4 [illegible] 9	2.2 2.7	2.0 [illegible].4	1.7 [illegible]	1.6 2.[illegible]
2½	¾	.176	3800	30.4	5.6 6.9	4.9 6.0	. .	. .	3 0	3.0 3.7	.7 .3	. .	
3	¾	.226	6200	36.5	6.9 8.4	6.1 7.5	. .	. .	2 1	3.8 4.6	.5 .3	. .	 3.4
3½	¾	.277	9300	42.7	8.2 10.0	7.2 8.8	. .	. .	0 1	4.6 5.6	.2 .[illegible]	. .	3 4 4 [illegible]
4	1	.302	11000	48.7	8.6 10.5	7.7 9.4	. .	. .	4 6	5.0 6.1	. .	. .	3 4
5	1	.403	19600	60.9	10.8 13.2	9.8 12.0	. 1 .	. .	1 7	6.5 8.0	. .	. .	4 5
6	1¼	.478	27700	73.1	12.2 14.9	11.2 13.7	[illegible] . .	. 1 .	2 1 0	7.5 9.2	. .	. .	5 7
7	1¼	.579	40600	85.4	14.1 17.2	13.0 15.9	. .	[illegible] . .	7 1 9	9.0 11.0	. 1 .	. .	6 8
8	1¼	.680	55900	97.8	15.9 19.4	14.7 18.0	. .	. .	[illegible] 2 7	10.3 2.6	. 1 .	. 1 .	7 9
9	1½	.755	69000	109.8	16.9 20.6	15.7 19.2	. .	. .	2 9	1.3 3.8	[illegible] . .	. 1 .	8 10
10	1½	.856	88600	122.0	18.5 22.6	17.3 21.1	. .	. .	5 5	2.6 5.4	. .	1 . 1 .	9 11
12	1½	1.057	135300	146.4	21.4 26.2	20.1 24.6	. .	. .	1 7	5.0 8.3	. .	1 . 1 .	11 14.4

6. $f_c=600$ $f_s=15,000$ $R=98$ $p=.0075$

Total Thickness of Slab, Inches.	Thickness of Concrete below Steel, Inches.	Required Area of Steel per Foot of Slab, Sq. In.	Moment of Resistance per Foot of Slab, In.-lbs.	Weight of Slab per Square Foot, Lbs.	50	75	100	150	200	250	300	400	500
2	¾	.112	1800	24.2	4.[illegible] 5.[illegible]	3.5 [illegible]	3.2 [illegible]	2.7 [illegible]	2.3 [illegible]	2.1 [illegible]	1.9 [illegible]	1.7 [illegible].1	1.5 1.8
2½	¾	.157	3600	30.4	5. 6.	. .	. .	. .	. .	. .	. .	.4 .9	[illegible] .
3	¾	.202	6000	36.5	6. 8.	. .	. .	. .	. .	. .	. .	.0 .7	. .
3½	¾	.247	8900	42.7	8. 9.	. .	. .	. .	. .	. .	. .	.7 .5	. .
4	1	.270	10600	48.7	8. 10.	. .	. .	. .	. .	. .	. .	.0 .9	. .
5	1	.360	18900	60.9	10. 13.	. 1 .	. 1 .	. .	. .	. .	. .	.2 .4	. .
6	1¼	.427	26700	73.1	12. [illegible]4.	[illegible] . .	[illegible] . .	. 1 .	. .	. .	. .	.1 .5	1[illegible]. .
7	1¼	.517	39100	85.5	13. [illegible]7.	. .	. .	[illegible] . .	. 1 .	. 1 .	. 1 .	.3 .9	. .
8	1¼	.607	53800	97.	15. 9.	. .	. .	. .	[illegible] . .	[illegible] . .	. 1 .	.5 1 .4	. .
9	1½	.675	66400	109.	6. 0.	. .	. .	. .	. .	. .	[illegible] . .	.4 1 .5	. .
10	1½	.765	85300	121.8	8. 2.	. .	. .	. .	. .	. .	. .	1 .4 1 .7	. 1 .
12	1½	.945	130200	146.2	21. 25.	. .	. .	. .	. .	. .	. .	1 .6 1 .4	[illegible] . .

TABLE No. 21 (*Continued*)—STRENGTH OF FLOOR-SLABS.

Bold-faced type, $M = \frac{1}{8}wl^2$; light-faced type, $M = \frac{1}{12}wl^2$.

7. $f_c = 600$ $f_s = 16,000$ $R = 95$ $p = .0068$

Total Thickness of Slab, Inches.	Thickness of Concrete below Steel, Inches.	Required Area of Steel per Foot of Slab, Sq. In.	Moment of Resistance per Foot of Slab, In.-lbs.	Weight of Slab per Square Foot, Lbs.	Span in Feet for Given Net Loads per Square Foot of Floor in Pounds.								
					50	75	100	150	200	250	300	400	500
2	¾	.101	1800	24.2	**4.0** 4.9	**3.5** 4.3	**3.1** 3.8	**2.6** 3.2	**2.3** 2.8	**2.1** 2.6	**1.9** 2.3	**1.7** 2.1	**1.5** 1.8
2½	¾	.142	3500	30.3	**5.4** 6.6	**4.7** 5.8	**4.2** 5.1	**3.6** 4.4	**3.2** 3.9	**2.9** 3.6	**2.7** 3.3	**2.3** 2.8	**2.1** 2.6
3	¾	.182	5800	36.4	**6.7** 8.2	**5.9** 7.2	**5.3** 6.5	**4.5** 5.5	**4.0** 4.9	**3.7** 4.5	**3.4** 4.2	**3.0** 3.7	**2.7** 3.3
3½	¾	.223	8600	42.6	**7.9** 9.7	**7.0** 8.6	**6.3** 7.7	**5.5** 6.7	**4.9** 6.0	**4.4** 5.4	**4.1** 5.0	**3.6** 4.4	**3.3** 4.0
4	1	.243	10300	48.6	**8.3** 10.2	**7.4** 9.0	**6.8** 8.3	**5.9** 7.2	**5.2** 6.4	**4.8** 5.9	**4.4** 5.4	**3.9** 4.8	**3.5** 4.3
5	1	.324	18300	60.8	**10.5** 12.8	**9.5** 11.6	**8.7** 10.[illegible]	[illegible]**.6** .3	**6.8** 8.3	**6.3** 7.7	[illegible] .	**5.1** 6.2	**4.7** 5.8
6	1¼	.385	25700	73.0	**11.8** 14.4	**10.7** 13.1	**9.** 12.	**.7** .6	**7.9** 9.7	**7.3** 8.9	. .	**6.0** 7.4	**5.5** 6.7
7	1¼	.466	37700	85.2	**13.6** 1[illegible].6	**12.5** 15.3	**11.** 14.	**.3** .6	**9.4** 11.5	**8.7** 10.6	. .	**7.2** 8.8	**6.5** 8.0
8	1¼	.547	52000	97.7	**1 .3** 1 .7	**14.2** 17.3	**13.** 16.	**.8** .4	**10.8** 3.2	**10.0** 1[illegible].2	. .	**8.3** 10.2	**7.6** 9.3
9	1½	.608	64200	109.4	**1 .4** 2 .0	**15.2** 18.6	**14.** 17.	**.8** .7	**1.8** 4.4	**1 .9** 1 .3	. .	**9.2** 11.2	**8.4** 10.3
10	1½	.689	82400	121.6	**1 .9** 2 .9	**16.7** 20.4	**15.** 19.	**.2** .3	**3.1** 6.0	**1 .2** 1 .9	. .	**10.2** 12.5	**9.4** 11.5
12	1½	.851	125800	146.2	**2 .6** 2 .2	**19.4** 23.7	**18.** 22.	**8** .5	**5.5** 9.0	**1 .5** 1 .7	. .	**12.4** 15.2	**11.4** 13.9

8. $f_c = 600$ $f_s = 18,000$ $R = 89$ $p = .0056$

Total Thickness of Slab, Inches.	Thickness of Concrete below Steel, Inches.	Required Area of Steel per Foot of Slab, Sq. In.	Moment of Resistance per Foot of Slab, In.-lbs.	Weight of Slab per Square Foot, Lbs.	50	75	100	150	200	250	300	400	500
2	¾	.083[illegible]	1700 33	24.1[illegible]	**3.9** [illegible]	**3.3** [illegible]	**3.0** [illegible]	**2.5** 3.1	**2.2** 2.7	**2.0** [illegible]	**1.8** [illegible]	**1.6** 2.0	**1.4** **2.0**
2½	¾	.11	00	30.	. .	. .	. .	**3.5** 4.3	**3.1** 3.8	. .	. .	**2.2** 2.7	. 2.4
3	¾	.15	5400	36.	. .	. .	. .	**4.4** 5.4	**3.9** 4.8	. .	. .	**2.9** 3.6	**2.6** [illegible]
3½	¾	.18	8100	42.	. .	. .	. .	**5.3** 6.5	**4.7** 5.8	. .	. .	**3.5** 4.3	. .
4	1	.200	9600	48.	. .	. .	. .	**5.7** 7.0	**5.1** 6.2	. .	. .	**3.8** 4.6	. .
5	1	.267[illegible]	17100 [illegible]	60.	1 . 1 .	. 1 .	. 1 .	**7.4** 9.0	**6.6** 8.1	. .	. .	**5.0** 6.1	. .
6	1¼	.31	100	72.	. .	. .	. 1 .	**8.5** 10.4	**7.7** 9.[illegible]	. .	. .	**5.8** 7.1	. .
7	1¼	.38	300	85.	. 1 .	. .	. .	**10.0** 2.2	**9.** 11.	. 1 .	. .	**7.0** 8.6	. .
8	1¼	.45	600	97.	1 . .	. .	. .	**1.4** 3.9	**10.** 2.	. 1 .	. 1 .	**8.1** 9.9	. .
9	1½	.500	60000 11	109. 1	. .	. .	. .	**2.4** 5.2	**1.** 3.	**1** . .	. 1 .	**8.9** 10.9	. .
10	1½	.567	7100 7	21.	1 . 2 .	. .	. .	**3.8** 6.9	**2.** 5.	. .	. .	**9.9** 12.1	. 1 .
12	1½	.700	00	45.	**2** . .	. .	. .	**6.3** 9.9	**5.** 8.	. .	. .	**12.0** 14.7	**1** . .

TABLE No. 21 (*Continued*)—STRENGTH OF FLOOR-SLABS.

Bold-faced type, $M=\frac{1}{8}wl^2$; light-faced type, $M=\frac{1}{12}wl^2$.

9. $f_c=700$ $f_s=14,000$ $R=129$ $p=.0107$

Total Thickness of Slab, Inches.	Thickness of Concrete below Steel, Inches.	Required Area of Steel per Foot of Slab, Sq. In.	Moment of Resistance per Foot of Slab, In.-lbs.	Weight of Slab per Square Foot, lbs.	Span in Feet for Given Net Loads per Square Foot of Floor in Pounds.								
					50	75	100	150	200	250	300	400	500
2	$\frac{3}{4}$	.161	2400	24.4	4.7	4.0	3.6	3.0	2.7	2.4	2.2	1.9	1.7
$2\frac{1}{2}$	$\frac{3}{4}$	.225	4700	30.6	[illegible]	[illegible]	[illegible]	[illegible]	[illegible]	[illegible]	[illegible]	[illegible]	[illegible]
3	$\frac{3}{4}$	.289	7800	36.7	[illegible]	[illegible]	[illegible]	[illegible]	[illegible]	[illegible]	[illegible]	[illegible]	[illegible]
$3\frac{1}{2}$	$\frac{3}{4}$	.354	11700	43.0	[illegible]	[illegible]	[illegible]	[illegible]	[illegible]	[illegible]	[illegible]	[illegible]	4.6
4	1	.386	13900	48.9	[illegible]	[illegible]	[illegible]	[illegible]	[illegible]	[illegible]	[illegible]	[illegible]	[illegible]
5	1	.514	24700	61.1	[illegible]	[illegible]	[illegible]	[illegible]	[illegible]	[illegible]	[illegible]	[illegible]	[illegible]
6	$1\frac{1}{4}$	.611	34800	73.5	[illegible]	[illegible]	[illegible]	[illegible]	[illegible]	[illegible]	[illegible]	[illegible]	[illegible]
7	$1\frac{1}{4}$	.739	51000	85.7	[illegible]	[illegible]	[illegible]	[illegible]	[illegible]	[illegible]	[illegible]	[illegible]	[illegible]
8	$1\frac{1}{4}$	.868	70300	98.1	[illegible]	[illegible]	[illegible]	[illegible]	[illegible]	[illegible]	[illegible]	[illegible]	[illegible]
9	$1\frac{1}{2}$	.964	86800	110.3	[illegible]	[illegible]	[illegible]	[illegible]	[illegible]	[illegible]	[illegible]	[illegible]	[illegible]
10	$1\frac{1}{2}$	1.093	111500	122.6	[illegible]	[illegible]	[illegible]	[illegible]	[illegible]	[illegible]	6.2	[illegible]	[illegible]
12	$1\frac{1}{2}$	1.350	170100	147.3	[illegible]	[illegible]	[illegible]	[illegible]	[illegible]	[illegible]	5.9	[illegible]	[illegible]

10. $f_c=700$ $f_s=15,000$ $R=124$ $p=.0096$

Total Thickness of Slab, Inches.	Thickness of Concrete below Steel, Inches.	Required Area of Steel per Foot of Slab, Sq. In.	Moment of Resistance per Foot of Slab, In.-lbs.	Weight of Slab per Square Foot, lbs.	50	75	100	150	200	250	300	400	500
2	$\frac{3}{4}$	.144	2300	24.4	4.6	4.0	3.5	3.0	2.6	2.4	2.2	1.9	1.7
$2\frac{1}{2}$	$\frac{3}{4}$	[illegible]	[illegible]	[illegible]	[illegible]	[illegible]	[illegible]	[illegible]	[illegible]	[illegible]	[illegible]	[illegible]	[illegible]
3	$\frac{3}{4}$	[illegible]	[illegible]	[illegible]	[illegible]	[illegible]	[illegible]	[illegible]	[illegible]	[illegible]	[illegible]	[illegible]	[illegible]
$3\frac{1}{2}$	$\frac{3}{4}$	[illegible]	[illegible]	[illegible]	[illegible]	[illegible]	[illegible]	[illegible]	[illegible]	[illegible]	[illegible]	[illegible]	[illegible]
4	1	[illegible]	[illegible]	[illegible]	[illegible]	[illegible]	[illegible]	[illegible]	[illegible]	[illegible]	[illegible]	[illegible]	4.0
5	1	.461	[illegible]	[illegible]	[illegible]	[illegible]	[illegible]	[illegible]	[illegible]	[illegible]	[illegible]	[illegible]	[illegible]
6	$1\frac{1}{4}$	.548	[illegible]	[illegible]	[illegible]	[illegible]	[illegible]	[illegible]	[illegible]	[illegible]	[illegible]	[illegible]	[illegible]
7	$1\frac{1}{4}$	[illegible]	[illegible]	[illegible]	[illegible]	[illegible]	[illegible]	[illegible]	[illegible]	[illegible]	[illegible]	[illegible]	[illegible]
8	$1\frac{1}{4}$	[illegible]	[illegible]	[illegible]	[illegible]	[illegible]	[illegible]	[illegible]	[illegible]	[illegible]	[illegible]	[illegible]	[illegible]
9	$1\frac{1}{2}$	.865	[illegible]	[illegible]	[illegible]	[illegible]	[illegible]	[illegible]	[illegible]	[illegible]	[illegible]	[illegible]	[illegible]
10	$1\frac{1}{2}$	.980	[illegible]	[illegible]	[illegible]	[illegible]	[illegible]	[illegible]	[illegible]	[illegible]	[illegible]	[illegible]	[illegible]
12	$1\frac{1}{2}$	[illegible]	[illegible]	[illegible]	[illegible]	[illegible]	[illegible]	[illegible]	[illegible]	[illegible]	[illegible]	[illegible]	[illegible]

TABLE No. 21 (*Continued*)—STRENGTH OF FLOOR-SLABS.

Bold-faced type, $M=\frac{1}{8}wl^2$; light-faced type, $M=\frac{1}{12}wl^2$.

11. $f_c=700$ $f_s=16,000$ $R=120$ $p=.0087$

Total Thickness of Slab, Inches.	Thickness of Concrete below Steel, Inches.	Required Area of Steel per Foot of Slab, Sq. In.	Moment of Resistance per Foot of Slab, In.-lbs.	Weight of Slab per Square Foot, Lbs.	Span in Feet for Given Net Loads per Square Foot of Floor in Pounds. 50	75	100	150	200	250	300	400	500
2	¾	.130	2300	24.3	**4.5** 5.5	**3.9** 4.8	**3.5** 4.3	**2.9** 3.6	**2.6** 3.2	**2.3** 2.8	**2.1** 2.6	**1.9** 2.3	**1.7** 2.1
2½	¾	.182	4400	30.5	**6.1** 7.5	**5.3** 6.5	**4.8** 5.9	**4.0** 4.9	**3.6** 4.4	**3.2** 3.9	**3.0** 3.7	**2.6** 3.2	**2.4** 2.9
3	¾	.234	7300	36.6	**7.5** 9.2	**6.6** 8.1	**6.0** 7.4	**5.1** 6.2	**4.5** 5.5	**4.1** 5.0	**3.8** 4.6	**3.3** 4.0	**3.0** 3.7
3½	¾	.286	10900	42.9	**8.9** 10.9	**7.9** 9.7	**7.2** 8.8	**6.2** 7.6	**5.5** 6.7	**5.0** 6.1	**4.6** 5.6	**4.0** 4.9	**3.7** 4.5
4	1	.312	13000	48.8	**9.3** 11.3	**8.4** 10.3	**7.6** 9.3	**6.6** 8.1	**5.9** 7.2	**5.4** 6.6	**5.0** 6.1	**4.4** 5.4	**4.0** 4.9
5	1	.416	23100	61.0	**11.8** 14.4	**10.6** 12.9	**9.8** 12.0	**8.5** 10.4	**7.7** 9.4	**7.0** 8.6	**6.5** 8.0	**5.8** 7.1	**5.2** 6.4
6	1¼	.494	32600	73.2	**13.3** 16.2	**12.1** 14.8	**11.2** 13.7	**9.9** 12.1	**8.9** 10.9	**8.2** 10.0	**7.6** 9.3	**6.8** 8.3	**6.2** 7.6
7	1¼	.598	47700	85.4	**15.4** 18.8	**14.1** 17.2	**13.1** 16.0	**11.6** 14.2	**10.6** 12.9	**9.8** 12.0	**9.1** 11.1	**8.1** 9.9	**7.4** 9.0
8	1¼	.702	65800	97.7	**17.3** 21.1	**15.9** 19.4	**14.9** 18.2	**13.3** 16.2	**12.1** 14.8	**11.2** 13.7	**10.5** 12.8	**9.4** 11.5	**8.6** 10.5
9	1½	.780	81200	109.9	**18.4** 23.0	**17.1** 20.9	**16.1** 19.7	**14.4** 17.6	**13.2** 16.1	**12.3** 15.0	**11.5** 14.1	**10.3** 12.6	**9.4** 11.5
10	1½	.884	104300	122.1	**20.1** 24.6	**18.8** 23.0	**17.7** 21.6	**16.0** 19.5	**14.7** 18.0	**13.7** 16.8	**12.9** 15.8	**11.6** 14.2	**10.6** 12.9
12	1½	1.092	159200	146.6	**23.2** 28.3	**21.9** 26.8	**20.8** 25.5	**18.9** 23.1	**17.5** 21.4	**16.4** 20.0	**15.4** 18.8	**14.0** 17.0	**12.8** 15.7

12. $f_c=700$ $f_s=18,000$ $R=113$ $p=.0072$

Total Thickness of Slab, Inches.	Thickness of Concrete below Steel, Inches.	Required Area of Steel per Foot of Slab, Sq. In.	Moment of Resistance per Foot of Slab, In.-lbs.	Weight of Slab per Square Foot, Lbs.	50	75	100	150	200	250	300	400	500
2	¾	.107	2100	24.3	**4.4** 5.4	**3.8** 4.6	**3.4** 4.2	**2.8** 3.4	**2.5** 3.1	**2.3**	**2.1**	**1.8**	**1.6**
2½	¾	.150	4200	30.4	**5.9** 7.2	**5.1** 6.2	**4.6** 5.6	**3.9** 4.8	**3.5** 4.3	.	.	.	.
3	¾	.193	6900	36.4	**7.3** 8.9	**6.4** 7.8	**5.8** 7.1	**5.0** 6.1	**4.4** 5.4	.	.	.	.
3½	¾	.236	10300	42.7	**8.6** 10.5	**7.6** 9.3	**6.9** 8.4	**6.0** 7.4	**5.3** 6.5	.	.	.	.
4	1	.258	12200	48.6	**9.1** 11.1	**8.1** 9.9	**7.4** 9.0	**6.4** 7.8	**5.7** 7.0	.	.	.	.
5	1	.344	21700	60.8	**11.4** 13.9	**10.3** 12.6	**9.5** 11.6	**8.3** 10.2	**7.5** 9.2	.	.	.	.
6	1¼	.408	30600	72.9	**12.8** 15.7	**11.7** 14.3	**10.9** 13.3	**9.6** 11.7	**8.6** 10.5	.	.	.	.
7	1¼	.494	44800	85.4	**14.8** 18.1	**13.6** 16.6	**12.7** 15.5	**11.3** 13.8	**10.2** 12.5	.	.	.	.
8	1¼	.580	61800	97.9	**16.7** 20.4	**15.4** 18.8	**14.5** 17.7	**12.9** 15.8	**11.7** 14.3	.	.	.	.
9	1½	.644	76300	109.6	**17.9** 21.9	**16.6** 20.3	**15.6** 19.1	**14.0** 17.1	**12.8** 15.7	.	.	.	.
10	1½	.730	98000	121.8	**19.5** 23.8	**18.2** 22.2	**17.2** 21.0	**15.5** 19.0	**14.3** 17.5	.	.	.	.
12	1½	.902	149500	146.3	**22.6** 27.6	**21.2** 25.9	**20.1** 24.5	**18.4** 22.5	**17.0** 20.8	.	.	.	.

CHAPTER VII.

BUILDING CONSTRUCTION.

153. Division of the Subject.—The various elements of building construction relating to reinforced-concrete design may be grouped under the following heads: (1) Beams forming a continuous surface, as floor- and roof-slabs; (2) Floor-beams and girders; (3) Columns; (4) Footings; (5) Walls and partitions. In the discussion of these various elements consideration will be given to the determination of stresses, the design of the members, and the arrangement of connective details.

154. General Arrangement of Concrete Floors.—Two general types of floors may be considered: (1) that in which the floor-slab is supported on steel beams, and (2) that in which concrete beams are used, the floor of the entire structure being of a monolithic character. In the former case the steel skeleton consists of columns, girders, and cross-beams, the beams being spaced commonly about 6 feet apart. The floor-slab is supported mainly by the cross-beams. The same variety of arrangements is used in the case of all-concrete structures, the cross-beams being spaced usually from four to six feet apart. The cross-beams may in this case be entirely omitted, giving span lengths of 15 to 20 feet. Sometimes, also, the cross-beams are inserted only at columns, forming a nearly square panel of the floor-slab, which is then considered as supported on four sides.

155. Stresses in Continuous Beams.—Since floor-slabs and beams are commonly designed to act as continuous beams it is important to investigate the possible stresses under such conditions, although exact calculation is impracticable and unnecessary. In the case of floor-slabs there is usually a large number of consecutive spans and the loading producing the theoretical maximum moments at various points would involve unreasonable assumptions as to position of live loads. Sufficiently exact analysis may be arrived at by considering certain simple cases.

Moments in Beams of Two and Three Spans.—Inasmuch as the conditions for a theoretical maximum are more likely to occur in beams of two or three spans than where the number of spans is large, an exact analysis will be made of maximum moments at all points for the beam of two spans and for the beam of three spans. The spans will be assumed equal and the beam considered as continuous but freely supported at all points. Assume the dead load to be a uniformly distributed load, $=w$ per lineal foot, and the live load to be also a uniform load, $=p$ per lineal foot, but distributed over such portions of the beam as to cause a maximum moment at the given section. The maximum moment is readily calculated by means of the usual continuous girder formulas. The calculation will not be repeated here. The results are shown graphically to scale in Figs. 71 and 72. The dotted lines relate to dead-load effects and the dashed lines to live-load effects. The span length and the load per foot are assumed equal to unity. If l=span length, the true moments will be found by multiplying the proper ordinates by wl^2 or by pl^2 respectively.

The coefficients of wl^2 and pl^2 for the maximum positive and negative moments for the two beams are as follows:

		Maximum near Center of Span. (+)	Maximum at Support. (−)
Beams of two spans (Fig. 77):			
Dead load		.070	.125
Live load		.095	.125
Beam of three spans (Fig. 72):			
Dead load	1st span	.080	.100
	2d span	.025	
Live load	1st span	.100	.117
	2d span	.075	

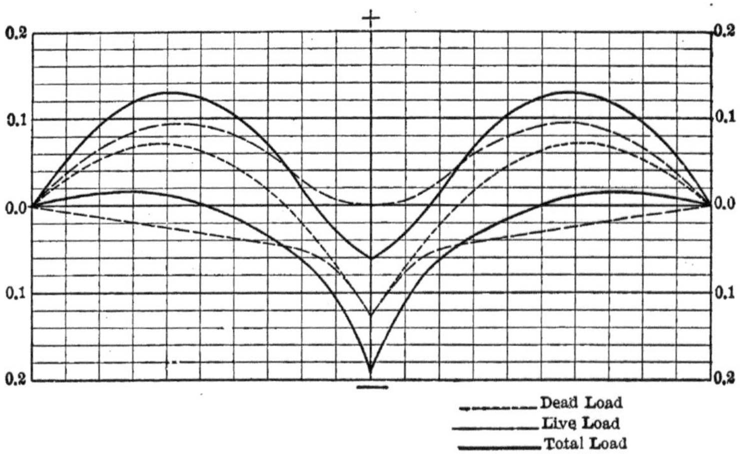

Fig. 71.—Moments in Beams of Two Spans.

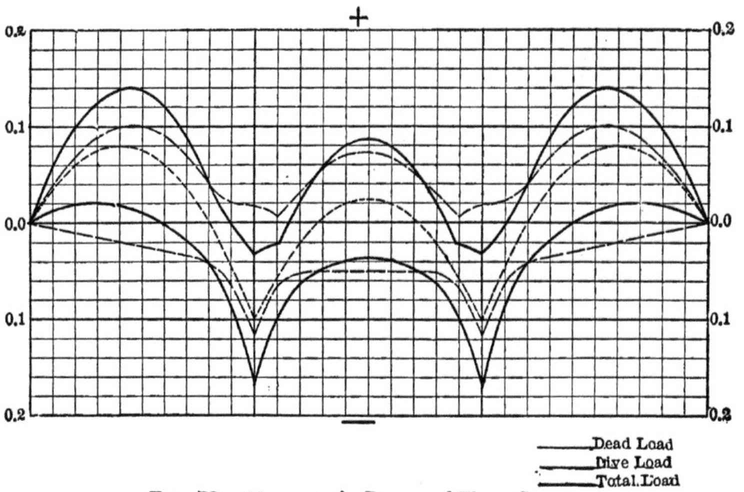

Fig. 72.—Moments in Beams of Three Spans.

If the dead and live loads are combined into a single unit for the purposes of calculation, the proper coefficient for $(w+p)$ will depend on the relation of dead and live load. If, for example, the dead load is one-half the live load, then there results, for the first case,

$$\text{Maximum positive moment} = .087(w+p)l^2,$$
$$\text{Maximum negative moment} = \tfrac{1}{8}(w+p)l^2,$$

and for the second case

$$\text{Maximum positive moment} = .093(w+p)l^2,$$
$$\text{Maximum negative moment} = .111(w+p)l^2.$$

In Figs. 71 and 72 the full lines represent the maximum moments throughout the beam for the condition that $w=\frac{1}{2}p$; these lines are particularly useful in showing the relative distances from the supports over which positive and negative moments may occur.

Moments in Beams of Several Spans.—In the case of several spans it will be practically correct in calculating positive

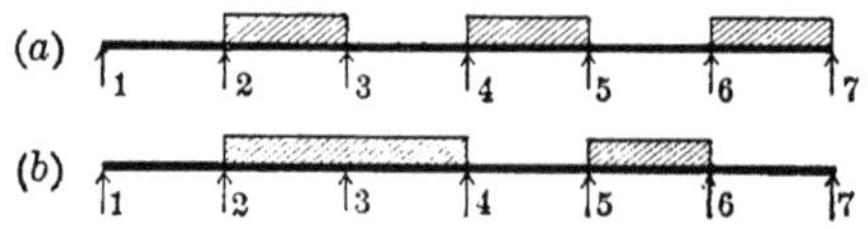

FIG. 72*a*.

moments to consider that the maximum moment at the center of the span is the maximum desired. (Strictly the maximum is generally not quite at the center.) The loading required for maximum live-load moments is illustrated in Fig. 72*a*, which shows in (*a*) the loading for maximum positive moment in spans 2–3, 4–5, and 6–7, and in (*b*) the loading for maximum negative moment at support No. 3. For the former case each alternate span is loaded and, for the latter, the two adjoining spans are loaded and then each alternate span. Calculations have been made for each span and each support for 4, 5, 6, and 7 spans, and given in Table 21*A*. It is found in general that for all spans and supports, except the end span and support adjacent thereto, the maximum positive and negative moments do not vary greatly for the different spans, but for these end spans and supports they are considerably larger than for intermediate spans. The results are, accordingly, arranged in two groups in the table. For the intermediate spans the greatest value for the several spans is given. The table also includes the previous results for two and three spans.

TABLE NO. 21*A*.

COEFFCIENTS FOR MAXIMUM MOMENTS IN CONTINUOUS BEAMS.

No. of Spans.	Intermediate Spans and Supports.				End Span and 2d Support.			
	At Center (+)		At Support (−)		At Center (+)		At Support (−)	
	Dead.	Live.	Dead.	Live.	Dead.	Live.	Dead.	Live.
Two	....	....	....	....	.070	.095	.125	.125
Three...	.025	.075	....	....	.080	.100	.100	.117
Four ...	.036	.081	.071	.107	.071	.098	.107	.120 (.115)
Five	.046	.086	.079	.111 (.106)	.072	.099	.105	.120 (.116)
Six	.043	.084	.086	.116 (.106)	.072	.099	.106	.120 (.116)
Seven ..	.044	.084	.085	.114 (.106)	.072	.099	.106	.120 (.116)

The quantities in parentheses are the coefficients for live-load moments over supports where the two adjoining spans only are loaded. The effect of loading each alternate span in addition to these two spans is seen to be small; and considering that such a loading would be extremely improbable, and also the fact that a comparatively small amount of load on the other spans would neutralize this effect, it is apparent that the quantities in parentheses may be taken as reasonable maximum values. The two-span beam should preferably be treated as a special case.

Finally, leaving out of account the two-span beam, the following values may be taken as reasonable maximum values for beams of any number of spans.

	Intermediate Spans.		End Spans.	
	At Center.	At Support.	At Center.	At Support.
Dead load moments	.045	.085	.075	.105
Live load moments	.085	.105	.100	.115

Practical Working Coefficients for Moments.—It is generally convenient to adopt some simple fraction, such as $\frac{1}{8}$, $\frac{1}{10}$, or $\frac{1}{12}$ for the coefficient for both dead and live loads for ordinary calculations, and if the same coefficient can be used for both dead and live load it is desirable to do so. The live load will generally range from two to five times the dead load. The average coefficients for the combined loads, for various ratios of live to dead load, using the separate values above given, are as follows:

Ratio of Live : Dead.	Intermediate Spans.		End Spans.	
	At Center.	At Support.	At Center.	At Support
2:1	.072	.098	.092	.112
3:1	.075	.100	.094	.112
4:1	.077	.101	.095	.113
5:1	.078	.102	.096	.113

It will be seen from this table that for ordinary proportions a single coefficient may well be used for both dead and live loads.

Before adopting final values consideration should be given to certain modifying influences. The beams and slabs are not freely supported as assumed, but are, to a considerable extent, fixed at the supports. This tends to reduce the maximum moments. The supports, also, are of considerable width, so that if the span lengths be taken center to center, the negative moment at the edge of the support is considerably less than the caculated maximum. Thus if the width of support is $\frac{1}{12}$th the span length the negative moment at edge of support is about 25% less than at center of support. The slab also is greatly strengthened by the adjoining floor structure, as explained in Art. 156, and is also much simpler in design than the beam, and hence need not be so liberally proportioned. Furthermore, it is generally convenient to use the same amount of steel over the support as at the center, so that the moment at support will govern the design. Considering all these elements the following coefficients are proposed for both dead and live loads, and for both positive and negative moments:

For slabs of medium or short span:

Intermediate and end spans . . . . $\frac{1}{12}$

For beams and for slabs of long span:

Intermediate spans . . . . $\frac{1}{12}$
End spans . . . . $\frac{1}{10}$

A value of $\frac{1}{10}$ is commonly used throughout, but this is unnecessarily large for most slab spans, although it might well be used for beams if it is desired to have a fixed value for all spans.

Remarks.—The foregoing calculations assume uniform moment of inertia and therefore that about the same amount of steel is used for negative as for positive moments. The effect of variation in moment of inertia is discussed in Art. 166. Equal spans are also assumed. Where span lengths vary greatly, special calculations should be made, but the use of $\frac{1}{10}$ for the general coefficient will provide ample strength in all ordinary cases. This would require an actual resisting moment at each support of only about $\frac{1}{40}pl^2$ in order that the center moment be reduced to $\frac{1}{10}pl^2$. Cases of heavy concentrated loads must be given special consideration.

Shears. The maximum shears near supports are not greatly affected by moving loads. For intermediate spans the maximum end shear may be taken at one-half of the span load; for end spans the shear near the second support will be approximately six-tenths of a span load.

156. Effect of Rigid Supports on the Resisting Moment. If a flat slab is held between unyielding supports, such as fixed I-beams, a strength, or resisting moment, will be developed in the slab even though there be no steel reinforcement. Failure cannot take place without the crushing of the concrete either at the center or at the support. For short spans this resisting moment (the so-called "arch action") is about as great as will exist in the slab if reinforced and simply supported at the ends. In the case of a flat reinforced slab such rigid supports likewise add considerably to the strength of the slab, giving the effect of partial continuity.

In practice, the supports of slabs of short span length, whether consisting of I-beams or of concrete beams of which the slab is a part, are rendered very rigid by reason of the action of the adjoining floor-panels. Even where the slabs are simply supported on the tops of steel beams the adjoining slabs prevent to some extent lateral motion, rendering all such spans partially continuous. The strengthening effect of rigid supports is, therefore, especially great in the case of narrow floor-spans and where there is a large number of consecutive unbroken panels. Under such conditions reinforcement against negative moment is hardly necessary. For long spans and for spans on the outside of a system the effect is small.

157. Slabs Reinforced in Two Directions.—If the panel between beams is square, or nearly so, the slab may advantageously be reinforced in both directions. The exact analysis of stresses in such a case is difficult, if not impossible, as the effect of the more or less rigid supports is especially important and the problem is otherwise difficult of exact treatment.

The following solution for square and rectangular slabs will serve to show, approximately, the relation of the loads carried by the two systems of reinforcement. The results are certainly safe and do not vary much from rules of practice, but point to a somewhat more economical use of material.

158. *Square Slabs.*—In this case the reinforcement should be of equal amount in the two directions. It may be calculated on the assumption that one half the load is carried by each system of reinforcement. The concrete is proportioned for only one system, or one-half the load, as the stresses due to the two systems are at right angles to each other and it is assumed that the stresses in one direction do not weaken the concrete with respect to stresses in the other direction. The loading on each system is usually assumed to be uniformly distributed, resulting in an equal spacing of rods throughout the beam. This assumption is, however, far from the truth, and while giving safe results it is desirable to consider a more exact analysis of the problem which will show that the rods should be spaced closer at the center than at the edge.

In Fig 73, $ABCD$ represents a square slab supported on all sides and loaded with a uniform load w per unit area. Consider the relative amounts of load carried by the system parallel to aa' and the system parallel to mm'. At the centre O, and at all points on the diagonal lines AD and CB, it follows from symmetry that the loading is equally distributed on the two systems and is equal to $w/2$. At point E the proportion of the load carried by the system aa' will be much greater than that carried by the system mm', since for given loads the beam element along aa' will deflect much less at point E than will the element along mm'. In general, therefore, as we approach the support BD the proportion of load carried by the system aa' increases, reaching a value of w at the extreme end a'. The distribution of load on aa' may then be roughly represented by the ordinates from AB to the curved line aOa' of Fig. 74. Consider now the load along a line bb'. At points F and F' the load will be $w/2$; at point G' it will be less than $w/2$, being the same as the load on the system mm' at G. It will be shown in Fig. 74 by the ordinate GH from Oa' to the line aa'. At the end b' the load will be w. The curve of distribution will then be somewhat as represented by the line $aG'a'$ in Fig. 74, in which $G'K=GH$.

Assuming the curve aOa' to be a parabola it is found that the centre bending moment along the line aa', for a beam one foot wide, will be $\frac{7}{48}\,(w/2)l^2$ instead of $\frac{6}{48}\,(w/2)l^2$, as results

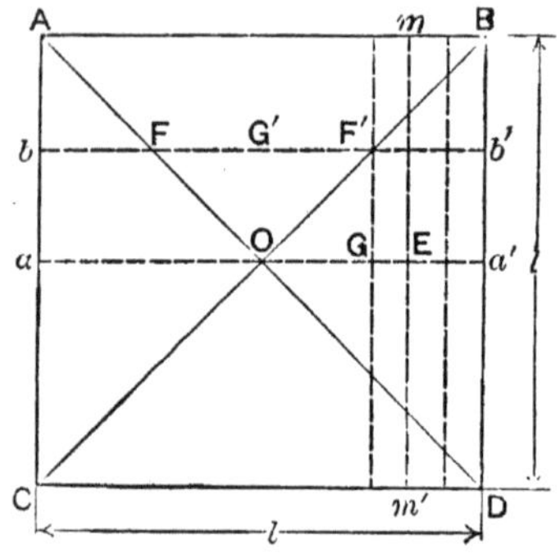

FIG. 73.

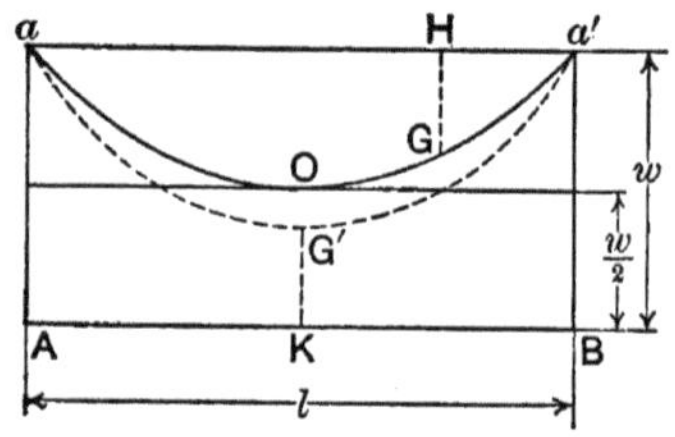

FIG. 74.

from the usual assumption. The spacing of the rods at the centre may then be determined on this basis. At points intermediate between the centre and the edge, the rods might well be spaced so that the number per foot would vary from the required number at the centre to zero at the edge, following the law of the parabola. If N represents the total number required on the ordinary assumption of equal spacing, then $\frac{7}{6}N\times\frac{2}{3}$, or $\frac{7}{9}N$, would represent the more correct number when spaced as here calculated. Practically as good results will be secured if the rods are spaced uniformly at the usual spacing, determined by the formula $M=\frac{1}{8}(w/2)l^2$, for the centre half of the slab, then gradually reduce the number per foot to the edge of the slab, using one-half as many rods for the remaining two quarters. The total number used would then be $\frac{3}{4}N$ instead of $\frac{7}{9}N$ as above determined, but the strength would be ample. If the slabs are continuous, then

$\frac{1}{10}$ or $\frac{1}{12}$ should be substituted for $\frac{1}{8}$ in the formula for M, as may be permissible.

159. *Rectangular Slabs of Greater Length than Breadth.*— As a slab becomes oblong in form the relative amount of load carried by the longitudinal system becomes rapidly less. Fig. 75 represents an oblong slab of length l and breadth b. Consider a central strip one foot wide along the line aa' and also along the line mm'. Suppose the rods to be spaced equally in the two directions so that the moments of inertia of the strips are equal. Let w=load per foot on the strip mm' and w'=load per foot on aa'. The deflection of a beam uniformly loaded is proportional to wl^4, hence, since the deflections of the two beams are equal, we have $wl^4=w'b^4$ or $w:w'=b^4:l^4$. That is, the amount of the load carried (per square foot) by the two systems is inversely proportional to the fourth power of the respective dimensions. For points nearer the ends of the slab the proportion carried by the longitudinal system will be greater, but in any case the longitudinal rods will be much under-stressed.

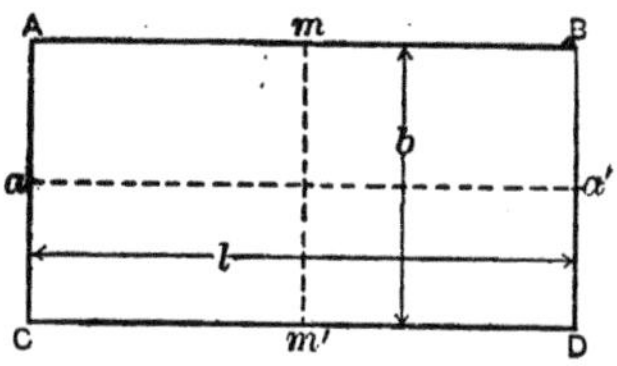

FIG. 75.

In accordance with this theory the proportion of the total load carried by the transverse system for various ratios of $l:b$ is as follows:

Ratio $l:b$	1	1.1	1.2	1.3	1.4	1.5
Proportion of load carried by transverse system	0.50	0.59	0.67	0.75	0.80	0.83

In the report of the French Commission the factors recommended give greater weight to the strengthening effect of a double system. They are as follows:

Ratio $l:b$	1	1.1	1.2	1.3	1.4	1.5
Proportion of load carried by transverse system	0.33	0.42	0.50	0.58	0.65	0.72
Proportion of load carried by longitudinal system	0.33	0.26	0.20	0.15	0.12	0.08

It is evident in any case that if the length is as much as 25% more than the breadth the working stresses in the longitudinal rods will be much less than in the transverse rods, and that, in general, it is not economical to reinforce long and narrow panels in two directions.

From this discussion it is evident that longitudinal reinforcement should not be used to carry load in oblong panels where the length exceeds the breadth by more than 15 to 20 %. An excess of 25% would seem to be about the practical limit. Whatever steel is placed in the longitudinal direction is used uneconomically.

160. Reinforcement to Prevent Cracks.—While longitudinal reinforcement is of little value in carrying loads, a small amount is nevertheless often desirable in preventing cracks and in binding the entire structure together. For a close beam spacing such reinforcement is hardly necessary, as the beam reinforcement itself thoroughly ties the structure longitudinally along the beam lines. For wide beam spacing it is more important. Just what amount of steel is needed is a matter of experience. The use of ¼-inch or ⅜-inch rods spaced about two feet apart is common practice. If a metal fabric is used for oblong panels, the longitudinal metal should be proportioned in accordance with the principles discussed in this and the preceding articles.

161. Floor-slabs Supported on Steel Beams.—Many "systems" have been developed of this type of construction, differing from each other in form of steel used, position of the concrete relative to the beam, use of curved or flat slabs, use of various kinds of hollow tile in connection with the concrete, etc. Sufficient examples only will be given to illustrate the principles involved; further information regarding the many systems can readily be had from trade catalogues.

Fig. 76 shows the floor placed directly on the tops of the beams. The reinforcement may be small rods or a meshwork of expanded metal or woven fabric. If reinforced as shown, the slab must be calculated as a simple beam, there being no reinforcement against negative moment over the support. For spans of considerable length some reinforcement for negative moments is desirable to secure economy and to

FIG. 76.

prevent cracks in the upper surface, although the lateral rigidity due to adjoining panels is of much assistance, as explained in Art. 156.

Fig. 77 represents a slab constructed after Hennebique's

FIG. 77.

system, to be supported by walls or steel beams. Small rods are used for reinforcement, every alternate rod being bent up and stirrups of flat steel looped on the straight rods. This is a very effective design to secure strength against shear or diagonal tensile stresses, but except where the floor-load is very heavy special shear reinforcement is hardly needed in floor-slabs. Fig. 78 shows a more common design of non-

FIG. 78.

continuous slab, the concrete being supported on the lower flange and the entire beam surrounded.

Fig. 79 shows a standard form of construction in which

FIG. 79.

the slab is practically continuous. The reinforcing material may be rods or a metal fabric continuous over several spans.

Figs. 80 and 81 show two forms in which a bar is hooked around the beam flange.

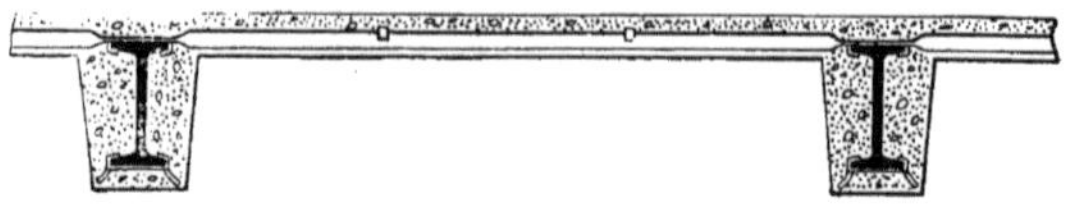

FIG. 80.

FIG. 81.

Many other forms are employed, some using a concrete arch with more or less reinforcement. In some, also, the concrete slab is brought down somewhat below the beam, giving a plane surface on the under side.

162. Floor-slabs in All-concrete Construction.—Where concrete beams are used the slab and beam are usually built simultaneously, giving a monolithic structure. The slab thus constitutes part of the beam, but to be effective these two parts must be well tied together. Where cross-beams are used the span of the slab will commonly range from 4 to 6 feet in length. For such short spans a reinforcement of rods or metal meshwork near the bottom only will be effective as explained in Art. 156. This reinforcement, if of rods, should be laid with lapped and broken joints to give continuity and to prevent the localization of contraction cracks in undesirable places (Fig. 82). The

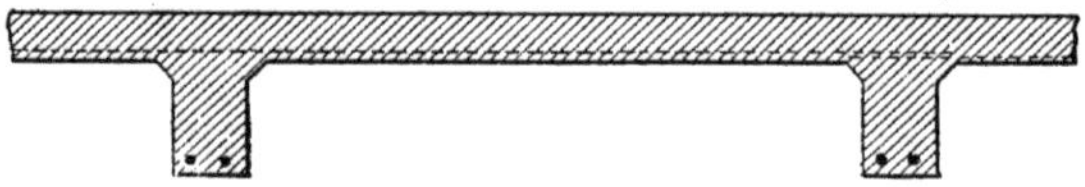

FIG. 82.

beam, if well bonded to the slab, will make a very rigid support comparable to the I-beam.

In the case of spans longer than 5 to 6 feet it becomes desirable to reinforce against negative moment. This may readily be done by bending up a part or all of the rods and extending the bent ends beyond the beam. Fig. 83 illustrates two arrangements of this kind. In either case the amount of steel at the

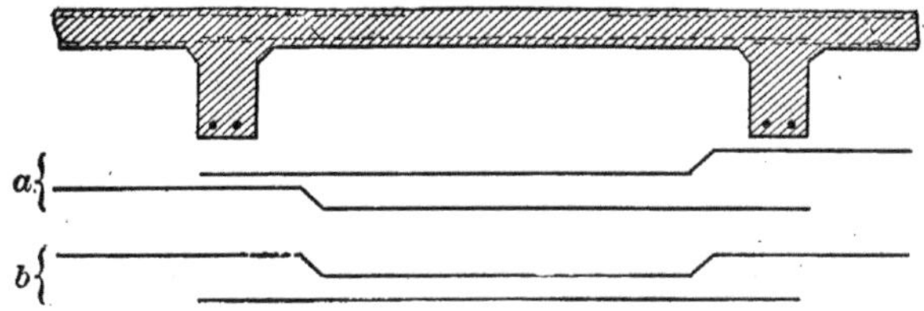

FIG. 83.

top above the beam is the same as at the bottom in the centre of the slab. The result may also be arrived at by using separate straight rods, as shown in Fig. 84. The plan of bent

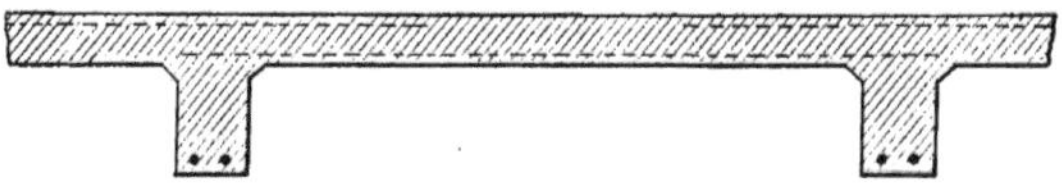

FIG. 84.

rods has a slight advantage as it reinforces somewhat against shearing failures, but this is not usually important in slabs. For very heavy loads, however, it becomes of importance, and the same care should be used as in the design of large beams.

Fig. 85 shows the Hennebique bent-rod and stirrup system applied to long-span slabs.

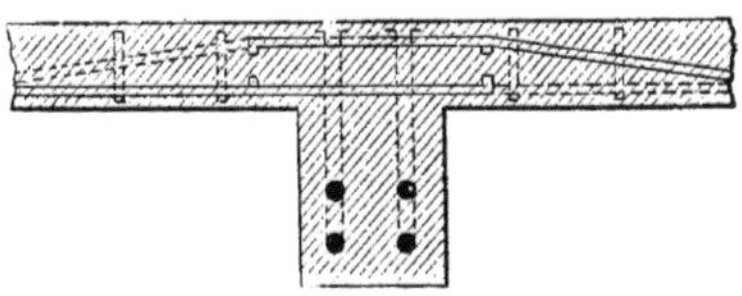

FIG. 85

The length of span over which negative moment is likely to exist may be estimated from Fig 72. It is seen that in the centre span of a three-span girder, where the dead load is one-half the live load, negative moment may, under extreme conditions, occur entirely across the beam. For long spans a top reinforcement at least to the third point will be desirable, but for short spans a less extensive reinforcement will be sufficient. The effect of a less amount of steel is discussed in Art. 165. Other examples of slab construction are shown in Art. 168.

163. Beams and Girders.—*Economical Arrangement.*—The arrangement of columns, girders, and beams is determined according to the same principles as in steel construction. The spacing of columns and girders will be determined largely by architectural considerations. The best spacing of cross-beams will differ in different cases. Where the spacing of girders is not too great (12 to 15 feet) and where cross-beams are not needed to secure lateral stiffness, it will be a question of omitting all cross-beams, of inserting them only at columns so as to form a square or nearly square panel, or of spacing them at closer intervals of 4 to 8 feet, using two or more to a girder-panel. The preceding analysis shows that double reinforcement will not be economical for oblong panels. Cross-beams, if used, should therefore be arranged to give very nearly square panels or else be spaced much more closely, designing the reinforcement so as to carry the entire load to the beams and thence to the girders.

If not otherwise needed, the use of cross-beams to secure square panels effects little if any saving. The amount of concrete will be less, but the amount of steel required will be more, and the extra beam will be more costly per unit volume than the slab. However, for the sake of lateral stiffness it will usually be desirable to place cross-beams at columns.

Where close spacing of beams is adopted the best arrangement depends upon the loading and the working stresses, as well as upon the cost of the material and forms. Heavy loads and low stresses call for large weights of concrete and tend to require the use of the material more in the form of deep ribs or beams, as the deeper the beam the greater its moment of resistance for a given volume. If cross-beams are used, a spacing greater than 10 or 12 feet or less than 4 or 5 feet will seldom be economical. Architectural considerations will often govern, and frequently building regulations relative to ratio of span to depth will control.

164. *Distribution of Floor-loads to Beams.*—Where the floor-slab is reinforced in one direction only the load will practically all be transmitted to the corresponding beams, but at the ends of the panels a small part will be transferred directly to the girder. This may be neglected in the calculations. In the case of reinforcement in two directions, unless the panel is nearly square, the load may still be assumed as all transferred to the side beams. If the panels are square, or nearly so, the distribution may be assumed in accordance with the discussion of Art. 158. Thus the load brought to point a' (Fig. 74) will be one-half of the area below the curve aOa',

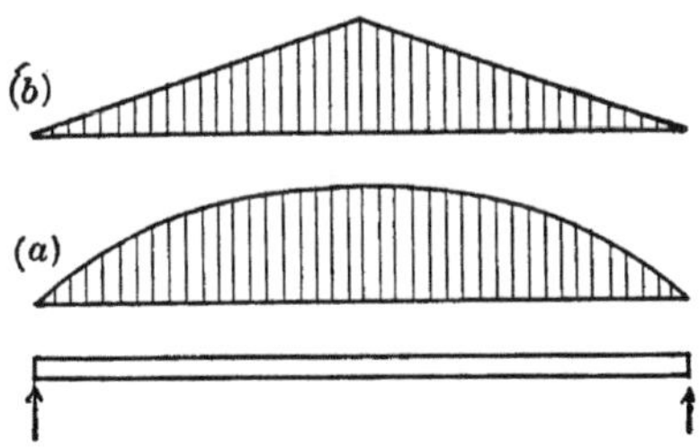

Fig. 86.

and the load brought to b' will be one-half the area below the curve $bG'b'$, etc. The distribution along the beam will then follow some such law as represented by the shaded area in Fig. 86 (a), the total load being necessarily $\frac{1}{4}wl$, where w = floor-load per square foot. It will be sufficiently accurate to assume this curve a parabola. The centre bending moment in the beam, assumed as a simple beam, will then be equal to

$$M = \frac{5}{128}wl^2.$$

A distribution of load as represented in Fig. 86 (b), as is sometimes assumed, gives a centre moment equal to $\frac{1}{24}wl^2$, a value about 7% higher than the above. A uniform distribution gives a moment equal to $\frac{1}{32}wl^2$, a value 20% lower.

165. *Design of Cross-beams.*—In the design of beams the chief features are the determination of the cross-section, the amount of steel and its make-up, provision for shearing stress, provision for negative bending moment and connections with slabs, other beams, and columns. The proportions of the beam, whether considered as a rectangular beam or as a T-beam, will be determined by considerations discussed in Chapter V. Ratios of depth to width greater than 2 or 2½ are seldom used. Requirements of head-room, space for rods, and shearing strength will limit the possible variations in proportions to a comparatively narrow range. Deep beams are economical of concrete but cost more for forms than do shallow beams.

If the beam may be calculated as a T-beam, the width of slab which may be counted on as a part of the beam is an important question. Specifications usually allow a width of six to ten times the thickness of the slab, but not to exceed the width between beams. As regards *strength* it would be very difficult to secure so thorough a reinforcement of web as to make it possible to crush a flange as much as four times the width of the web; the excessive shearing stresses in the web would cause failure. As regards *stiffness*, which controls the position of the neutral axis, the width of the slab to be counted as part of the beam may and should be taken relatively great. The width of flange being known, the design of the T-beam consists chiefly in the design of the web and the calculation of the steel cross-section. It will be only in the case of large girders that the compressive stress in the concrete will be a determining factor. Usually there is a large excess of material.

If the beam is to be considered as continuous over supports, the moment of resistance at the support must also be investigated. At this point the tension side is uppermost and the effective beam is now a *rectangular* beam. The maximum moment is about the same as at the centre, thus requiring about the same amount of steel at the top as is required in the centre of the span at the bottom. The maximum compression in the concrete will be greater than in the centre and will probably determine the size of beam required unless special provision is made for these stresses. This may be done by increasing the depth of the beam near the end, as shown in Fig. 87, or by the use of compressive reinforcement. Such reinforcement may be provided to a considerable extent by merely continuing the horizontal steel sufficiently to give the necessary bond strength (see Art. 123). If the horizontal steel near the end amounts to as much as 1% of the rectangular section, then considering both of two adjoining beams there would be available about 2% of compressive reinforcement. By Plate XII, p. 286, this amount would reduce the compressive concrete stresses about 42%. This would usually be sufficient. Inasmuch as a slight excess of stress at this

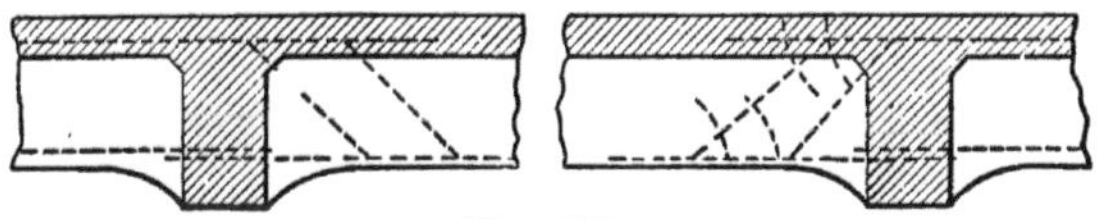

Fig. 87.

point does not in any way endanger the structure, merely increasing somewhat the positive moment on the beam, it would seem to be proper to permit the use of a higher working stress than at the center. An increase of 10–15% would be entirely safe. The necessary top steel at the end may be provided, as in the slab, by bending up a portion of the lower rods, or by using separate short rods, or by both methods combined. To provide thoroughly for negative moment the upper reinforcement should extend to about the third point, and in some cases still farther. Various arrangements of bent-up rods are illustrated in the examples cited in Art. 168.

It has been assumed in the determination of positive and negative bending moments in Art. 155 that the moment of inertia of the beam is uniform throughout. As there shown, the resulting maximum moments at center and support are not greatly different and for all practical purposes may be taken as equal, so that if fully reinforced the amount of steel and the moments of inertia will accord with the assumption. It is the practice of some designers, however, to consider the beam primarily as a simple beam and design it to carry all, or nearly all, of the load as such. A relatively small amount of steel is then placed in the top of the beam over the support, mainly to prevent objectionable cracks, but which is also in some cases counted upon to carry a portion of the moment. It is therefore of considerable importance to determine the actual moments and stresses which occur at the centre and support in such a case.

This problem has been concisely analyzed by Mr. P. E. Stevens,* and the following results are from his paper. He assumed a uniform moment of inertia, $=I_0$, for that portion

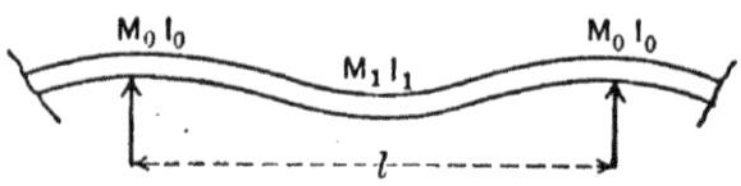

FIG. 87*a*.

over the support in which negative moments exist, and another moment of inertia, $=I_1$, for that portion of the central part of the span in which positive moments exist (see Fig. 87*a*). Let M_0=bending moment at support and M_1=bending moment at center. Then for a uniformly distributed load the ratios of these bending moments for various ratios of $I_0:I_1$ are as given in the table on page 321.

The table also gives in the third and fourth columns the values of these moments expressed as percentages of the centre moment, $\frac{1}{8}wl^2$, for a simple beam. In the last column are given the ratios of the corresponding unit stresses in the steel, f_0 and f_1, assuming the moments of inertia to be proportional to the amount of steel used. If, for example, the moment of inertia is the same throughout, the moment at the centre is $\times\frac{1}{8}wl^2$ and at the end is $\frac{2}{3}\times\frac{1}{8}wl^2$, as is well known. The

* Trans. Am. Soc. C. E., Vol. LX, 1908, p. 496.

MOMENTS AND STRESSES IN BEAMS WITH VARIABLE MOMENT OF INERTIA

Ratio, $I_0:I_1$.	Ratio, $M_0:M_1$.	$\frac{M_1}{\frac{1}{8}wl^2}$	$\frac{M_0}{\frac{1}{8}wl^2}$	Ratio of Stresses, $f_0:f_1$.
$\frac{1}{5}$	0.8	.55	.45	4.1
$\frac{1}{4}$	0.9	.52	.48	3.7
$\frac{1}{3}$	1.1	.48	.52	3.2
$\frac{1}{2}$	1.4	.42	.58	2.8
1	2.0	.33$\frac{1}{3}$	.66$\frac{2}{3}$	2.0
1$\frac{1}{2}$	2.6	.28	.72	1.7
2	3.0	.25	.75	1.5
6	5.8	.15	.85	1.0

amount of steel should then be based upon the end moment and made uniform. Again suppose the amount of steel at the support be made one-half that at the center and that the center be designed for $\frac{1}{10}wl^2 = .8 \times \frac{1}{8}wl^2$. The actual moment will be $\frac{42}{80}$ of the assumed moment and the stress at the center will be the same proportion of the assumed working stress. At the end the fiber stress will be $\frac{42}{80} \times 2.8 = 1.5$ times the assumed working stress. If a small amount of steel be placed at the end, such that $I_0/I_1 = \frac{1}{5}$, and the full bending moment provided for at the center, the stress at the center will be 55% of the working stress and at the end will be $4.1 \times .55 =$ 2.25 times the working stress.

The treatment of girders is the same as described for beams, it being especially important that the reinforcement pass well through the column.

The arrangement of shear or web reinforcement for beams and girders is of great importance, as it is in these forms where the web tensile stresses will be high. At points where the allowable shearing stress in the concrete is exceeded steel must be added in some form to carry a part of the stress, as explained in Art. 125. Where bent-up rods are used, as in Fig. 87, these rods aid greatly in carrying shear, and where not spaced too widely may be counted on to add perhaps 50% to the strength of the web. For thorough web reinforcement the stirrup is usually employed, or some form of bent bar closely spaced. This reinforcement may be calculated as explained in Art. 125, not too much reliance being placed on one or two bent rods. Web reinforcement will usually be needed only for the end quarter or third of the beam. Near the support, where the moment is negative, the tendency is for diagonal cracks to start at the top, while farther along the cracks tend to start at the bottom, as shown in Fig. 87. Stirrups at points of negative moment should loop about the upper bars, and at points of positive moment should loop about the lower bars. A correct appreciation of the diagonal stresses in such continuous beams is important.

The beam should be well bonded to the slab, especially near the end where the differential stresses between the two parts are large. This is well accomplished by means of the bent rods brought up as high as possible, and by means of the slab reinforcement which crosses the beam. Along the centre of the beam the matter is not of so great importance, but it is better to provide such bond by some form of vertical rein-

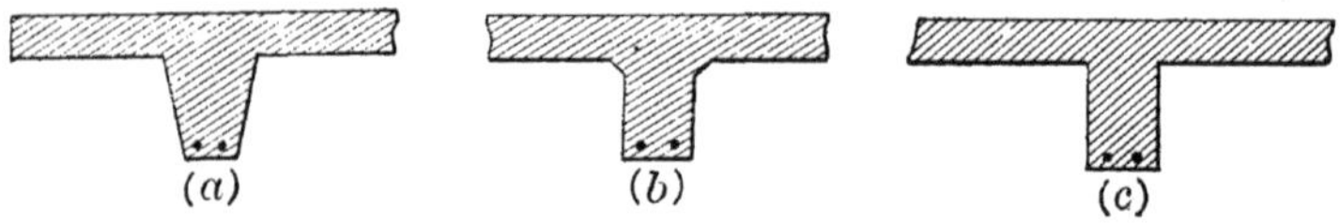

FIG. 88.

forcement, such as stirrups, extending up into the slab at occasional intervals. This is of especial importance in the case of girders where the main slab reinforcement runs parallel to the beam. A good bond is also more necessary the thinner the sections. Sections shown in Fig. 88, (*a*) and (*b*), are more favorable than such a section as in Fig. 88 (*c*). Sharp reentrant angles in such a brittle material as concrete are points of weakness, and where they exist a steel bond is desirable.

166. Columns.—There is little to be said here relative to column design. Much difference of opinion still exists as to the use of large or small quantities of steel and methods of calculation. A conservative course should be pursued in this matter, as the *columns* and *beams* in a reinforced structure are the vital parts of the structure. Working stresses in columns such as 700 or 800 lbs/in² should not be employed. Where large areas of steel are used, and figured at ordinary working stresses, such steel skeleton should not rely upon the concrete for rigidity. Concrete may, however, be relied upon to transmit loads from girders to columns. Where small areas of steel are used the rods should be well lapped at the floor-level, and those from the lower columns should extend upwards the full depth of the connecting beams. The rods should be well banded together by steel bands or large wire so as to hold all parts in position and to strengthen the column circumferentially. Unless such banding is spaced very closely it should not be counted upon, however, as "hooping." Brackets under all connecting girders are serviceable in stiffening the frame as well as in decreasing the stress in the girders. Rods of connecting girders should pass well through the columns.

167. *Eccentric Loads on Columns.*—Where loads are applied on free brackets or cantilevers the load is definitely eccentric, and the moment due to the same can readily be calculated. Moments are also caused in columns by unevenly loaded panels through the rigid beam connections. Assuming the beams rigidly fixed at the ends, a panel load on one without a load on the corresponding one on the opposite side will cause a bending moment in the beam at the column equal to $\frac{1}{12}pl^2$, where p = live load per lineal foot of beam. This moment is resisted mainly by the column and the members attached to it in the same plane as the loaded beam, and in proportion to their moments of inertia divided by their lengths.* If the two beams are about as rigid as the column, then the moment in the column above and below the floor will be about one-fourth of the given moment, $=\frac{1}{48}pl^2$. This indicates, roughly, what

* See Johnson's Framed Structures, Art. 154.

may be expected from unequally loaded floors. In the lower stories of a high building such a moment would be of little consequence, but in the upper floors it might add a large percentage to the column stress.

168. Examples of Floor and Column Design.—The following examples have been selected from published designs as representing good practice and as illustrating more or less specifically various features of design.

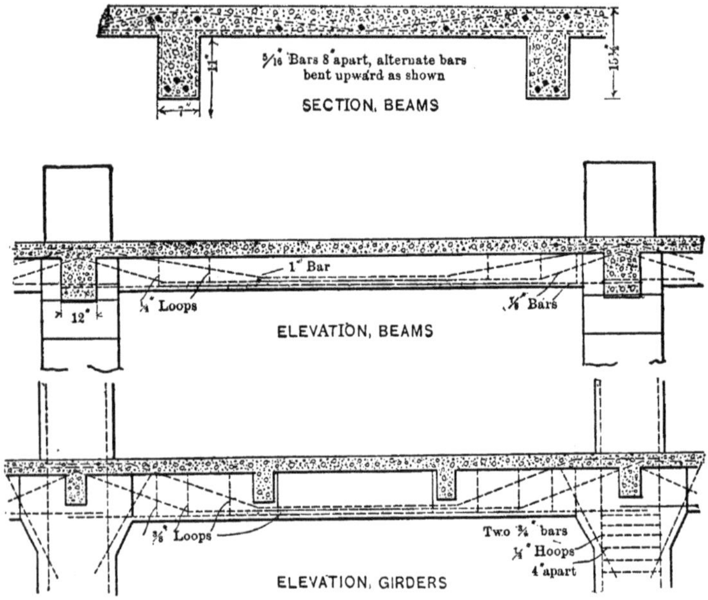

FIG. 89.—Details of the Robert Gair Factory, Brooklyn.

Fig. 89 illustrates the details of the Robert Gair factory, Brooklyn.* The reinforcement of the columns varies from eight $1\frac{1}{8}$-in. round rods at the base to four $\frac{3}{4}$-in. rods at the top. In the lower stories the bars are threaded and connected by sleeves. The rods are connected by hoops spaced from 4 to 10 in. apart. The girders are about 16 ft. apart, and the

* Eng. Record, Vol. 51, 1905, p. 279.

beams about one-third of this distance. Features of design to be noted are the brackets on the columns, bent rods and stirrups in the beams, bent rods in the slabs, and longitudinal reinforcement by the use of $\frac{5}{16}$-inch bars. The stirrups are rather widely spaced. The Ransome bar was used except in the columns.

Fig. 90 shows the details of the Park Square Garage, Boston.* The slabs are reinforced with expanded metal brought

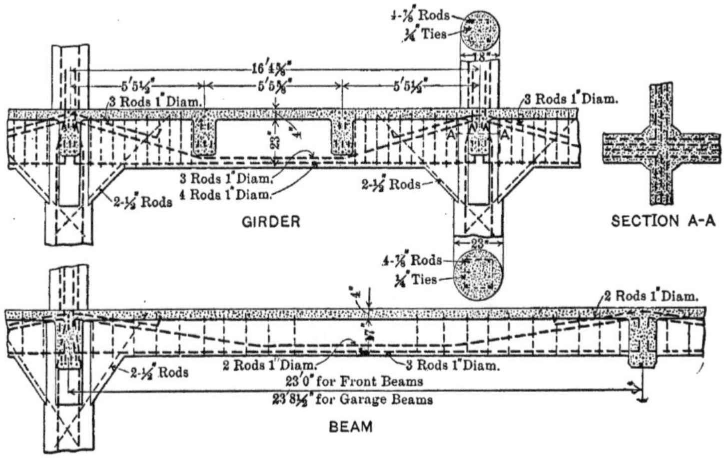

FIG. 90.—Details of the Park Square Garage, Boston.

near to the top surface over the beams. They were calculated as continuous girders. Beams and girders were calculated as T sections and figured on the basis of 375 lbs/in² compressive stress, and 30 lbs/in² shearing stress with no web reinforcement, and 100 lbs/in² with such reinforcement. The column reinforcement consisted of round rods from $2\frac{3}{4}$ in. to $\frac{7}{8}$ in. in diameter. They were banded by $\frac{1}{4}$ in. bands. The concrete used in the columns was $1:1\frac{1}{2}:3$; for the lower parts of the beams, $1:2\frac{1}{2}:4$; and for the upper parts of the beams and for slabs $1:2\frac{1}{2}:5$. The close spacing of the stirrups is noteworthy.

---

* Eng. Record, Vol. 52, 1905, p. 373.

Fig. 91 shows the construction for the Thompson and Morris factory, Brooklyn. Corrugated bars are used throughout.

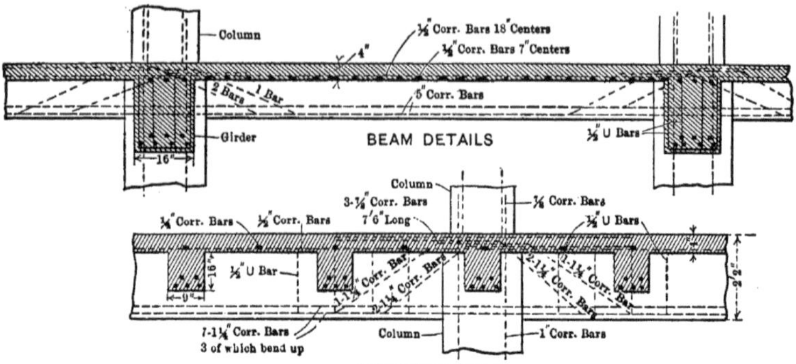

FIG. 91.—Details of the Thompson and Morris Factory, Brooklyn.

The girders are spaced 12 to 15 ft. apart and the beams 3 ft. 9 in. apart, four to a panel. The heavy reinforcement for negative moment should be noted.

Fig. 92 shows details of the Citizens' National Bank Building, Los Angeles, Cal.* Large girders connect the columns in both directions, forming panels 17 ft. by 22 ft. These panels are then subdivided into four smaller ones by cross-beams in both directions, a somewhat peculiar arrangement used probably for architectural effect. The slabs are reinforced both

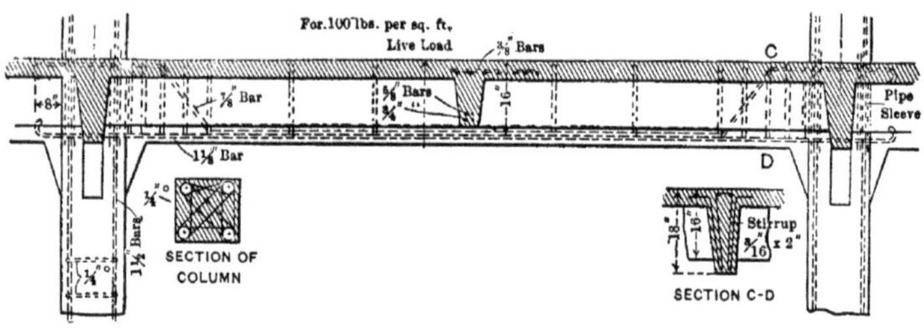

FIG. 92.—Details of the Citizens' National Bank Building, Los Angeles.

ways by $\frac{3}{8}$-in. twisted bars, $4\frac{1}{2}$ in. apart. The stirrups are flat bands $\frac{3}{16}'' \times 2''$ and spaced 18 ins. apart except near the end, as shown. They are looped about the rods in a very effective

* Eng. News, Vol. 56, 1906, p. 16

manner. Note the sleeve-splice for the column bars. The beams and columns are made of 1:2:3 concrete.

Fig. 93 shows a typical girder constructed with the Kahn bar illustrated in Fig. 7, Art. 33. By using inverted bars over the supports negative moment can be provided for, and at the same time additional shear reinforcement.

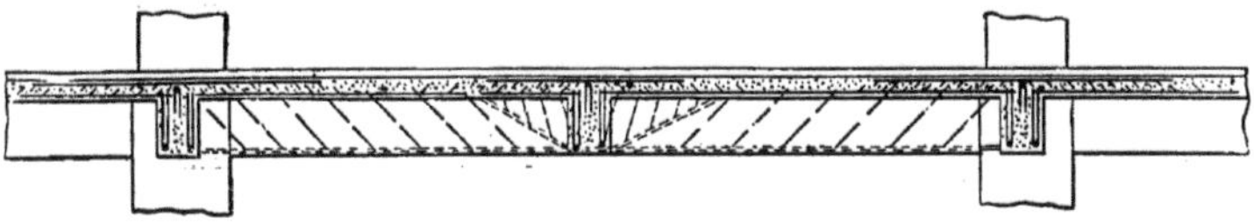

FIG. 93.—Reinforcement with the Kahn Bar.

In executing work a practical difficulty of considerable importance is that of placing and keeping all bars in their proper position until the concrete is in place. Very considerable labor is required in wiring bars in position, or in providing other means of support, and careful supervision is necessary during construction to see that they remain in place. To avoid these difficulties various arrangements have been devised for fastening together all, or a part, of the rods of a single span into a group which can be handled as a unit, giving rise to the so-called "unit frame". These units are obviously not so adaptable to a great variety of conditions as single independent bars, but their advantages are considerable and they are being used to quite an extent. Fig. 94 illustrates one such type of

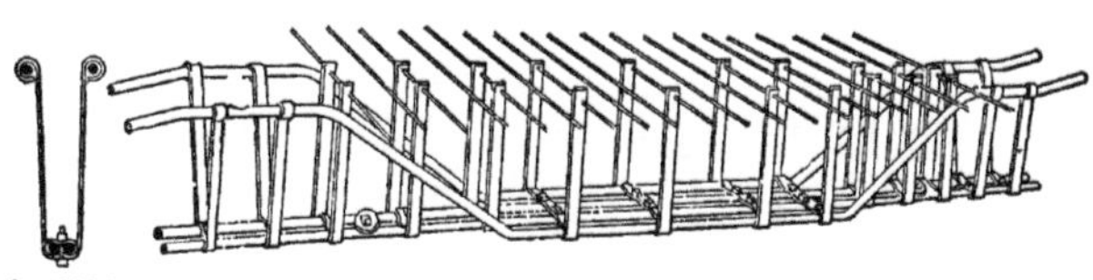

FIG. 94.—Unit Frame.

construction manufactured by the Unit Concrete Steel Frame Co. of Philadelphia, and has been used in several buildings. Some of the transverse slab rods pass through the upper ends of the stirrups as shown.

Fig. 95 illustrates a kind of unit reinforcement on the Bertine system and used in the warehouse of the Bush Terminal Co., Brooklyn.* Round rods are used and tied together by round steel stirrups.

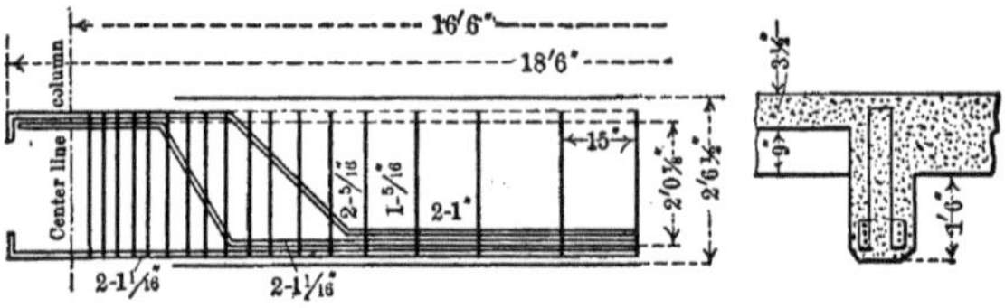

Fig. 95.—Unit Frame.

In all the examples here given it is to be noted that the differences of detail refer almost entirely to the method of caring for the shearing stresses and in handling the reinforcing members. The beam and slab arrangement is used in all.

A system of construction quite radically different from the foregoing is shown in Fig. 96, called the "mushroom" system, devised by Mr. C. A. P. Turner. No beams or ribs are used, the loads being transmitted from floor-slab directly to the column. The reinforcement is essentially radial and the column is enlarged at the top to increase the circumference at the line of maximum stress in the slab. The floor is of uniform thickness throughout.

To a certain extent this type of construction follows the natural lines of stress more closely than the rectangular ribbed-panel type; it is best adapted to large areas with few large openings.

The analysis of stresses in this system may be made approximately by the application of the method given in Art. 150, Chapter VI, and Plates X and XI. In applying this method to a continuous floor like the "mushroom" system, an estimate must first be made of the position of the "line" of inflexion with reference to the column. Noting that the point of inflexion of a beam fixed at the ends and uniformly

* Eng. Record, Vol. 53, 1906, p. 36.

loaded is about one-fifth the span length from the end, a sufficiently close estimate of the "line" of inflexion can be made. It will evidently be nearer the column than if the support were a continuous wall. Having estimated the line of inflexion the area within may be treated roughly as a circular plate loaded with the given uniform load on its area and a vertical load along its periphery equal to the remaining part of the load

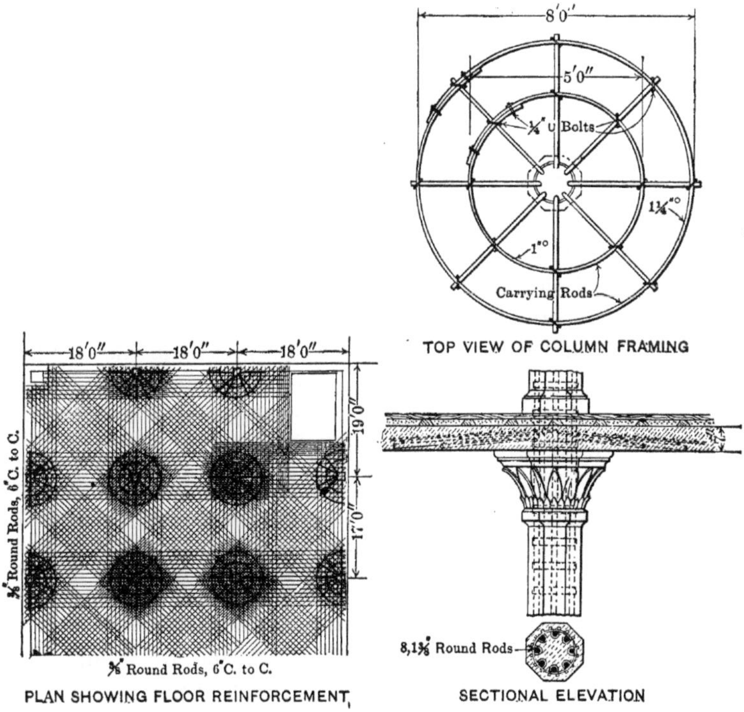

FIG. 96.—The "Mushroom" System.

tributary to the column. The diagrams then apply directly. Thus, suppose the columns are spaced 16 ft. apart and are 20 ins. in diameter at their upper ends. Suppose the load to be 150 lbs/ft^2 over the entire area. With columns 16 ft. apart, the diagonal spacing will be about 22.5 ft. The line of inflexion will probably not be less than 3 ft. nor more than 4 ft. from the column centre. Call it 3.5 ft. The area of this circle will be 38.5 sq. ft. The area of the entire square tributary to the column is $16 \times 16 = 256$ sq. ft. Hence the total load applied along the periphery will be $(256-38.5) \times 150 = 32,600$ lbs., which will be equal to 1480 lbs. per lineal ft. From Plates X and XI the value of M_1 and M_2 are found to be: (*a*) for the direct load of 150 lbs/ft^2, $M_1 = 280$ ft.-lbs., $M_2 = 1040$ ft.-lbs.; (*b*) for the peripheral load of 1480 lbs/ft, $M_1 = 2960$ ft.-lbs., $M_2 = 8650$ ft.-lbs. If r_1 had been assumed at 4 ft. the values of M_2 would have been 1500 and 9180 ft.-lbs., respectively. An increase in column diameter to 30 ins. would reduce the moments M_2 to about 980 and 6600 ft.-lbs. respectively, assuming a value of r_1 of 4 ft.

In the illustrations shown the columns have been mainly reinforced by longitudinal rods. Various types of banded or hooped columns are used more or less, but usually in connection with longitudinal reinforcement. From the discussion of Chapter IV it would seem that large amounts of hooped reinforcement should not be counted upon too greatly in the strength of the column. In some forms the columns are banded with spirally wound hooping, as in the Considère column, in others flat steel is used in riveted or welded hoops. Expanded metal is also used by wrapping around longitudinal bars.

169. Footings.—The problem of the design of footings is in general the same as that of floors. On account of the heavy concentrated loads and the large unit upward pressures of the earth against the footings the beam construction will be relatively heavy. The beams will be short and deep and will require special attention to provide against excessive shearing and bond stresses. For single footings of ordinary size a single symmetrical slab is most convenient. For larger footings and for footings carrying more than one column, a combination of beam and slab, similar to floor construction, is often most economical.

It is difficult to calculate accurately the stresses in a square footing, but assumptions may be made which will simplify the problem and give results well on the safe side (see Fig. 97).

As a general principle the pressures should be carried as directly as possible from the extremities to the centre. Two sets of main reinforcing rods aa' and bb' will then be used as shown in the figure. The reinforcing of the remaining corners can best be done by sets of diagonal rods dd'. If these cannot cover the area, then a few short cross-rods may be used. Reinforced in this way the total pressure on the area $ABCD$ may be assumed to be carried to the line BC, where the bending moment and shear will be a maximum. Figured as a free cantilever the resulting stresses will be higher than actually exist. If the entire square be reinforced by rods in

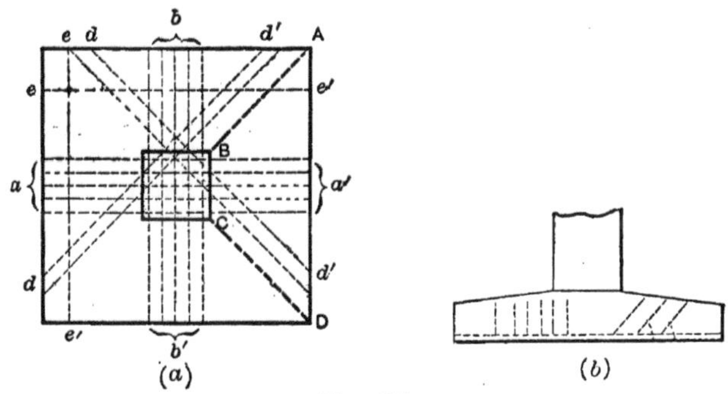

FIG. 97.

two directions only, as ee', then a considerable part of such rods in the corners of the square are ineffective.

The method of analysis of Art. 150, Chapter VI, may also be applied to this problem.

In the case of cantilever beams such as in footings the maximum shearing stress is near the centre where the maximum moment occurs. Shear cracks tend to form on the dotted curved lines, Fig. 97 (*b*). Bent rods, if used, must be bent up just outside the column, and not at the end of the beam, and stirrups must be spaced closely at this point. The beam being short it may require special attention to bond stress.

For large individual footings a beam and slab may be economical. To secure the benefit of a T section and to give a flat upper surface the beam may be placed under the slab as shown in Fig. 98. This arrangement requires some attention as to connection of slab to beam, as the upward pressure against the slab tends to pull it away from the beam. The use of an extra horizontal rod in the top of the main beam

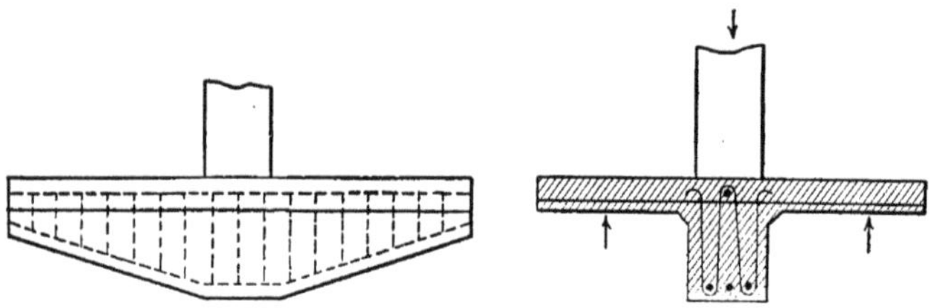

FIG. 98.

bonded by stirrups will give a thoroughly good anchorage for the transverse rods of the slab. For still larger areas a system of girders and beams may be adopted constituting a floor reversed as to loads.

170. Walls and Partitions.—The reinforcing of these parts is largely for the purpose of preventing cracks or of localizing them to desired lines. Where lateral pressures occur, of course the beam action must be considered. Walls are usually 3–6 inches thick and reinforced both ways with $\frac{1}{4}$- to $\frac{1}{2}$-in. rods, spaced about 2 feet apart.

CHAPTER VIII.

ARCHES.

171. Advantages of the Reinforced Arch.—If the loads on an arch were all fixed loads, it would be possible in any case to construct an arch ring so that the resultant pressure at all sections would intersect the centre of gravity of the section. The compressive stress at any section would then be uniformly distributed over the section, and the arch would be proportioned only for this uniform compression. The "line of pressure" would lie at the axis of the arch throughout. If, however, the arch ring is not made to fit the "line of pressure", or if part of the load is a live load, then the resultant pressure will not in general coincide with the axis of the arch. There will exist both bending and direct compression. If the resultant pressure and its position are known, the analysis of the stresses at any section is made in accordance with the method explained in Arts. 80–85, Chapter III.

In ordinary masonry or concrete arches tensile stresses are not permissible. The ring must therefore be designed so that the line of pressure will not pass outside the middle third. In reinforced arches this limitation does not exist. The arch rib is a beam, and if properly reinforced it may carry heavy bending moments involving tensile stresses in the steel.

Theoretically the gain in economy by the use of steel in a concrete arch is not great. If the pressure line does not depart from the middle third, the steel reinforces only in compression and in this respect is not as economical as concrete. If the line of pressure deviates farther from the centre, resulting in tensile stresses in the steel, the conditions are such that those stresses must be provided for by the use of the steel at very low working values. That is to say, the direct compression in the arch is so large a factor that the limiting stresses in the concrete will always result in very small unit tensile stresses in the steel where any tension exists at all.

Practically the value of reinforcement is very considerable. It renders an arch a much more secure and reliable structure, it greatly aids in preventing cracks due to any slight settlement, and by furnishing a form of construction of greater reliability makes possible the use of working stresses in the concrete considerably higher than is usual in plain masonry. Furthermore, in long-span arches where the dead load constitutes by far the larger part of the load, any possible increase in average working stress counts greatly towards economy. It affects not only the arch but the abutments and foundations.

172. Methods of Reinforcement.—The reinforcement of arches is arranged in various ways. Since the arch is a beam subject to either positive or negative bending moments it is essential that it should be reinforced on both sides, but the shearing stresses due to beam action are relatively small, so that little is needed in the way of web reinforcement. The arch is also subjected to heavy compression, so that it is desirable that the inner and outer reinforcement be tied together, somewhat as in a column, although in this case the necessity therefor is much less.

A large proportion of the arches which have been constructed have been built according to some one of the various "systems" that have been devised. The most important of these systems are the Monier and the Melan. In the Monier system, invented about 1865, the reinforcement consists of wire netting, one net being placed near the intrados and one near the extrados. The longitudinal wires are made smaller than those following the arch ring, as they serve only to aid in equalizing the load and in preventing cracks. A large number of bridges have been built in Europe on this system.

In the Melan type, invented about 1890, the steel is in

the form of ribs of rolled I sections, or of built-up lattice girders, which are spaced two to three feet apart. The flanges constitute the principal reinforcement, but the web enables the steel frame to be self-supporting and to carry shearing stresses, and in the open lattice type it furnishes a good bond with the concrete. The Melan arch has been built extensively in this country, largely under the direction of Mr. Edwin Thacher.

Many arches are now being constructed in which reinforcing bars of any satisfactory form are employed without reference to any particular system, being used in accordance with the requirements of the case. The problem of reinforcement is quite as simple as in a beam, after the moments and thrusts in the arch have been found.

ANALYSIS OF THE ARCH.

173. General Method of Procedure.—The method of analysis presented here is based on the elastic theory and is of general application to arches of variable moment of inertia and loaded in any manner. It is mainly an algebraic method, although certain simple graphical aids may be used advantageously. It necessarily assumes that a preliminary design has been made by the aid of approximate or empirical rules or by reference to the proportions of existing arches. This arch is then exactly analyzed and the results used in correcting the design; the corrected design may then in turn be analyzed if it departs too greatly from the one first assumed. A discussion of the various rules for thickness of crown and form of arch will not be entered upon here. For this information the reader is referred to the various treatises on the arch, and especially to those of Professor Cain and Professor M. A. Howe. The work of Professor Howe on "Symmetrical Masonry Arches" * contains a very useful table of data of existing masonry and reinforced-concrete arches.

* New York, 1906.

The analysis of an arch consists in the determination of the forces acting at any section, usually expressed as the *thrust*, the *shear* and the *bending moment*, at such section. The thrust is here taken to be the component of the resultant parallel to the arch axis at the given point, and the shear is the component at right angles to such axis. The thrust causes simple compressive stresses; the shear causes stresses similar to those produced by the vertical shear in a simple beam.

The method of procedure will be to determine, first, the thrust, shear, and bending moment at the crown. These being known, the values of similar quantities for any other section can readily be determined. A length of arch of one unit will be considered.

174. Thrust, Shear, and Moment at the Crown (H_0, V_0, M_0).

Notation. (See Fig. 100.)

Let H_0 = thrust at the crown;

V_0 = shear at the crown;

M_0 = bending moment at the crown, assumed as positive when causing compression in the upper fibres;

N, V, and M = thrust, shear, and moment at any other section;

R = resultant pressure at any section = resultant of N and V;

δs = length of a division of the arch ring measured along the arch axis;

n = number of divisions in one-half of the arch;

I = moment of inertia of any section = $I_{\text{concrete}} + nI_{\text{steel}}$ (see p. 92);

P = any load on the arch;

x, y = co-ordinates of any point on the arch axis referred to the crown as origin, and all to be considered as positive in sign;

m = bending moment at any point in the cantilever, Fig. 100, due to external loads.

Let AB, Fig. 99, be a symmetrical arch loaded in any manner with loads P_1, P_2, etc. Divide the arch into an even number of divisions (ten to twenty usually), making the divisions of such a length that the ratio $\delta s:I$ will be constant. This may be done by trial or by the more direct method explained in Art. 178. Mark the centre point of each division and num-

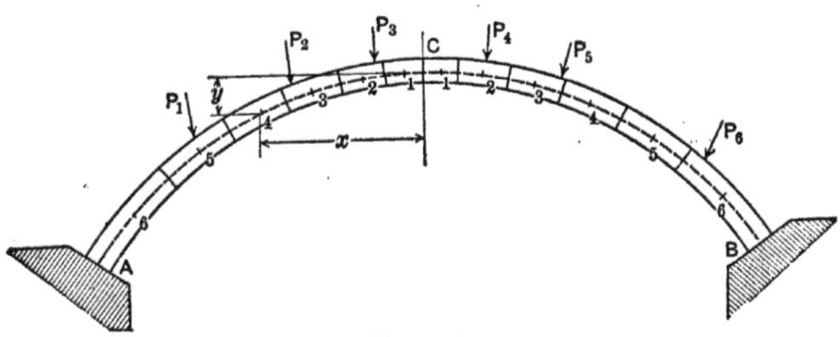

FIG. 99.

ber the points as shown. Consider the arch to be cut at the crown and each half to act as a cantilever subjected to exactly the same forces as exist in the arch itself, that is, the given loads, the reactions, and the stresses at the crown, represented

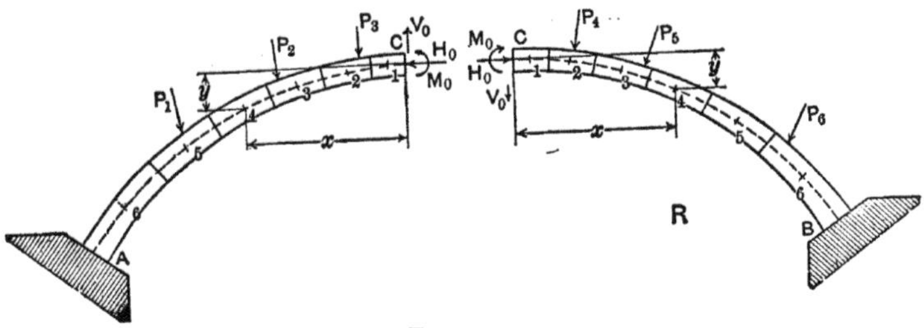

FIG. 100.

by H_0, V_0, and M_0 (Fig. 100). H_0 is applied at the gravity axis.

Let m=bending moment at any point, 1, 2, 3, etc., due to the given external loads P (considering the arch as two cantilevers). Denote by m_R the moments in the right half and by m_L those in the left half of the arch. All the moments m

will be negative. The values of H_0, V_0, and M_0 will then be given by the following equations:

$$H_0=\frac{n\Sigma my-\Sigma m\Sigma y}{2[(\Sigma y)^2-n\Sigma y^2]}, \quad \ldots \ldots \quad (1)$$

$$V_0=\frac{\Sigma(m_R-m_L)x}{2\Sigma x^2}, \quad \ldots \ldots \ldots \quad (2)$$

$$M_0=-\frac{\Sigma m+2H_0\Sigma y}{2n}. \quad \ldots \ldots \quad (3)$$

In these equations the summations Σy, Σy^2, and Σx^2 are for one-half of the arch only; the summation Σm is for the entire arch and is equal to $\Sigma m_R+\Sigma m_L$; the summation $\Sigma(m_R-m_L)x$ is a summation of the products $(m_R-m_L)x$, in which m_R and m_L are the bending moments at corresponding points in the right and left halves which have equal abscissas x; and the summation Σmy is for the entire arch, but since symmetrical points have equal y's this quantity may be calculated as $\Sigma(m_R+m_L)y$. A positive result for V_0 indicates action as shown in Fig. 100. All quantities are readily calculated Distances should be scaled and quantities tabulated.*

* *Demonstration.*—Consider the left-hand cantilever of Fig. 100. Under the forces acting the point C will deflect and the tangent to the axis at this point will change direction (the abutment at A being fixed). Let Δy, Δx, and $\Delta\phi$ be, respectively, the vertical and horizontal components of this motion and the change in angle of the tangent. Then according to the principles relating to curved beams [1] we have the values

$$\Delta y=\Sigma Mx\frac{\delta s}{EI}, \quad \Delta x=\Sigma My\frac{\delta s}{EI}, \quad \text{and} \quad \Delta\phi=\Sigma M\frac{\delta s}{EI}, \quad \ldots \quad (a)$$

in which the various quantities have the same significance as in Art. 174.

In like manner, referring to the right cantilever, let $\Delta y'$, $\Delta x'$, and $\Delta\phi'$ represent the components of the movement of C and the change of angle of the tangent. These may be expressed in terms similar to Eq. (a).

Now evidently

$$\Delta y=\Delta y', \quad \Delta x=-\Delta x', \quad \text{and} \quad \Delta\phi=-\Delta\phi'. \quad \ldots \ldots \quad (b)$$

Furthermore, since $\delta s/I$ is constant and likewise E, the quantity $\delta s/EI$ may be placed outside the summation sign.

[1] See Church's Mechanics, or Johnson's Framed Structures, p. 236.

175. Thrust, Shear, and Moment at any Section.—The values of H_0, V_0, and M_0 having been found, the total bending moment at any section, 1, 2, etc., is

$$M=m+M_0+H_0y\pm V_0x. \quad . \quad . \quad . \quad . \quad (4)$$

The plus sign is to be used for the left half and the minus sign for the right half of the arch.

The resultant pressure, R, at any section of the arch is equal in magnitude to the combined resultant of the external loads between the crown and the section in question, and the forces H_0 and V_0. These resultants can best be found graphically. The thrust, N, is the component of the resultant, R, parallel to the arch axis and the shear, V, is the component perpendicular to this axis.

176. Partial Graphical Calculation.—Where the loads are vertical the calculation of the quantities m can be advantageously performed by means of an equilibrium polygon as follows:

Let AB, Fig. 101, represent the arch axis. The load line is a-c-b. Select any convenient pole O on a horizontal line through

---

Using the subscript L to denote left side and R to denote right side we then derive the relations

$$\left.\begin{aligned}\Sigma M_L x &= \Sigma M_R x,\\ \Sigma M_L y &= -\Sigma M_R y,\\ \Sigma M_L &= -\Sigma M_R.\end{aligned}\right\} \quad . \quad . \quad . \quad . \quad . \quad . \quad . \quad . \quad . \quad (c)$$

The moment M may in general be expressed in terms of known and unknown quantities thus:

$$M_L=m_L+M_0+H_0y+V_0x \text{ for the left side}$$

and

$$M_R=m_R+M_0+H_0y-V_0x \text{ for the right side.}$$

Hence, substituting in (c) and combining terms, and noting that ΣM_0 for one half is equal to nM_0, we have

$$\Sigma m_L x-\Sigma m_R x+2V_0\Sigma x^2=0, \quad . \quad . \quad . \quad . \quad . \quad . \quad . \quad . \quad . \quad (d)$$
$$\Sigma m_L y+\Sigma m_R y+2M_0\Sigma y+2H_0\Sigma y^2=0, \quad . \quad . \quad . \quad . \quad . \quad . \quad (e)$$
$$\Sigma m_L+\Sigma m_R+2nM_0+2H_0\Sigma y=0, \quad . \quad . \quad . \quad . \quad . \quad . \quad (f)$$

From Eq. (d) is obtained Eq. (2), p. 268; and from Eqs. (e) and (f) are obtained Eqs. (1) and (3), noting that $\Sigma m_L+\Sigma m_R=\Sigma m$, and $\Sigma m_L y+\Sigma m_R y=\Sigma my$.

the point c, at the junction of loads P_3 and P_4, the loads adjacent to the crown C. Construct the equilibrium polygon $efgh$, producing to i and k the segment fg between loads P_3 and P_4. Drop verticals from the points 1, 2, 3, etc. The desired bending moments m, at the several points, will then be equal to the corresponding intercepts z_2, z_3, etc., on these verticals between the equilibrium polygon and the line ik, multiplied by the pole distance H; or in general $m=Hz$.

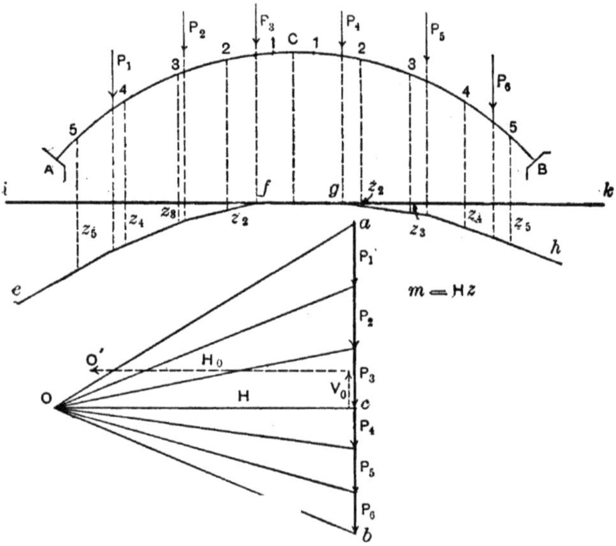

FIG. 101.

Finally, after the values of H_0, V_0, and M_0 are found by Eqs. (1), (2), and (3), the true equilibrium polygon can be drawn, if desired, and values of thrust, shear, and moment at various points determined graphically. The true pole is found by laying off V_0 and H_0 from the point c. The distance of the equilibrium polygon above or below the axis at the crown is equal to M_0/H_0. It lies above the axis if the result is positive and below if negative. The equilibrium polygon is then drawn from the crown each way towards the ends.

Where the loads are inclined at various angles it is still possible to use a graphical process for getting values of m, but there is little to be gained in such a case. After the values of H_0, V_0, and M_0 are found, however, it will be expedient to draw a final equilibrium polygon, or line of pressure, as explained above.

177. General Observations.—The method of analysis just described is brief, general, and easily followed. The arithmetical calculations are not longer than those required in the usual graphical process, while the graphical aids here suggested are of the simplest character.

The loads and their points of application have been considered apart from the divisions of the arch ring, as the two things are in no wise related. Where no spandrel arches are used and the entire load is applied continuously along the arch ring, the load may for convenience be divided to correspond with the arch divisions and applied at the center points, 1, 2, 3, etc. This division is, however, of no importance, the only requirement being a sufficiently small subdivision of the arch ring and of the load so that the errors of approximation will be negligible. Where spandrel arches are used, the live load and a large part of the dead load will be applied at the centers of the arch piers. The weight of the main arch ring may also be considered as concentrated at these same points.

If calculations are to be made for more than one loading it will be noted that the denominators of the values for H_0, V_0, and M_0 do not change. The quantities involving m are the only ones requiring recalculation, and if the load on but one-half of the arch is changed, then the values of m for that half only need be recalculated. In the case of a symmetrical loading, or a load on one-half only, the calculation of m is also necessary for one-half the arch only. For symmetrical loads, $V_0=0$.

178. Division of Arch Ring to give Constant $\delta s/I$.—In most cases the depth of the arch ring increases from crown towards springing line, giving a variable moment of inertia. Considering the concrete only, the moment of inertia will increase as d^3 so that a comparatively small change in depth will cause a large change in moment of inertia. To maintain $\delta s/I$ constant, the value of δs will therefore be much greater near the springing line than at the crown, and hence to secure the desired accuracy the length of division at the crown will need to be made fairly short. The value of $\delta s/I$ to adopt so that there will be no fractional division may be determined as follows:

Let $i=\frac{1}{I}$;

i_a = mean value of i;

s = half length of the arch ring measured along the axis;

n = desired number of divisions in one-half the arch.

Calculate first the mean value of i for the half arch ring by determining several values at equal intervals along the arch. Then the desired value of $\delta s/I$ is

$$\frac{\delta s}{I}=\frac{si_a}{n}. \qquad (5)$$

The value of $\delta s/I$ being known, the proper length of δs for any part of the arch ring can readily be determined. Beginning at the crown, the length of the first division is determined, then the second, third, etc., to the end. The length of a division not being exactly known beforehand, the value of I for that division will not be exactly known, but the necessary adjustment is very simple.

In determining the value of I the steel reinforcement must be duly considered.

179. Temperature Stresses.—For a rise of temperature of t degrees the increase in span-length, if the arch be not restrained, would be equal to ctl, where c = coefficient of expansion and l=span. The restraint of the abutments induces a thrust H_0 at the crown given by the equation

$$H_0=\frac{EI}{\delta s}\cdot\frac{ctln}{2[n\Sigma y^2-(\Sigma y)^2]}.\ ^*\quad\ldots\ldots\quad(6)$$

The summations refer to one-half the arch.

Having determined H_0, then

$$M_0=-\frac{H_0\Sigma y}{n}.\ ^*\quad\ldots\ldots\quad(7)$$

The bending moment at any point is

$$M=M_0+H_0y.\quad\ldots\ldots\quad(8)$$

The thrust and shear at any point in the arch are found by resolving H_0 parallel and normal to the arch axis at that point. Graphically, the true equilibrium polygon is a horizontal line drawn a distance below the crown equal to $M_0/H_0 = \Sigma y/n$.

180. Stresses Due to Shortening of Arch from Thrust.—A thrust throughout the arch producing an average stress on the concrete equal to f_c lbs/in² would shorten the arch span an amount equal to f_cl/E if unrestrained. This action develops horizontal reactions in the same manner as a *lowering* of tem-

* *Demonstration.*—For temperature stresses $\Delta\phi$ of Eq. (*a*), p. 338, is zero and Δx is equal to the change in length of the half-span, $=\frac{ctl}{2}$. We therefore have

$$\Sigma M_L y\frac{\delta s}{EI}=\frac{ctl}{2},$$

and

$$\Sigma M_L=0.$$

In this case, there being no external loads, $m=0$, and from symmetry, $V_0=0$, hence $M=M_0+H_0y$. Substituting this value of M in the above equations we have

$$M_0\Sigma y+H_0\Sigma y^2=\frac{ctl}{2}\cdot\frac{EI}{\delta s},$$

and

$$nM_0+H_0\Sigma y=0.$$

From these are readily derived Eqs. (6) and (7).

perature. The value of the resulting reactions, or the crown thrust, may then be found by substituting $f_c l/E$ for ctl of Eq. (6). There results

$$H_0 = -\frac{I}{\delta s} \cdot \frac{f_c l n}{2[n\Sigma y^2 - (\Sigma y)^2]}. \quad . \quad . \quad . \quad . \quad (9)$$

The moments at crown and elsewhere are given by Eqs. (7) and (8), using the value of H_0 from Eq. (9).

The thrusts and moments due to arch shortening will not usually be large. They may be applied as corrections to the thrusts and moments found before.

181. Deflection of the Crown.—The downward deflection of the crown under a load is given by Eq. (*a*), p. 268. It is

$$\Delta y = -\frac{\delta s}{EI} \Sigma Mx. \quad . \quad . \quad . \quad . \quad . \quad . \quad . \quad (10)$$

If M is not determined for all points, use the value of M from Eq. (4), deriving

$$\Delta y = -\frac{\delta s}{EI}[\Sigma mx + M_0 \Sigma x + H_0 \Sigma xy + V_0 \Sigma x^2]. \quad . \quad (11)$$

The summations are for one-half only.

The *rise* of crown due to an increase of temperature is obtained from Eq. (11) by substituting from Eqs. (6) and (7). There results

$$\Delta y = \frac{ctl}{2} \cdot \frac{n\Sigma xy - \Sigma x \Sigma y}{n\Sigma y^2 - (\Sigma y)^2}. \quad . \quad . \quad . \quad . \quad . \quad . \quad (12)$$

182. Unsymmetrical Arches.—If the arch is unsymmetrical the value of $\delta s/I$ should be made constant for the entire arch, and the number of divisions made even as before. The central point of the arch with reference to the *number* of divisions may then be taken as the crown, and the X-axis made tangent to the arch at this point. The two halves of the arch are now unlike and all terms resulting from substitution in Eq. (*c*), p. 269, must be retained. Explicit formulas for H_0,

V_0, and M_0 are very cumbersome, but the three equations derived from (c) are as follows:

$$(\Sigma_L x - \Sigma_R x)M_0 + (\Sigma_L xy - \Sigma_R xy)H_0 + \Sigma x^2 V_0 = \Sigma_R mx - \Sigma_L mx, \quad (13)$$

$$\Sigma y M_0 + \Sigma y^2 H_0 + (\Sigma_L xy - \Sigma_R xy)V_0 = -\Sigma my, \quad . \quad . \quad (14)$$

$$2nM_0 + H_0 \Sigma y + (\Sigma_L x - \Sigma_R x)V_0 = -\Sigma m. \quad . \quad . \quad (15)$$

Where no subscript is used the summation is for the entire arch. Numerical values of the coefficients of M_0, H_0, and V_0 should be obtained and the equations then solved.

183. Application of the Preceding Theory.—*Example 1.*—An arch ring will be assumed of the dimensions shown in Fig. 102. Span length with reference to the axis = 30 ft., rise = 8 ft. Thickness at crown = 1 ft., at springing lines = 1 ft. 6 in. For a small arch such as this great accuracy is not needed, hence a small number of divisions may be used. The number will be four for each half. These divisions are determined so that $\delta s/I$ is constant. The loads are applied at the centre-points 1, 2, 3, 4, and are assumed to be somewhat inclined (excepting loads P_4 and P_5), the several vertical and horizontal components being as given in the figure.

TABLE A.

CALCULATIONS FOR H_0, V_0, AND M_0.

1	2	3	4	5	6	7	8	9
Point.	x	y	x^2	y^2	m_L	m_R	$(m_L+m_R)y$	$(m_R-m_L)x$
1	1.55	.09	2.40	.01	0	0	0	0
2	4.90	.68	24.01	.46	− 13,840	− 8,640	− 15,300	+ 25,500
3	8.45	2.10	71.40	4.41	− 46,800	− 29,560	− 160,400	+ 145,700
4	12.85	5.35	165.12	28.62	−116,680	− 75,490	−1,028,100	+ 529,300
Σ		8.22	262.93	33.50	−291,010		−1,203,800	+ 700,500
Springing.	15.00	8.00			−180,820	−120,640		

Eq. (1) gives $H_0 = \dfrac{4(-1203800) - (-291010 \times 8.22)}{2[(8.22)^2 - 4 \times 33.50]} = +\ 18,230$ lbs.

Eq. (2) gives $V_0 = \dfrac{700500}{2 \times 262.93} = +\ 1,330$ lbs.

Eq. (3) gives $M_0 = -\dfrac{-291010 + 2 \times 18230 \times 8.22}{2 \times 4} = -1,090$ ft-lbs.

TABLE B.

BENDING MOMENTS, THRUSTS, AND ECCENTRIC DISTANCES.

1	2	3	4	5	6	7	8	9
Point.	H_0y	V_0x	Bending Moment M.		Thrusts.		Eccentric Distances.	
			Left.	Right.	Left.	Right.	Left.	Right.
1	1,640	2,060	+2,610	−1,520	18,450	18,640	+.14	−.08
2	12,400	6,530	+4,000	−3,860	19,580	19,310	+.21	−.20
3	38,300	11,260	+1,650	−3,620	22,050	20,770	+.07	−.17
4	97,500	17,120	−3,100	+3,850	28,800	24,970	−.11	+.15
Springing.	145,900	20,000	−16,070	+4,140	28,800	24,970	−.56	+.17

The calculations of the several quantities in the formulas for H_0, V_0, and M_0 (p. 338) are given in Table A. The co-ordinates x, y of the several points are given in cols. 2 and 3; then x^2 and y^2 in cols. 4 and 5; then in cols. 6 and 7 are given the quantities m_L and m_R, considering each half-arch a cantilever. These are readily calculated. Thus, on the left, for point 1, $m=0$; for point 2, $m=4130\times(4.90-1.55)=13{,}840$; for point 3, $m=4130\times(8.45-1.55)+5035\times(8.45-4.90)+310\times(2.10-.68)=46{,}800$; and for point 4, $m=4130\times(12.85-1.55)+5035\times(12.85-4.90)+5950\times(12.85-8.45)+310\times(5.35-.68)+725(5.35-2.10)=116{,}680$. The value of m at the springing line is also calculated and placed in this table for future use. The moments on the right are similarly found. All moments m are negative. In cols. 8 and 9 are then given the products $(m_L+m_R)y$ and $(m_R-m_L)x$.

Substituting in Eqs. (1), (2), and (3), p. 268, there are obtained the values for H_0, V_0, and M_0 given below the table.

The values of the bending moments, thrusts, and shears at any point may now be found either graphically or algebraically. The force-diagram method will be much the better for obtaining thrusts and shears; the moments may then be obtained either by constructing an equilibrium polygon or by the application of Eq. (4), p. 339.

In Fig. 102 the graphical construction is given. The load-line is *a-c-b*. The true pole is found by laying off $V_0 = +1330$ from point c (at the junction of the loads adjacent to the crown, P_4 and P_5); then $H_0 = 18{,}230$ horizontally to O. The

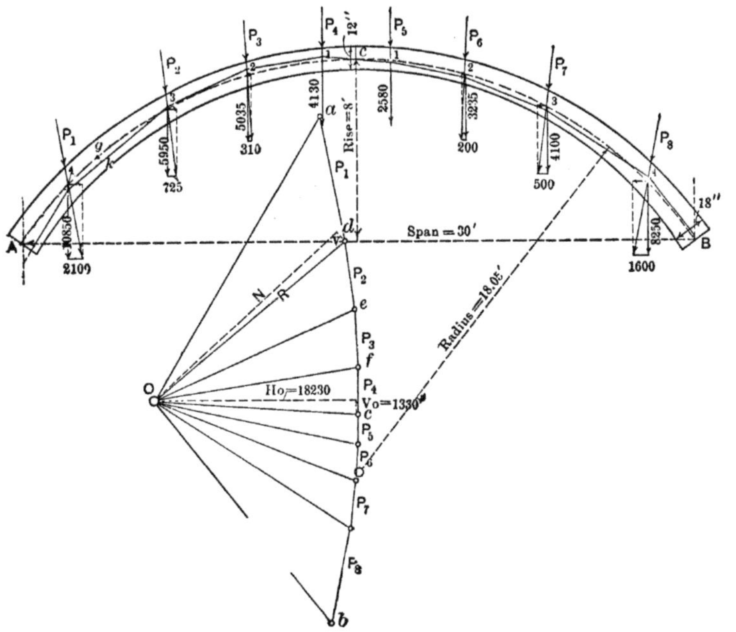

FIG. 102.

force diagram is drawn and then the equilibrium polygon, beginning at the crown and drawing the segment 1−1 at a distance *below* the crown equal to 1090/18230 = .06 ft. The resultant, R, acting at any section may be scaled from the force polygon, and the moment at any point will be equal to this resultant multiplied by the perpendicular distance at that point from the arch axis to the equilibrium polygon. For example, the bending moment at point g is equal to the force Od multiplied by the arm gk. The tangential component of the resultant R (the true thrust N) may be found by resolving the force R parallel and normal to the arch axis at the point in question. In most cases the thrust N may be taken as equal to R. The shear V will be the normal component of R; it will not usually require consideration.

Table B contains calculated values of moments and eccentric distances for points 1, 2, 3, 4 and the springing lines. The moments are calculated from the formula (Eq. (4)) $M=m+M_0+H_0y\pm V_0x$. The quantities m are obtained from Table A, cols. 6 and 7. The thrusts are scaled from the force polygon, being in each case the thrust on the abutment side of the point in question. The eccentric distances are equal to the moments divided by the thrusts; they are of use in calculating stresses in the arch. Obviously the bending moment at any other point, such as g, may be calculated in the same way as those here given.

184. *Example 2.* (Fig. 103.)—For another example an arch will be assumed of 100-ft. span and 20-ft. rise; thickness at crown = 30 in., thickness at springing line = 42 in. It will be assumed that the roadway is supported on spandrel piers 10 ft. apart, thus concentrating most of the load at points 10 ft. apart as shown; the weight of the arch ring will also be assumed as applied at these points. The loads as given represent an arch with live load on the left half only. The half arch is divided into ten divisions, making $\delta s/I$ constant. The loads in this case are vertical, so that the graphical method may be used to advantage in determining the cantilever moments m. The load line a-c-b is drawn and a pole O' selected on a horizontal line through c at the center of the crown load P_5. The pole distance is H. An equilibrium polygon, $efgh$, is then drawn, and the moments m will be equal to the intercepts, z, from this polygon to the horizontal line ik, multiplied by H. These moments, and the remainder of the calculations for H_0, V_0, and M_0 are given in Table C. The true pole O is then found as before and the correct equilibrium polygon drawn. The thrusts are then scaled from the force polygon, and the eccentric

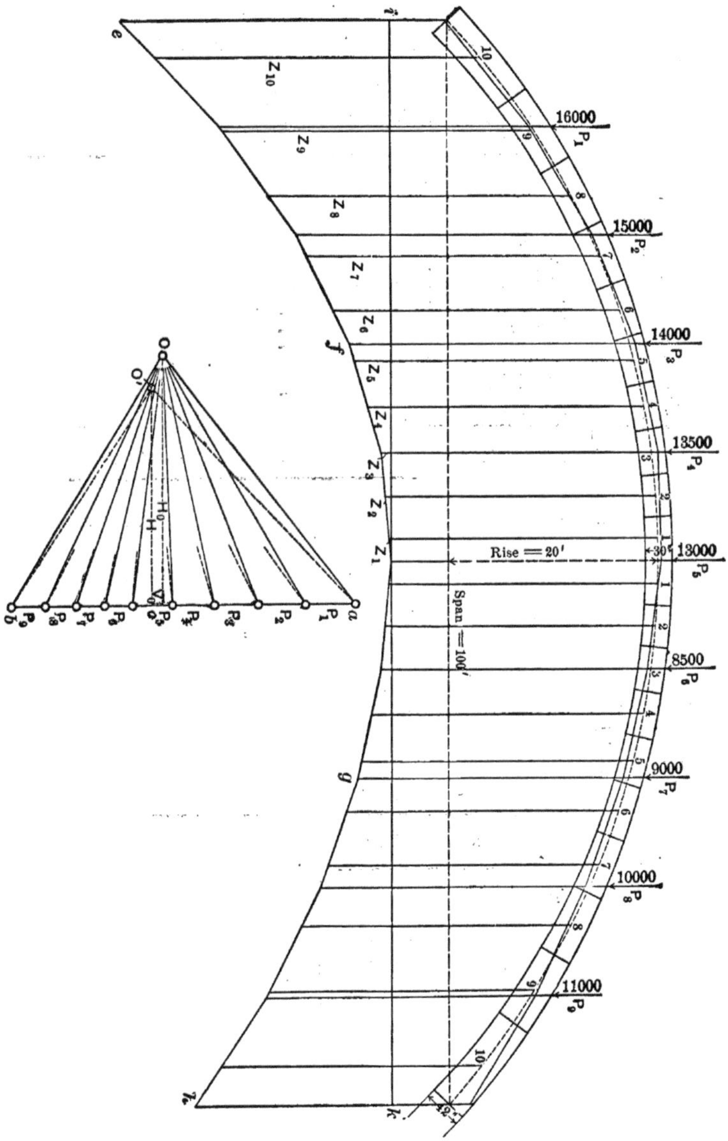
16000
15000
14000
13500
13000
8500
9000
10000
11000
Rise = 20'
Span = 100'

FIG. 103.

distances from the equilibrium polygon. These are given in Table D, together with resulting bending moments. The bending moments may also be calculated as done in Example 1.

TABLE C.

CALCULATIONS FOR H_0, V_0, AND M_0.

1	2	3	4	5	6	7	8	9
Point	x	y	x^2	y^2	m_L	m_R	$(m_L+m_R)y$	$(m_R-m_L)x$
1	2.07	.07	4.3	.00	− 13,500	− 13,500	− 2,000	——
2	5.96	.30	35.5	.09	− 38,700	− 38,700	− 23,000	——
3	10.00	.72	100.0	.52	− 65,000	− 65,000	− 94,000	..
4	14.18	1.40	201.1	1.96	− 148,600	− 127,700	− 387,000	+ 296,000
5	18.43	2.37	339.7	5.62	− 233,600	− 191,400	− 1,007,000	+ 777,000
6	23.06	3.77	531.8	14.21	− 369,000	− 288,400	− 2,479,000	+ 1,859 000
7	28.06	5.64	787.4	31.81	− 539,000	− 408,400	− 5,348,000	+ 3,665,000
8	33.60	8.23	1129.0	67.73	− 781,400	− 577,400	−11,183,000	+ 6,854,000
9	39.60	11.80	1568.2	139.24	−1,075,400	− 781,400	−21,910,000	+11,642,000
10	46.40	16.80	2153.0	282.24	−1,511,000	−1,083,000	−43,580,000	+19,859,000
Σ		51.10	6850.0	543.42	−8,350,100		−86,013,000	+44,952,000

$$H_0=\frac{10\times(-86013000)-(-8350100)\times 51.10}{2.[(51.10)^2-10\times 543.42]}=+76{,}760\text{ lbs.},$$

$$V_0=\frac{44952000}{2\times 6850}=+3{,}280\text{ lbs.}$$

$$M_0=-\frac{-8350100+2\times 76760\times 51.10}{2\times 10}=+25{,}260\text{ ft-lbs.}$$

TABLE D.

THRUSTS, ECCENTRIC DISTANCES, AND MOMENTS.

1	2	3	4	5	6	7
Point.	Thrusts.		Eccentric Distances.		Bending Moments.	
	Left.	Right	Left.	Right.	Left.	Right.
1	76,800	77,400	+ .31	+ .13	+23,700	+ 10,100
2	76,800	77,400	+ .38	− .13	+29,200	− 10,100
3	78,600	78,800	+ .61	− .22	+48,000	− 17,300
4	78,600	78,800	+ .39	− .53	+30,700	− 41,700
5	78,600	78,800	+ .44	− .57	+34,700	− 45,000
6	82,700	81,500	+ .26	− .61	+21,200	− 49,700
7	82,700	81,500	+ .14	− .52	+11 400	− 42,400
8	89,300	85,300	− .15	− .36	−13,400	− 30,700
9	89,300	85,300	− .16	+ .23	−14,300	+ 19,600
10	98,600	90,700	− .44	+ .88	−43,400	+ 80,000
Springing	98,600	90,700	− .21	+1.67	−20,700	+152,000

Temperature Stresses.—Suppose in Ex. 2 it is desired to know the thrust and bending moment at the crown due to a rise of temperature of 30°. Eqs. (6) and (7), Art. 179, will be used. Assume $E=2{,}000{,}000$ lbs/in² $=288{,}000{,}000$ lbs/ft². Suppose the value $\delta s/I$, in foot-units, is 3.1. Then from Eq. (6)

$$H_0=\frac{288{,}000{,}000}{3.1}\times\frac{.000006\times30\times100\times10}{2(10\times543-(51.1)^2)}=2970 \text{ lbs.},$$

$$M_0=-2970\times\frac{51.1}{10}=-15{,}200 \text{ ft-lbs.}$$

The equilibrium polygon is a horizontal line drawn a distance below the crown equal to $15200/2970=5.11$ ft. The moment at any point is equal to the thrust H_0 multiplied by the vertical distance from such point to this equilibrium polygon. At the springing line, $M=H_0\times(20-5.11)=2970\times14.89=44{,}200$ ft-lbs. This may also be calculated by Eq. (8).

Stresses Due to Shortening of Arch.—The modification of the thrust due to the compressive deformations of the arch ring is found by Eq. (9). The average compressive stress at any section is found by dividing the thrust at that section by the area of the transformed section of arch ring. This is nearly uniform throughout the arch and equal to about 150 lbs/in². Then

$$H_0=-\frac{1}{3.1}\times\frac{150\times144\times100\times10}{2[10\times543-(51.1)^2]}=1240 \text{ lbs.}$$

This thrust is equal to 42% of the thrust due to temperature change, already found. The resulting moments and stresses will then be 42% of those due to temperature change. They will be of opposite sign.

185. Maximum Stresses in the Arch Ring.—From the values of thrust, moment, and eccentric distance, as given in Tables B and D, the stresses in the concrete and steel can be found at any section of the arch, as explained in Chap. III and also in Art. 147, Chap. VI. The maximum value of fibre stress will be where the sum of the stresses due to thrust, N, and moment, M, is a maximum. This will not in general be where either the thrust or the moment is a maximum; but as the thrust varies slowly along the arch ring the maximum stress will occur very near to the point of maximum moment.

The position of live load causing maximum moment at any point will differ in arches of different proportions. In designing an arch it is sufficient generally to determine the maximum stresses at the crown, the haunch, and the springing line. This will require several different positions of the live load. For the crown the maximum positive moments are caused when a short length of the arch (one-fourth to one-third) at the center is loaded, and the maximum negative moments when the remaining portions are loaded. The maximum positive and negative moments at the haunch (about the $\frac{1}{4}$ point) are caused when the arch is loaded about two-thirds the span length and one-third the span length respectively. The same loading will give practically the maximum moments at the springing lines.

These conditions make it desirable to analyze the arch for various assumed loadings about as follows: full load; one-third of span loaded; two-thirds of span loaded; center third loaded; and end thirds loaded. In the case of large and important structures it may be found desirable to place the loads somewhat differently than here indicated. A complete and exact solution can readily be made by analyzing the arch for a load of unity at each load-point of one-half of the arch. Influence lines can then be drawn for moment or fibre stress and the exact maximum values determined.

185*a*. Example of Complete Analysis for Maximum Stresses.—To further illustrate the methods of analysis here described, and the use of influence lines in determining maximum stresses, a complete analysis will be made of the arch of Fig. 103, modified, for sake of illustration, by the addition of steel reinforcement amounting to 2 in^2 per foot of arch along both the extrados and the intrados, and placed 3 in. from the surface.

Calculation of Values of I and of δs.—The half length of the arch axis is found to be 55.17 ft. The depth at crown=2.5 ft., and at springing line=3.5 ft. The moment of inertia at any section=$I=I_c+15I_s$, where I_c and I_s=moment of inertia of the concrete and steel sections respectively. Following the procedure of Art. 178, the half arch will first be divided into a convenient number of equal divisions and the value of I determined at the center of each division. The reciprocal, i, is then found. It is convenient to use the same number of preliminary divisions as is desired for the final divisions. The results of these calculations are given in Table E. The calcu-

TABLE E.

DIVISIONS OF ARCH RING.

Properties of Preliminary Equal Divisions.						Properties of Final Divisions.			
No. of Division.	Depth, d	I_c	$15I_s$	$I=I_c+15I_s$	$i=\frac{1}{I}$	i	δs	I	d
1	2.55	1.38	.44	1.82	.549	.560	3.57	1.78	2.53
2	2.65	1.55	.48	2.03	.492	.520	3.85	1.92	2.59
3	2.75	1.73	.53	2.26	.442	.480	4.16	2.08	2.67
4	2.85	1.93	.57	2.50	.400	.440	4.53	2.26	2.75
5	2.95	2.14	.62	2.76	.362	.405	4 95	2.47	2.84
6	3.05	2.36	.68	3.04	.328	.370	5.42	2.71	2.93
7	3.15	2.61	.73	3.34	.300	.333	6.00	3.00	3.03
8	3.25	2.86	.79	3.65	.274	.300	6.67	3.33	3.15
9	3.35	3.13	.85	3.98	.251	.265	7.50	3.75	3.28
10	3.45	3.42	.91	4.33	.231	.235	8.51	4.26	3.42
					3.629		55.16		

$i_a=3.63/10=.363$. By Eq. (5) $\delta s \,.\, i=\frac{55.17\times .363}{10}=2.00.$

lations are made in foot-units. The first part of the table relates to the preliminary ten equal divisions, each=55.17/10 =5.517 ft. long. The resulting values of i are plotted in Fig. 103*a*. The line AB is 55.17 ft. long and is divided into ten equal divisions, 1, 2, 3, etc. At the centers of the several divisions the values of i are laid off as ordinates, i_1, i_2, i_3, etc.,

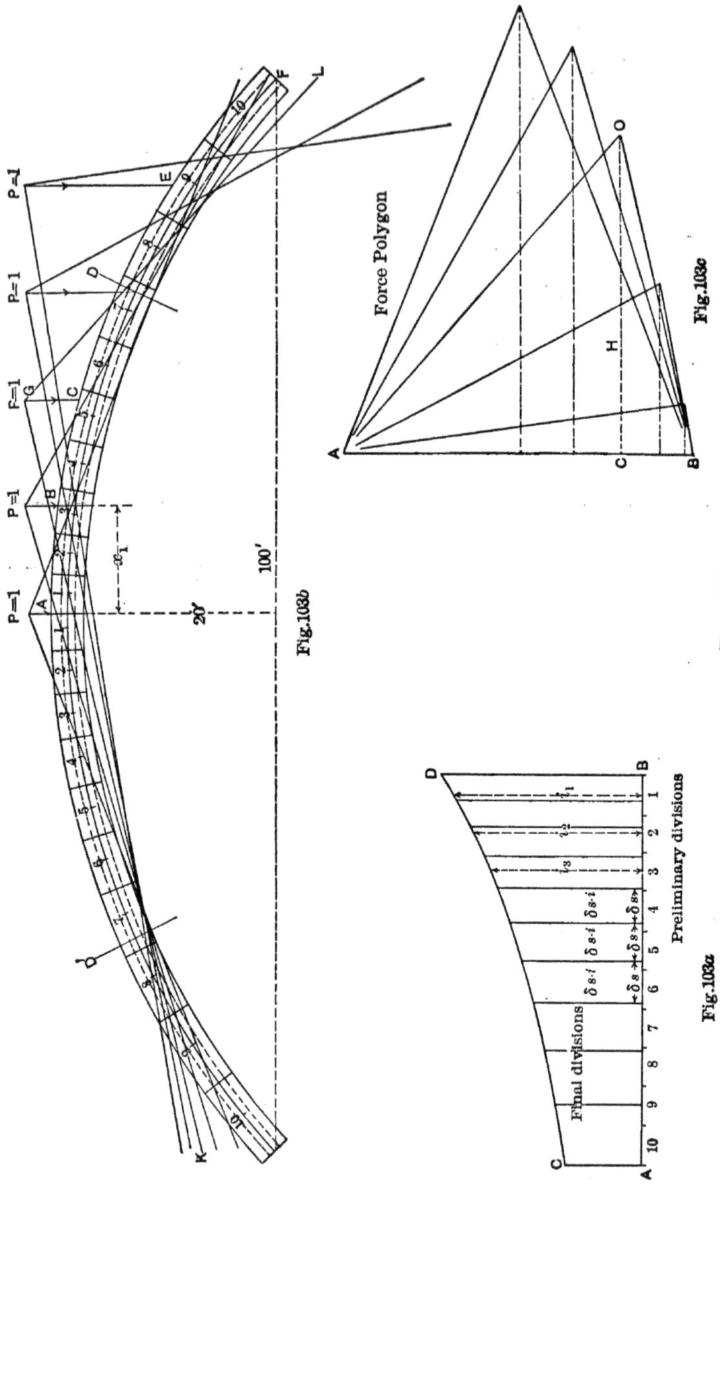
P=1
P=1
P=1
P=1
P=1
P=1
Fig.103b
100′
20′
Force Polygon
Fig.103c
Final divisions
Preliminary divisions
Fig.103a

FIG. 103a–c.

and the curve CD drawn. The area $ABCD=\frac{55.17}{10}\times\Sigma i$ $=55.17\times i_a$. This area is now to be divided into ten equal parts, each equal to $\delta s\cdot i$. Each of these parts will then be equal to $\frac{55.17\times i_a}{10}=2.00$, as given below Table E. Beginning at one end of the diagram, Fig. 103a, the several equal areas are then laid off, the values of i being scaled from the diagram, and δs being equal to $2.00/i$. These calculations are given in the latter part of Table E, where are also given the values of I and d for the center points of the final subdivisions. The diagram of Fig. 103a is not really necessary but is here introduced partly to make clear the methods employed in the calculations.

Calculation of H_0, V_0, and M_0 for a Load Unity at Each Load Point.—Fig. 103b shows the arch ring with the subdivisions made and the center points numbered 1, 2, 3, etc., as in Fig. 103. The load points considered are those on the right, A, B, C, D, and E. These are the center points of the supporting spandrel walls. [In the case of an arch with continuous loading, arbitrary load points may be selected sufficiently close to secure the desired degree of accuracy in the influence lines. Four of five such points will generally be sufficient.] The problem now is, first, to calculate H_0, V_0, and M_0, for a load unity placed successively at these several load points. The computations are performed in the same manner as in the preceding examples. Table F gives all the calculations. The quantities m are equal to $x-x_1$, where x_1 is the abscissa of the load point in question, measured from the crown. For the load point A the entire load is, for convenience, assumed as acting just to the right of the crown and hence as belonging to the right half. The shear V_0 is evidently equal to 0.5 in this case. The quantities m_L do not appear. For load points B, C, D, and E the calculations become progressively less numerous.

TABLE F.

CALCULATIONS FOR H_0, V_0, AND M_0.

Points.	x	y	x^2	y^2	Load at A; $x_1=0$.		
					m_R	m_Ry	m_Rx
1	1.78	0.02	3.2	.00	1.78	.0	3.2
2	5.49	0.29	30.1	.08	5.49	.8	30.1
3	9.47	0.62	89.7	.27	9.47	2.9	89.7
4	13.76	1.32	189.3	1.74	13.76	9.1	189.3
5	18.38	2.37	337.7	5.61	18.38	21.7	337.7
6	23.34	3.86	544.7	14.89	23.34	45.0	544.7
7	28.67	5.91	821.8	34.92	28.67	84.7	821.8
8	34.37	8.67	1181.5	75.10	34.37	148.9	1181.5
9	40.44	12.33	1635.5	151.93	40.44	249.3	1635.5
10	46.82	17.15	2192.5	294.08	46.82	401.5	2192.5
Σ	....	52.54	7026.0	578.62	−222.52	−1927.9	−7026.0

TABLE F—(*Continued*).

Points.	Load at B; $x_1=10.0$.			Load at C; $x_1=20.0$.		
	m_R	m_Ry	m_Rx	m_R	m_Ry	m_Rx
4	3.76	4.9	51.7			
5	8.38	19.8	154.1			
6	13.34	51.5	311.2	3.34	12.9	78.6
7	18.67	110.3	535.1	8.67	51.2	248.4
8	24.37	211.2	837.6	14.37	124.6	494.6
9	30.44	375.2	1231.3	20.44	252.0	826.6
10	36.82	631.5	1724.3	26.83	460.0	1255.2
Σ	−135.78	−1404.4	−4845.3	−73.65	−900.7	−2903.4

Points.	Load at D; $x_1=30.0$.			Load at E; $x_1=40.0$.		
	m_R	m_Ry	m_Rx	m_R	m_Ry	m_Rx
8	4.37	37.9	150.4			
9	10.44	128.7	422.5	0.44	5.4	17.8
10	16.83	288.5	787.9	6.83	117.0	31.96
Σ	−31.64	−455.1	−1360.8	− 7.27	−122.4	−337.4

From Table F we then have for the denominators of eqs. (1), (2), and (3) of Art. 174,

$$2[(\Sigma y)^2-n\Sigma y^2]=2[(52.54)^2-10\times 578.62]=-6051.4,$$

$$2\Sigma x^2=2\times 7026=14{,}052, \qquad 2n=20.$$

The values of H_0, V_0, and M_0 then result as follows:

Load at A:

$$H_0=\frac{10\times(-1927.9)-(-222.52)\times 52.54}{-6051.4}=+1.254;$$

$$V_0=\frac{-7206}{14{,}052}=-0.50;$$

$$M_0=-\frac{-222.52+2\times 1.254\times 52.54}{20}=+4.54.$$

Load at B:

$$H_0=\frac{10\times(-1404.4)-(-135.78)\times 52.54}{-6051.4}=+1.142;$$

$$V_0=\frac{-4845.3}{14{,}052}=-0.345;$$

$$M_0=-\frac{-135.78+2\times 1.142\times 52.54}{20}=+0.80.$$

Load at C:

$$H_0=\frac{10\times(-900.7)-(-73.65)\times 52.54}{-6051.4}=+0.849;$$

$$V_0=\frac{-2903.4}{14{,}052}=-0.207;$$

$$M_0=-\frac{-73.65+2\times 0.849\times 52.54}{20}=-0.78.$$

Load at D:

$$H_0=\frac{10\times(-455.1)-(-31.64)\times 52.54}{-6051.4}=+0.477;$$

$$V_0=\frac{-1360.8}{14{,}052}=-0.097;$$

$$M_0=-\frac{-31.64+2\times 0.477\times 52.54}{20}=-0.92.$$

Load at E:

$$H_0=\frac{10\times(-122.4)-(-7.27)\times 52.54}{-6051.4}=+0.139;$$

$$V_0=\frac{-337.4}{14{,}052}=-0.024;$$

$$M_0=-\frac{-7.27+2\times 0.139\times 52.54}{20}=-0.37.$$

Calculation of Moments and Thrusts at any Given Section of the Arch Due to Unit Loads at the Load Points.—Consider, for example, a load unity at C, Fig. 103*b*. The values of H_0, V_0, and M_0 being known, the moments, shears, and thrusts at any given section can now be found either graphically or analytically, as already explained. The graphical method is much the simpler. For this purpose draw the force polygon, Fig. 103*c*, the load AB being equal to unity, and lay off the shear $V_0=-.207$ from B, upwards to C, and then the thrust $H_0=.849$ to the right, fixing the pole O. Then from the center of the arch at the crown measure down a distance equal to $e=\frac{M_0}{H_0}=\frac{.78}{.849}=.92$ ft., fixing one point on the equilibrium polygon. Lines KG and GL drawn parallel to OB and AO, intersecting in the load vertical at G, complete the polygon. The other polygons are drawn in a like manner.

Having these polygons drawn, the bending moment at any section, due to any one of the unit loads, is equal to the vertical ordinate measured from the proper equilibrium polygon to the gravity axis of the arch, multiplied by the corresponding pole distance. From these several polygons the moments can therefore be found at any section for a unit load at any load point on either half of the arch, and the influence line drawn for such moment. The tangential thrusts can likewise be determined.

Influence Lines for Fiber Stresses at any Section.—Instead of constructing influence lines for moment and thrust it will be more direct to construct them at once for stress on extreme fiber.

From Art. 82 the stress on extreme fiber is

$$f_c=\frac{Mu}{I}+\frac{N}{A},$$

in which u is the distance from the neutral axis to the fiber in question and N is the thrust normal to the section. The bending moment is equal to the thrust times its eccentricity, or $M = Ne$.

If r = radius of gyration of the section, we have $Ar^2 = I$. The above expression for f_c may then be written in the form,

$$f = \frac{Neu}{I} + \frac{N\frac{r^2}{u} \cdot u}{I}$$

$$= \frac{N\left(e+\frac{r^2}{u}\right)u}{I}. \quad . \quad . \quad . \quad . \quad . \quad . \quad . \quad . \quad (16)$$

The quantity $\frac{r^2}{u}$ is a length, and $N\left(e+\frac{r^2}{u}\right)$ is a moment which may be written M'. Then we have

$$f = \frac{M'u}{I}. \quad . \quad . \quad . \quad . \quad . \quad . \quad . \quad . \quad . \quad (17)$$

This new moment, M', is equal to the thrust N, multiplied by the eccentricity e plus the additional distance $\frac{r^2}{u}$. We may then compute the value of $\frac{r^2}{u}$ for the section considered and take the center of moments at this distance above or below the neutral axis, according as the lower or upper fiber stress is desired. By so selecting the center of moments the fibre stress becomes equal to the moment multiplied by the usual factor $\frac{u}{I}$, and therefore varies with the moment. An influence line drawn for such moment will therefore serve as an influence line for fiber stress.

In this problem influence lines have thus been drawn for upper and lower fiber stress at sections taken at the various load points and at the springing line F. They are given in Fig. 103*d*. To explain further their construction consider the section at D, and suppose the arch is loaded with a unit load at C. The arch near D is shown to a larger scale in Fig. 103*e*. The force R is the resultant of the forces on the

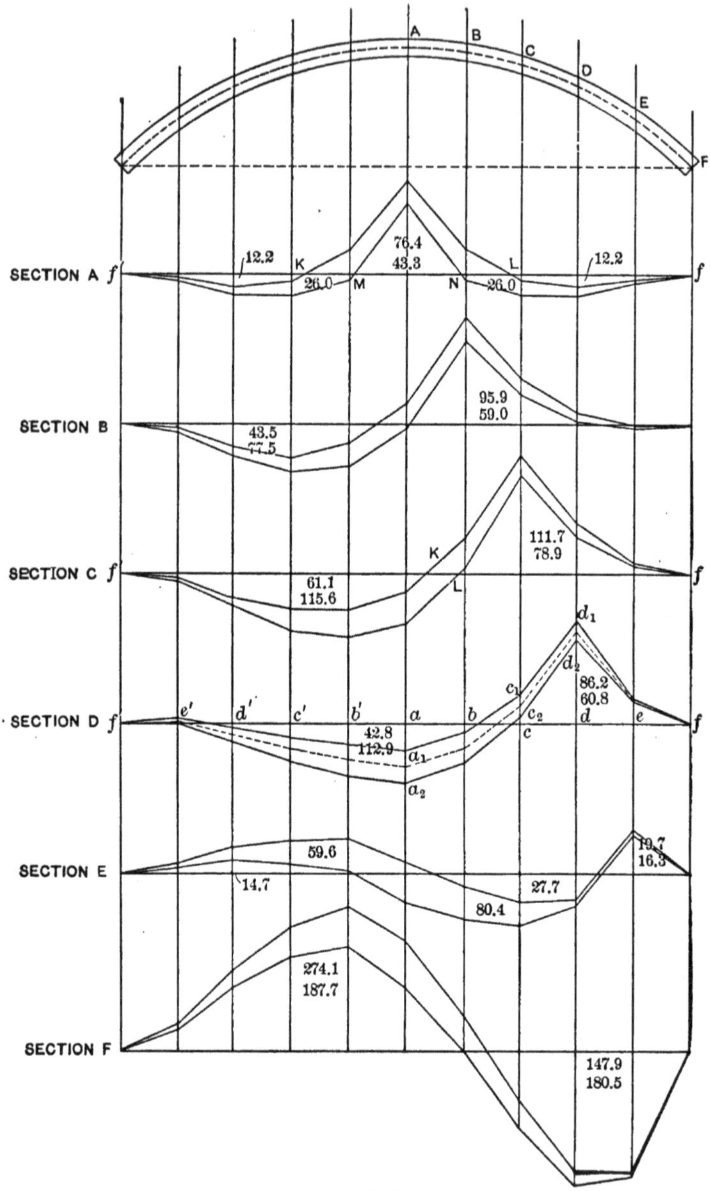

A
B
C
D
E
F
SECTION A
f
12.2
K
76.4
43.3
L
12.2
f
26.0
M
N
26.0
SECTION B
95.9
59.0
43.5
77.5
SECTION C
f
K
111.7
78.9
f
61.1
115.6
L
d_1
d_2
86.2
60.8
SECTION D
f
e′
d′
c′
b′
a
b
c_1
c_2
d
e
f
c
42.8
112.9
a_1
a_2
SECTION E
59.6
19.7
16.3
14.7
27.7
80.4
274.1
187.7
SECTION F
147.9
180.5

Fig. 103*d*.

left and acts on the line GL, Fig. 103b. In amount it is equal to AO in the force polygon. The neutral axis is at a, Fig. 103e, and the distances ab_1 and ab_2 are the values of r^2/u calculated for this section. The point b_1 is therefore the center of moments for M' for upper fiber stress and b_2 is the center of moment for M' for the lower fiber stress. (The points b_1 and b_2 are at the edges of the "kern" of the section. In a simple rectangular section they are at the edges of the middle third and in a truss or plate girder, where the flange takes all the moment, they are at the flange centers.)

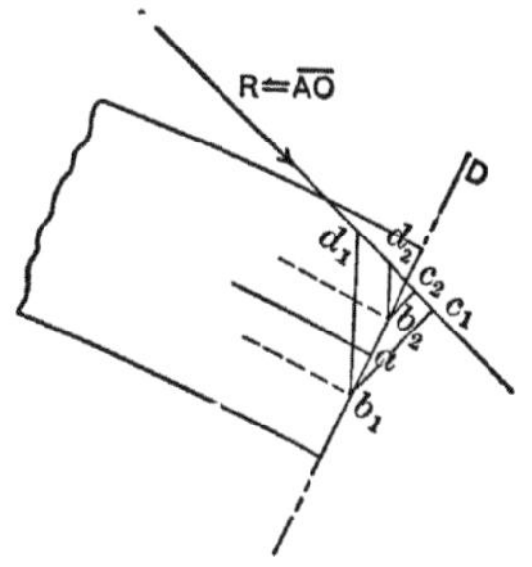

FIG. 103e.

These values of M' are given graphically by $R \times b_1c_1$ and $R \times b_2c_2$, or by the more convenient products $H \times b_1d_1$ and $H \times b_2d_2$. These moments are calculated and plotted in Fig. 103d at point C, under the load unity. They are the ordinates cc_1 and cc_2, and are plotted as positive ordinates since they represent positive bending moments. In the same manner values of the moments are determined at section D for unit loads at A, B, D, and E, and plotted at corresponding points in the diagram. For loads on the left half of the structure the values are found by considering the section D to be on the left half of the arch at D', and measuring the several ordinates to the reaction lines as drawn for loads on the right. The complete influence lines are given in the diagram. The upper line $f'a_1d_2f$ is for upper fiber stress and the lower line $f'a_2d_2f$ is for lower fiber stress. The dotted line (midway between, for symmetrical sections) is the influence line for the usual bending moment, center of moments at the neutral axis. Since positive ordinates indicate positive moments it follows that for the upper fibers positive ordinates indicate compression and for the lower fibers positive ordinates indicate tension. Negative ordinates indicate respectively tension on the upper fibers and compression on the lower. The actual fiber stress at section D due to a unit load at any point is now given by the ordinate to the influence line at the point, multiplied by $\frac{u}{I}$ for this section.

In Table G are given values of u, I, r^2, and $\frac{r^2}{u}$ for each of the sections $A \ldots F$. The distances r^2/u are shown by the dotted lines in the arch ring, Fig. 103*b*.

TABLE G.

PROPERTIES OF SECTIONS $A \ldots F$.

Section.	Depth, d.	$u=\frac{1}{2}d$.	I	A (A_c+15A_s)	r^2	$\frac{r^2}{u}$
A	2.50	1.25	1.70	2.92	.58	.46
B	2.78	1.39	2.10	3.20	.65	.47
C	3.00	1.50	2.54	3.42	.74	.49
D	3.18	1.59	3.06	3.60	.85	.53
E	3.35	1.67	3.70	3.77	.98	.59
F	3.50	1.75	4.46	3.92	1.14	.65

Where the arch is continuously loaded the influence lines should be drawn as smooth curves through the points determined by the ordinates at the load-points selected.

Maximum Fiber Stress at Any Section.—Having the influence lines drawn for several sections, as in Fig. 103*d*, the maximum fiber stress at the various sections can readily be determined for any given loading, either uniform or concentrated. In the figure the various positive and negative areas above and below the axis have been measured and are written within the respective areas. The upper figure refers in all cases to the area between the axis and the upper curve. For uniform loads these areas, multiplied by the load per foot, give the respective moments. For concentrated loads the moments are found by summing the products of the several loads times the corresponding ordinates. The maximum values can readily be determined by trial. The influence lines show very clearly the extent and general position of loads for maximum stresses.

For example, at the crown, section A, the moment for upper fiber stress, due to a uniform dead load of, say, 800 pounds per foot, $=[76.4-(2\times12.2)]\times800=41{,}600$ ft.-lbs. The stress $f_c=M'\frac{u}{I}=41{,}600\times\frac{1.25}{1.70}=20{,}700$ lbs/ft², $=330$ lbs/in². For a uniform live load of 500 lbs. per foot, the maximum stress is caused when the load extends from K to L, and the stress for such load $=76.4\times500\times\frac{1.25}{1.70}\div144=195$ lbs/in².

Again, at section C, the dead load upper fiber stress $=(111.7-61.1)\times800\times\frac{1.50}{2.54}\div144=166$ lbs/in²; and the maximum live load stress $=111.7\times500\times\frac{1.50}{2.54}\div144=230$ lbs/in², the load extending from K to f. For the maximum lower fiber stress the load extends from f' to L and the stress $=237$ lbs/in²; the dead load stress $=120$ lbs/in². Othe rstresses are found in a similar manner.

The influence lines show that in arches of proportions such as here considered, the loading for maximum stresses is about as noted in Art. 185. For the crown, the center third should be loaded; for the haunch (sections C and D) the load should extend over about two-thirds or one-third the span length, and about the same for the section at the springing line.

Case in which the Resultant Stress is Tensile.—If the resultant stress is tensile at any section and the tension in the concrete is to be neglected then the formulas for fiber stress of Art. 83 must be used, which makes it necessary to plot the influence lines for the true bending moment and to determine the thrust as well as the moment. The diagram, Plate XIV, p. 288, can then be used.

186. Illustrative Examples of Arch Design.—Fig. 104 shows a longitudinal section and part plan of a bridge at Grand Rapids, Mich.* It consists of five spans of lengths from 79 to 87 feet. The reinforcement is composed of 1¼-in. Thacher bars spaced 14 in. apart near both the extrados and intrados. Each pair is connected by ⅝-in. connecting-rods spaced 4 ft. apart.

* Eng. News, Vol. LII, 1904, p. 490.

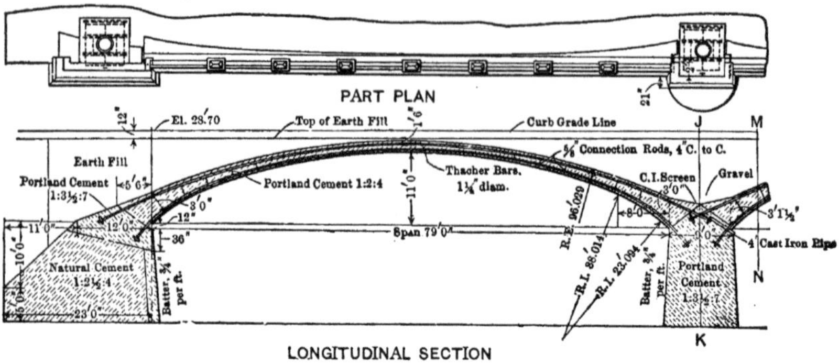

FIG. 104.—Arch Bridge at Grand Rapids, Mich.

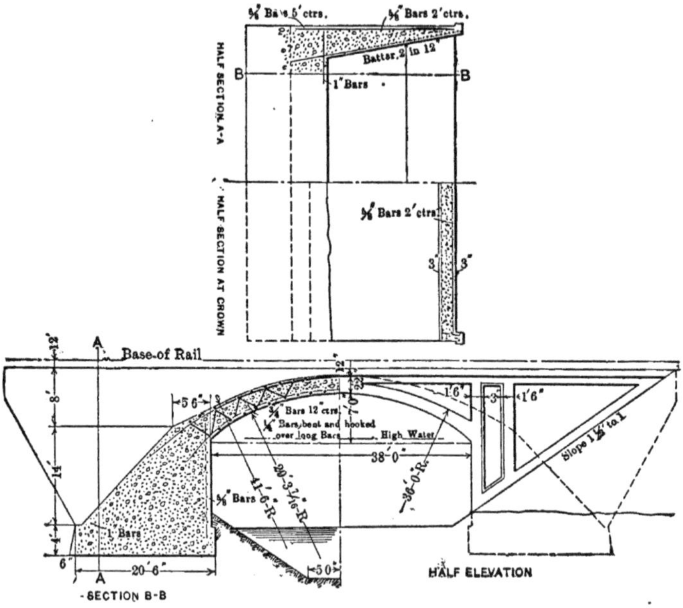

FIG. 105.—Arch Bridge on the Chicago & Eastern Illinois R.R.

Fig. 105 illustrates a 38-ft. span reinforced arch on the C. & E. I. R.R.* Johnson corrugated bars were used.

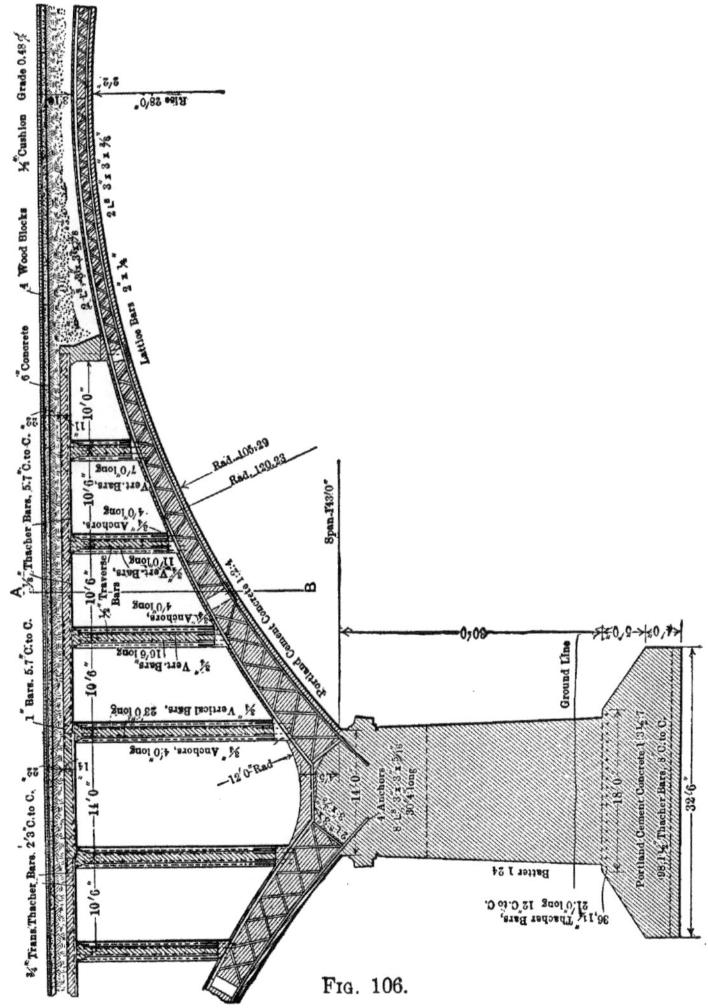

FIG. 106.

Figs. 106 and 107 show the details of a design for a concrete viaduct at Milwaukee, Wis.† The design was made by the

* R.R. Gaz., April 13, 1906, p. 390.

† Eng. News, Vol. LVII, 1907, p. 178.

Concrete-Steel Engineering Co., of New York City, and is a typical Melan arch. The reinforcement consists of built-up ribs of 3″×3″×⅜″ angles connected by lattice bars 2″×¼″.

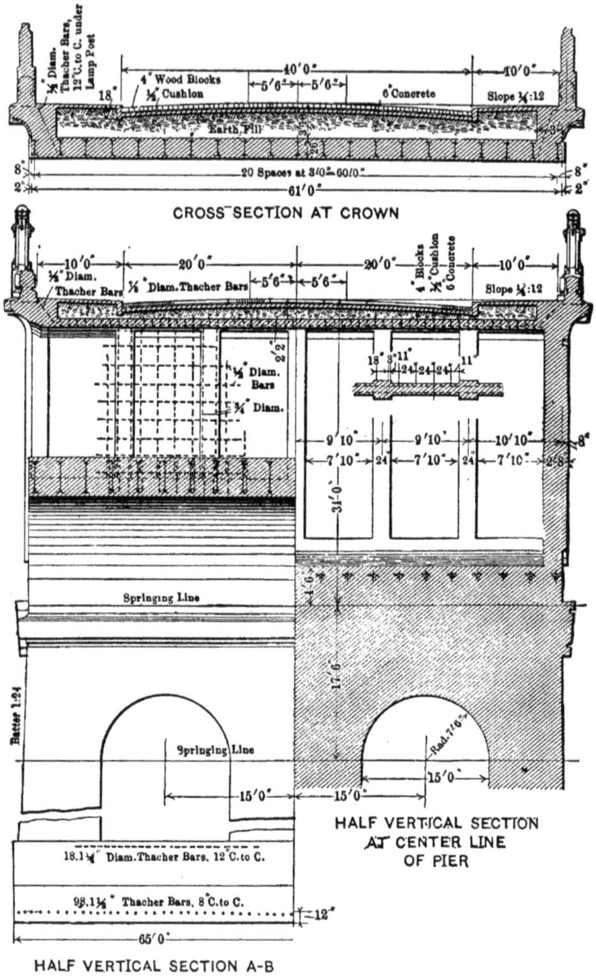

Fig. 107.

The ribs are spaced 3 ft. apart. These ribs are designed to carry the entire bending moment at a stress of 18,000 lbs/in². The stress in the concrete was limited to 500 lbs/in² in com-

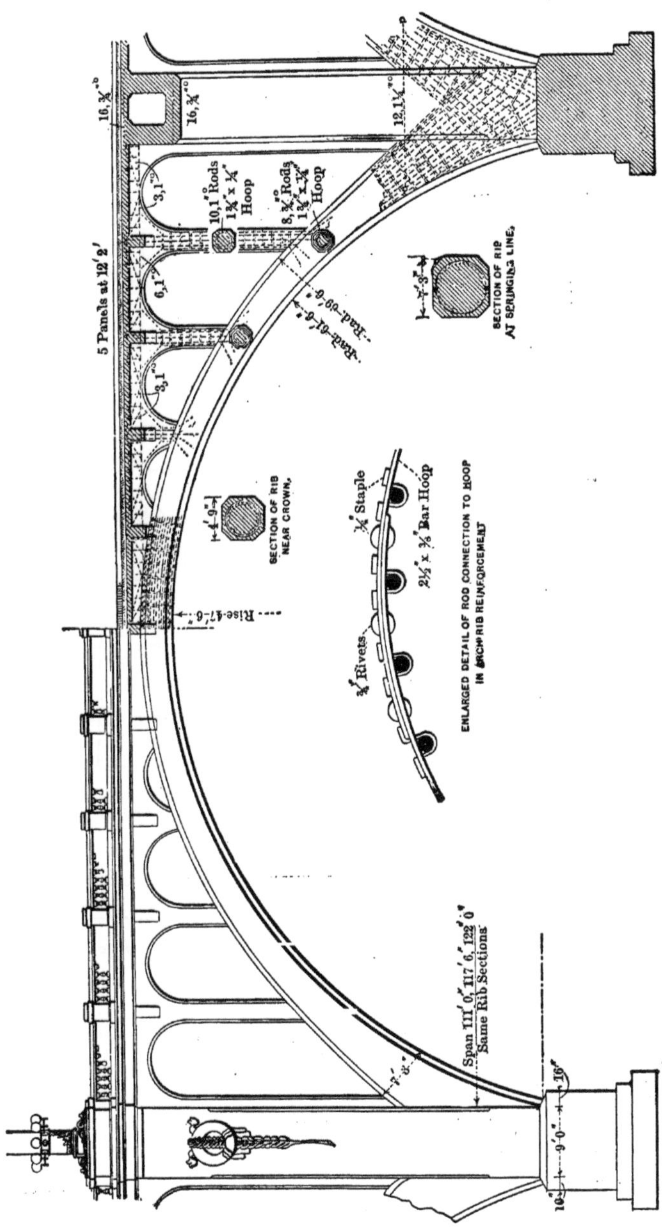

5 Panels at 12' 2'
Rise 41' 6"
Span 111' 0", 117' 6", 122' 0"
Same Rib Sections
SECTION OF RIB NEAR CROWN.
SECTION OF RIB AT SPRINGING LINE.
ENLARGED DETAIL OF ROD CONNECTION TO HOOP IN ARCH RIB REINFORCEMENT

FIG. 108.

pression, or 600 lbs/in^2 including temperature stresses. The roadway is supported over a considerable portion of the span length by means of a reinforced floor carried on vertical walls.

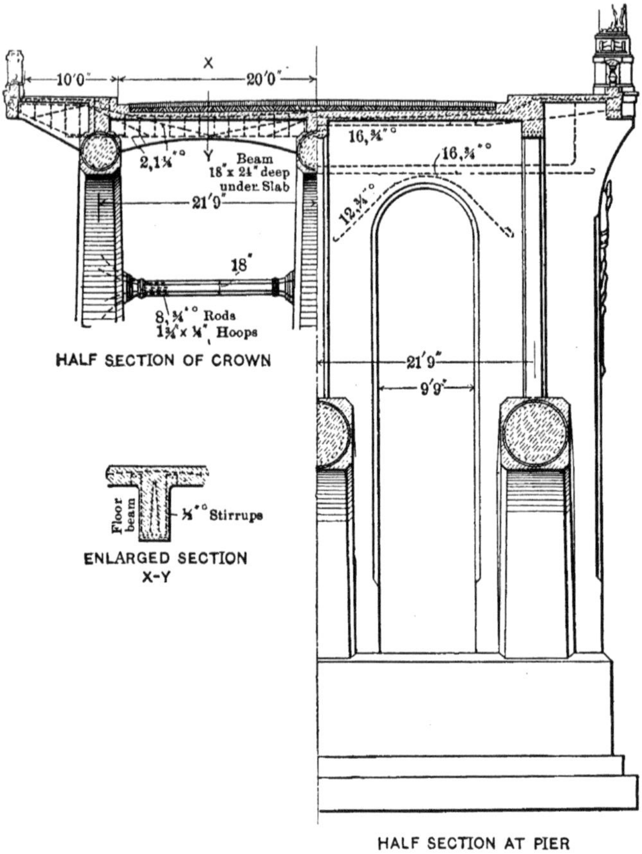

FIG. 109.

The viaduct contains eight spans of the dimensions shown in the illustration.

Figs. 108 and 109 illustrate another design for the same viaduct mentioned in the preceding paragraph.* This design was

---

* Eng. News, Vol. LVII, 1907, p. 178.

submitted by Mr. C. A. P. Turner, of Minneapolis, Minn. In this case the arch is composed of three ribs 4 ft. 9 in. square at the crown and 7 ft. 3 in. square at the pier. The rib reinforcement is composed of longitudinal rods arranged in a circle and connected at frequent intervals by bands of $2\frac{1}{8}'' \times \frac{3}{8}''$ metal. The stirrups and bent bars in the floor-beams and slabs give a very effective reinforcement.

CHAPTER IX.

RETAINING-WALLS AND DAMS.

187. Advantages of Reinforced Concrete.—Retaining-walls, dams, bridge abutments, and the like constitute a class of structures in which the outside forces acting are mainly horizontal, and in which, therefore, the question of stability is largely a question of safety against overturning. Where ordinary masonry is used in these structures the weight of the material must be depended upon to balance the overturning forces; for though the structure be anchored to the foundation no tensile stresses can be allowed in the masonry. As a consequence of these limitations the maximum compressive stresses in such structures are not high, except in extreme cases, so that generally the dimensions are determined by the weight of the material. The application of reinforced concrete in such cases enables the design to be so modified as to utilize the weight of the material to be retained as part of the resisting weight and to calculate the sections to develop more nearly the full strength of the concrete. A very considerable gain in economy therefore results.

RETAINING-WALLS.

188. Method of Determining Stability.—No attempt will be made here to present the various mathematical theories of earth pressure. Unless the results obtained from such theories are carefully controlled by the results of experience they are apt to be very misleading. Probably the most satisfactory way to design a reinforced concrete retaining-wall, as regards stability against overturning, is to proportion it so that it will be, as nearly as possible, equivalent to a solid masonry wall of such a section as is known to have given satisfactory results under the given conditions. Rules of practice as to solid masonry walls have long been established. They represent the accumulated experience of many engineers and are based upon data obtained from many failures as well as from successful designs. Until experience is had directly with the reinforced type of wall its stability may, therefore, well be determined by comparison with the older form of construction. The analysis given here will consequently be limited to a convenient method of comparison of the two types. It may be said in passing that good construction requires quite as much attention to the earth filling itself and to its drainage as to the design and construction of the wall.

In dimensioning a reinforced concrete wall which will possess stability equal to that of a given solid wall, it will be convenient to determine the equivalent fluid pressure under which the solid wall will be stable and then apply this pressure to the reinforced type of wall. The basis of the calculation of this fluid pressure will be to determine the weight per cubic foot of a fluid which will exert such a pressure against the solid wall as to cause the resultant of all forces above the base to intersect the base at the edge of the middle third. If, then, the reinforced wall be designed so that it will be equally stable against this pressure, it will be practically equivalent to the solid wall.

It will be seen that this method is very simple and adapts itself readily to the utilization of present rules of practice. If desired, the theory of earth pressure may of course be directly applied to the problem.

189. Equivalent Fluid Pressure for Ordinary Masonry Walls.—Two forms of wall will be considered (Fig. 110). Form (*a*) is the more common form of wall. A small batter is usually given to the front face, and the back face is sloped in an irregular line, the width of the top being as narrow as circumstances may warrant. Such a wall will be stable when the width of the base is made from one-third to one-half the height, four-tenths being a common rule of practice. Form (*b*) is used for relatively low walls. Its width may be a little less than that of form (*a*) for equal stability. While the calculations here given apply only to the two forms as represented in Fig. **110**, the results will be but little different for walls similar in form but which vary considerably therefrom.

Form (*a*).—The height is h and the bottom width l. The batter of the front face will be taken at 1:12, and the top width at 1/6 of the bottom width. The weight of the masonry will be assumed at 150, and that of the earth filling at 100 lbs/ft³. It will be assumed that the fluid pressure acts against a vertical plane FC; the stability of the entire volume to the left of this plane, including the weight of the earth, will be determined. Let W_1 denote the weight of masonry per lineal foot, and W_2 the weight of the earth filling to the left of FC. Let P denote the resultant fluid pressure acting at a distance $\frac{1}{3}h$ above the base. Let p denote the weight per cubic foot of such fluid. Then $P=\frac{1}{2}ph^2$.

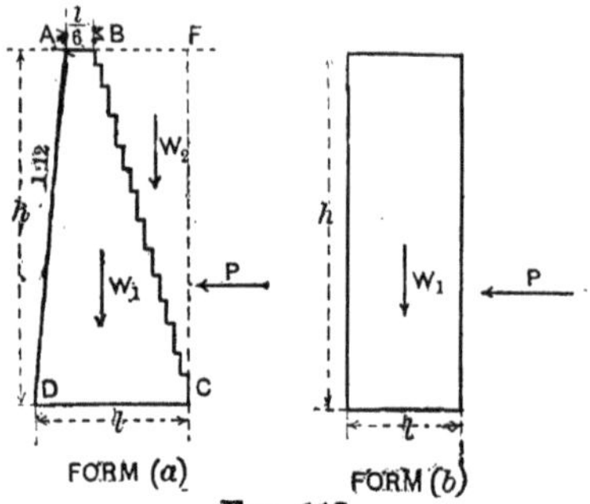

FIG. 110.

Assume that the resultant pressure due to the weight of the wall W_1, the weight of the earth W_2, and the pressure P, intersects the base at the edge of the middle third. Equating moments about this point we derive the relation

$$p=132\frac{l^2}{h^2}. \quad \quad (1)$$

Form (*b*).—In this form the only forces to be considered are the weight W_1 and the pressure P. Equating these as before there results

$$p=150\frac{l^2}{h^2}. \quad \quad (2)$$

If the top width in form (*a*) be made zero the effect on the result would be to change the coefficient in Eq. (1) from 132 to 127, thus showing that a considerable variation in top width has little effect on the result.

Substituting various values of l/h in Eqs. (1) and (2) we have the following values of p, or equivalent fluid weight, under which the wall is stable as above assumed.

	l/h	p
Form (*a*)	1/3	14.7 lbs/ft^3
	4/10	21.1 "
	1/2	33.0 "
Form (*b*)	1/4	9.4
	1/3	16.7
	4/10	24.0

According to these calculations a fluid weight of 20 to 25 lbs/ft^3 may be taken as a basis of design to secure stability equivalent to the ordinary wall, assuming the resultant pressures to cut the edge of the middle third and counting the weight of earth vertically above the back slope as part of the resisting load. It is to be noted that the pressures herein determined are not necessarily the actual earth pressures; the results are to be used only as a means of securing stability of reinforced walls approximately equal to that of solid walls of known proportions.

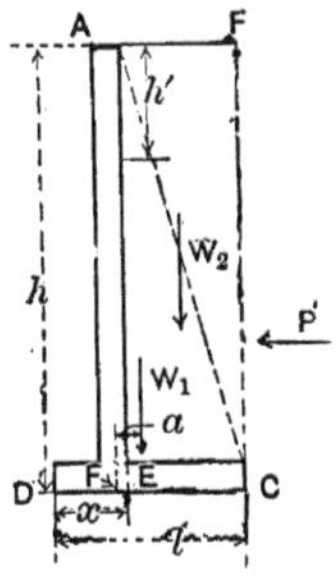

FIG. 111.

190. Stability of Reinforced Concrete Walls.—Fig. 111 represents in outline the usual type of reinforced wall. It consists of a vertical wall *AE* attached to a floor *DC*. For low walls the upright part *AE* may act simply as a cantilever; and likewise the parts *EC* and *ED*. For larger walls the part *AE* is tied to *EC* at intervals by back walls *ACE* in the form of narrow transverse walls with tension reinforcement. The projecting portion *ED* may still act as a cantilever, or it, also, may be connected to the vertical wall AE by means of buttresses. In either case the earth pressures act in essentially the same manner and the necessary width of base is found in the same way.

Let l=width of base;

x=distance from toe to back of wall AE;

h=height;

p=equivalent fluid weight as determined in Art. 189;

w_2=weight of earth filling per cubic foot;

W_1=weight of masonry per lineal foot;

W_2=weight per lineal foot of earth above the floor EC;

a=lever-arm of W_1 about point F, the edge of the middle third;

P=total fluid pressure$=\frac{1}{2}ph^2$.

Then equating moments about the point F we have

$$W_1a+W_2\left(\tfrac{2}{3}l-\frac{l-x}{2}\right)=\frac{Ph}{3}, \quad . \quad . \quad . \quad . \quad . \quad (3)$$

or

$$W_1a+w_2h(l-x)\left(\tfrac{2}{3}l-\frac{l-x}{2}\right)=\frac{ph^3}{6}. \quad . \quad . \quad . \quad . \quad (4)$$

If the wall AE is placed well towards the front the moment of the masonry will be small. Neglecting this term and putting $x=kl$ we may solve for l, getting

$$l=h\sqrt{\frac{p}{w_2(1+2k-3k^2)}}. \quad . \quad . \quad . \quad . \quad . \quad (5)$$

This is a minimum for $k=\frac{1}{3}$, that is, for $x=\frac{1}{3}l$. With this value of k we have

$$l=.87\sqrt{\frac{p}{w}}.h. \quad . \quad . \quad . \quad . \quad . \quad . \quad . \quad . \quad (6)$$

For $w=100$

$$l=.087\sqrt{p}.h. \quad . \quad . \quad . \quad . \quad . \quad . \quad . \quad (7)$$

If, for example, the value of p be taken at 21.1, corresponding to a value of $l/h=4/10$ for a solid wall, the value of l is equal to $.087 \times \sqrt{21.1} \times h = .4h$, or the same as the width of the solid wall.

As it may be desirable to use a smaller or larger value of x than $\frac{1}{3}l$, Table No. 22 has been prepared giving the values of l/h for various values of x/l and various values of p. An examination of the table shows plainly that the length of the projection x makes very little difference in the required total length of base. However, with x made very small or very large the weight of the wall should be taken into account. A further fact brought out by the table and by the table of Art. 189 is that the stability of the reinforced wall is about the same as a solid wall of form (*a*) shown in Fig. 110 and having the same base length.

Table No. 22.

PROPORTIONS OF REINFORCED-CONCRETE RETAINING-WALLS.
(See Fig. 111.)

Values of l/h for Different Values of p and for $w_2 = 100$ (Eq. (5)).

Values of $k = x/l$	Values of Equivalent Fluid Weight p. Pounds per Cubic Foot.			
	15	20	25	33
.5	.35	.40	.45	.51
.33	.34	.39	.43	.50
.25	.34	.39	.44	.50
.20	.34	.40	.44	.51
.15	.35	.40	.45	.52
.10	.36	.41	.46	.53
0	.39	.45	.50	.57

The resultant forces acting upon the three parts of the wall AE, DE, and EC must be determined. On the wall AE the force may be taken as a horizontal force equal to $P, = \frac{1}{2}ph^2$, and applied a distance $\frac{1}{3}h$ above the base. The resultant force acting on any length h' from the top is likewise $\frac{1}{2}ph'^2$ and applied a distance $\frac{2}{3}h'$ below the top. The pressure on the foundation will equal the total weight $W_1 + W_2$ and will be applied a distance $\frac{1}{3}l$ from point D. The average unit pressure will be

$(W_1+W_2)/l$, and the maximum pressure at D will be twice this value.

The upward pressure under the cantilever DE will vary from a maximum of $2\frac{W_1+W_2}{l}$ at D to a value under the point E of $2\frac{W_1+W_2}{l}\times\frac{l-x}{l}$. This is a "trapezoid" of pressure, and where x is large the centre of gravity of the trapezoid may be found and the resultant applied at this point. Usually it will be accurate enough to assume the pressure on DE as uniformly distributed at an average value and applied at the centre of the projection outside of the vertical wall.

The upward pressure on the floor EC varies from the value above given at E, to zero at C. It varies uniformly between these points. The downward pressure is the weight of the earth above the floor, $=W_2$. This may be assumed as uniformly distributed and equal to w_2h per unit area at all points. The total downward pressure on EC will be greater than the upward pressure unless x is very small.

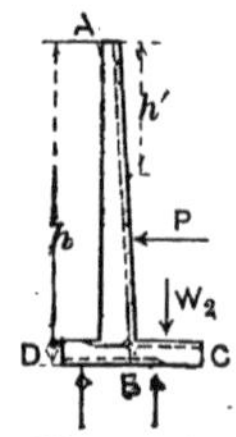

FIG. 112.

191. Design of Wall.—In discussing the design it will be necessary to consider two forms: (1) the cantilever wall without back tie-walls as in Fig. 112, and the wall provided with such back walls as in Fig. 113.

The form of Fig. 112 is adapted to heights of about 12 to 18 feet. For high walls the form of Fig. 113 will be more economical.

Form (*a*). (Fig. 112.)—The maximum moment in the upright portion AE is $P\frac{h}{3}=\frac{ph^3}{6}$. At any distance h' below the top the moment is $\frac{ph'^3}{6}$. Only a portion of the reinforcing-rods need be carried up the full height. The shear at the bottom is $P=\frac{ph^2}{2}$. This will be very small and will require no special attention. The reinforcing-rods of a cantilever beam have their maximum stress at the end of the beam, hence special care must be given to secure an effective bond or anchorage. In the figure the vertical rods have an insufficient length below the point of maximum moment to develop their full strength, and therefore they should be anchored in a substantial manner. This may be done by screw-ends and nuts, or by looping the rods around anchor-bars near the bottom of the floor *DC*.

The cantilever *DE* must be treated in the same manner as the upright cantilever. The pressures will be much heavier and the shear and bond stress may need attention. The reinforcement should extend far enough beyond *E* for bond strength.

The cantilever *EC* is acted upon by an upward and a downward force as shown in the figure. The maximum moment will be at *E* and will be negative. It is provided for by reinforcement as shown.

To secure maximum economy each one of the cantilevers may be tapered towards the end to a minimum practicable thickness. The bending moments at various sections in a cantilever beam uniformly loaded vary as the squares of the distances from the free end. The resisting moments vary approximately as the squares of the depths of the beam. Hence a beam tapering uniformly to zero depth at the end would be of the necessary depth at all points. The moments in the vertical beam *AE* vary as the cubes of the distances below the top, so that a straight taper will in this case give a beam whose weakest point will be at the bottom. At the top point *A* some form of coping is usually added, of a width according to the requirements of the case.

To prevent unsightly cracks a certain amount of longitudinal reinforcement is necessary. The amount required per square foot of cross-section will be less the heavier the wall, as temperature changes will be less in such a wall. On the basis of the discussion in Chap. V, Art. 142, the percentage required may be placed at about 0.4% as a maximum for thin walls, to perhaps one-half of this for heavy walls. High elastic-limit material is advantageous for this purpose.

Form (*b*). (Fig. 113.)—So far as the external pressures are concerned they have been explained in Art. 189, and are practically the same as in the previous case considered. The loads or pressures on the concrete are, however, carried quite differently. The toe *DE* is the same as in form (*a*) and reinforced in the same way. The pressure against the longitudinal wall *AE* is carried laterally for the most part and given over to the inclined

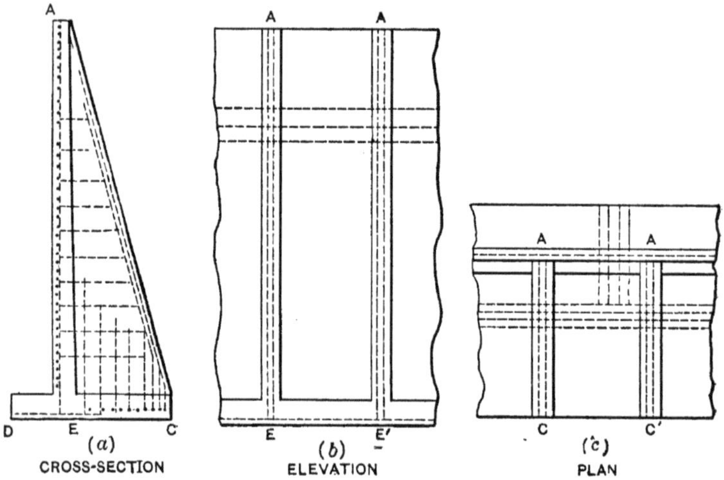

FIG. 113.

back walls. The wall *AE* must therefore be designed as a slab supported along the lines *AE* and *A'E'* (Fig. 113 (*b*)), and subjected to a pressure per square foot at any point a distance h' below the top equal to ph'. Near the bottom, the load on *AE* is transmitted more or less to the floor *EC*. The wall should therefore be bonded to the floor with a small amount of vertical reinforcement, which may well extend to the top to prevent cracks, although under ordinary conditions the wall *AE* is under some vertical pressure.

The floor *EC* is subjected to both upward and downward pressures, the latter exceeding the former towards the end *C*, and possibly throughout, as previously explained. This floor is supported by the back wall *AEC* and is therefore reinforced longitudinally as a floor-slab in accordance with the resultant pressure at any point. Here, again, it is well to bond the floor to the wall *AE* by extending the transverse reinforcement of the toe *DE* into the portion *EC*.

The back wall *ACE* acts as a cantilever beam anchored to the floor. It is also a T-beam, the flange being the longitudinal wall *AE*. The tension along the edge *AC* is carried by rods near this edge, whose stress at any point is found with sufficient accuracy by an equation of moments taken about the center of the front wall. The maximum stress will be at the bottom, if the wall is made with a straight profile. At the connection of the wall *AEC* to the floor, it is to be noted that the floor load is transferred to the wall along the line *EC*, but mainly near the end *C*. The main tension-rods in *AC* should therefore be distributed somewhat at their lower ends and well anchored to the reinforcing-rods of the floor *EC*. A few additional vertical rods should also be put in to insure thorough bonding of floor to wall. These will also carry a part of the tension in the back wall, but will not be as efficient as the rods nearer the outside edge. It is desirable, likewise, to bond the vertical wall *AE* to the back wall with short horizontal rods as shown. The slabs formed by the walls *AE* and the floor *EC* are continuous over supports, and if the span is long should be provided with some reinforcement for negative moments at these supports.

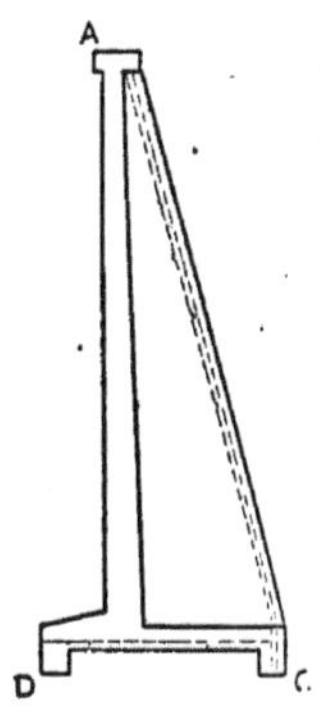

FIG. 114.

Fig. 114 shows some additional features of design which have been used. A longitudinal beam is built at *C* and the floor is thus supported on all four edges. The main rods along *AC* are then anchored into the beam.

A horizontal beam may also be made of the coping at *A*, thus giving some support to the wall *AB* along its upper edge.

A projection may be necessary at the toe D, or elsewhere, in order to increase the resistance against forward sliding. The beam C aids in this respect.

192. Illustrative Examples.—Fig. 115 shows the form of retaining-wall used on the Great Northern R.R. at Seattle,

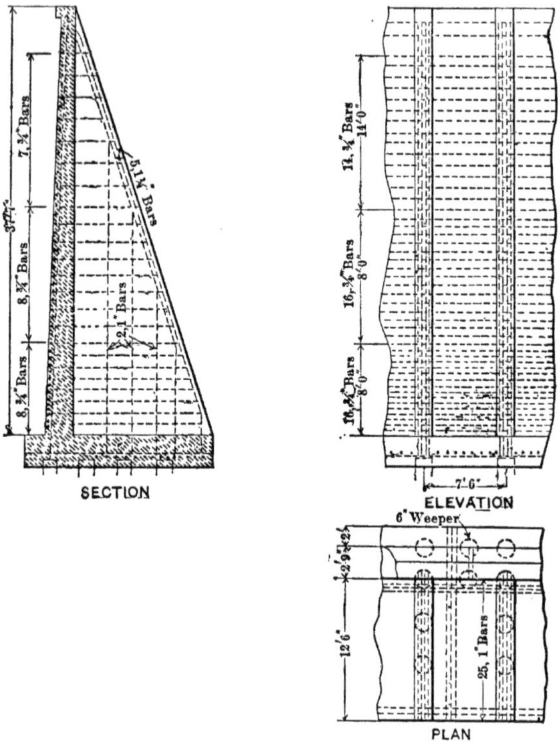

FIG. 115.—Retaining-wall, Great Northern Railway.

Wash.* This is a good illustration of the second type above discussed. An estimate by Mr. C. F. Graff of the amounts of material per lineal foot required in reinforced and plain con-

* Eng. News, Vol. LIII, 1905, p. 262.

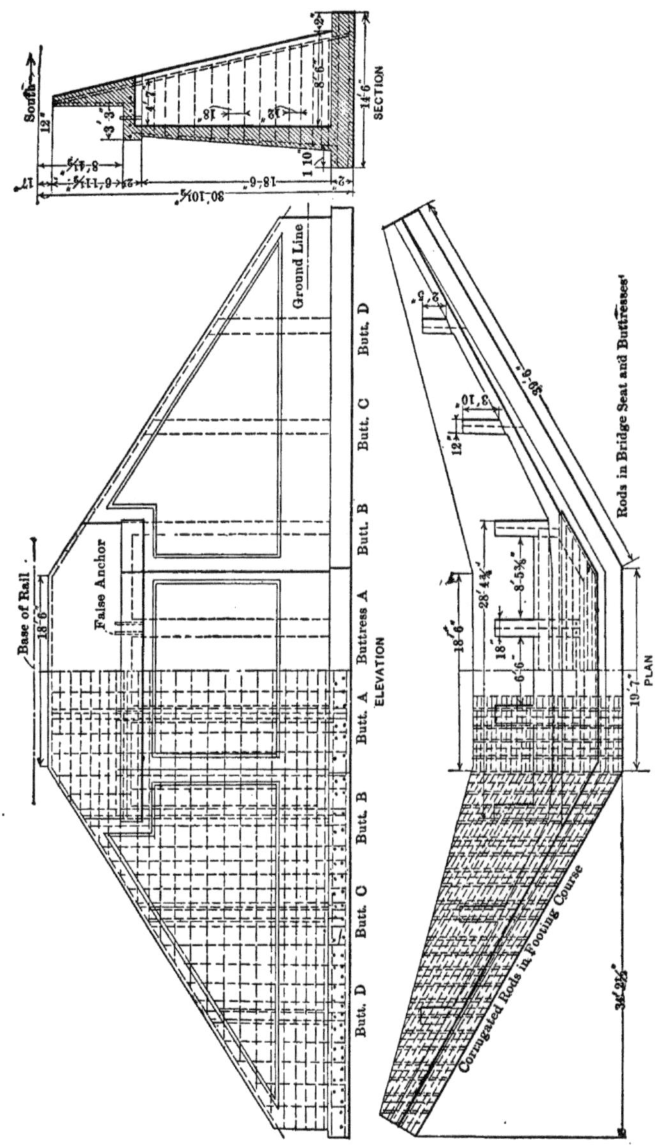

FIG. 116.—Standard Abutment, Wabash Railroad.

crete walls, made in connection with the design of Fig. 115, gave the following results:

Height of Wall, Feet.	Amount of Concrete per Lineal Foot.		Saving: Per Cent of Reinforced Wall.
	Plain Wall, Cubic Feet.	Reinforced Wall, Cubic Feet.	
40	396.4	218	45
30	226	127.8	43.3
20	110	69.9	36.4
10	44	34.9	20.4

The steel was included by adding its concrete equivalent.

Fig. 116 illustrates a standard form of abutment used by the Wabash R.R. Co.*

193. Retaining-walls Supported at the Top.—Frequently a retaining-wall may be supported at the top. In such a case it is designed as a simple beam supported at the top and bottom; or vertical ribs or beams may thus be calculated and the slab reinforced horizontally and supported by these ribs.

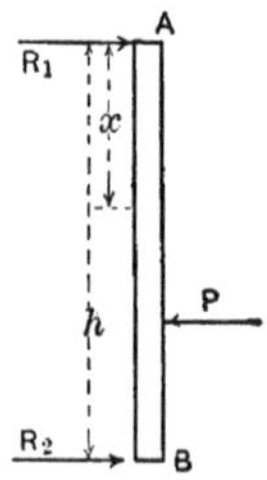

FIG. 117.

A wall AB (Fig. 117) acted upon by a pressure uniformly varying from zero at the top to a maximum at the bottom will be subjected to a bending moment whose maximum value will be determined. Let the pressure be that due to a fluid weighing p lbs/ft³. Then $P=\frac{1}{2}ph^2$, $R_1=\frac{1}{3}P=\frac{1}{6}ph^2$, $R_2=\frac{1}{3}ph^2$.

The bending moment M at a distance x below

$$A=R_1x-\tfrac{1}{6}px^3=p/6\,(h^2x-x^3).$$

This is a maximum for $x=h\sqrt{\frac{1}{3}}=.58h$. The maximum moment is then

$$M=.064ph^3. \quad . \quad . \quad . \quad . \quad . \quad . \quad . \quad . \quad (8)$$

If the pressure is water pressure, as in a reservoir, the value of the maximum moment becomes equal to

$$M=4h^3, \quad . \quad . \quad . \quad . \quad . \quad . \quad . \quad . \quad . \quad (9)$$

where the units are the foot and pound. For an earth retaining-wall with $p=20$, then $M=1.3h^3$, etc.

* Ry. Rev., Vol. XLV, 1905, p. 523.

DAMS.

194. The dam is a form of retaining-wall, but is subject to somewhat different conditions as to pressures. For this case a form of wall as shown in Fig. 118 is poorly adapted, owing to the fact that the water pressure will probably penetrate beneath the floor *DC* and exert an upward force nearly equal to the downward pressure, thus destroying the usefulness of the floor *EC*. To obviate these objections the wall *AE* must be brought back to the point *C*. Increased stability will then be secured by making it inclined. In this position it will naturally be supported by transverse walls or buttresses, resting on a floor *DC*, or directly on the foundation material, as shown in Fig. 119. The water pressure on the floor may then be relieved by

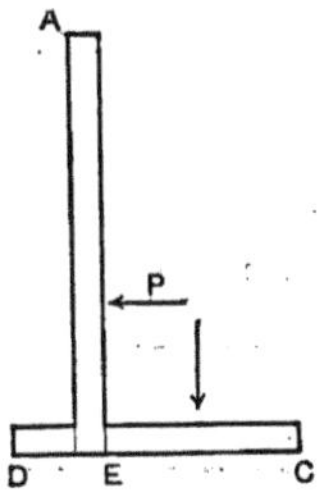

FIG. 118.

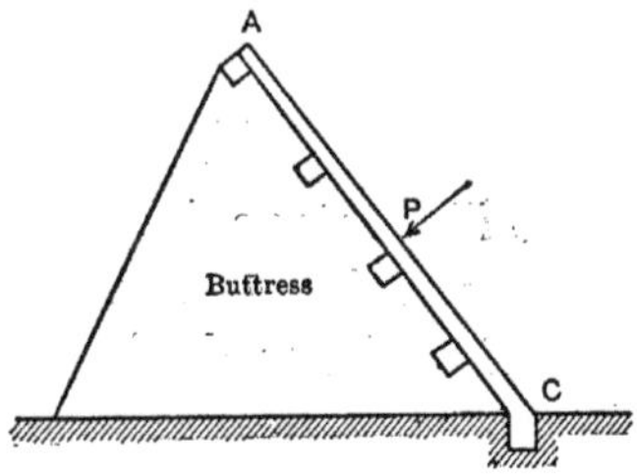

FIG. 119.

drain-openings allowing free exit for seepage-water. Thus built it forms a stable and efficient type of dam. Its design as to stresses and sections is simple and obvious. The wall or floor *AC* may be supported directly on the cross-walls and reinforced with longitudinal rods, or longitudinal beams may be used as shown and the slab supported on these. The pressure on the foundation is determined by considering the resultant of water pressure and weight of dam. The buttresses or cross-walls are subjected only to compressive stresses. Ample longitudinal reinforcement should be provided to thoroughly bind the structure together. Dams are often subjected to dynamic loads as well as static pressures, and sections must be provided more liberally than in many other structures.

The form shown in Fig. 119 is not suited to act as a spillway except for low falls. For a spillway the down-stream edge of the buttresses is also covered with a floor which may be curved in the usual manner.

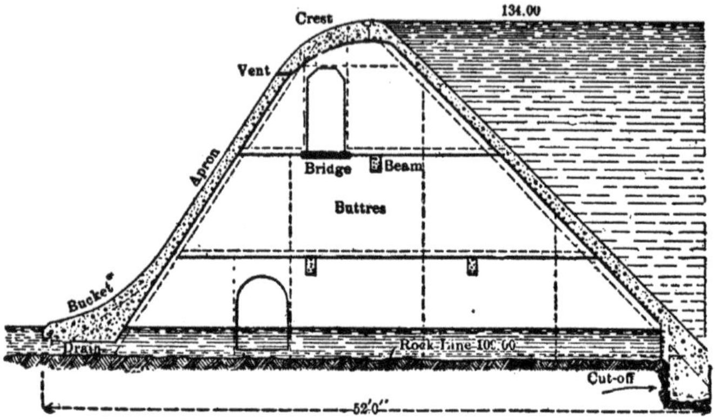

Fig. 120.—Dam at Schuylerville, N. Y.

Fig. 120 illustrates a dam of this type built at Schuylerville, N. Y., by the Ambursen Hydraulic Construction Co.* A footway is provided for in the interior. The design as to strength is obvious.

* Eng. News, Vol. LIII, p. 448.

CHAPTER X.

MISCELLANEOUS STRUCTURES.

GIRDER BRIDGES AND CULVERTS.

195. For short spans, the girder bridge or box culvert is likely to be a more economical form than the arch, owing to the less rigid requirements for foundations and abutments. For purposes of analysis this type of structure may be divided roughly into three classes: (1) Simple spans in which the girder rests upon independent abutments or piers; (2) concrete trestles or bridges in which the girders, abutments and piers form a monolithic structure; and (3) pipe culverts and box culverts built as square or rectangular pipes.

196. The Simple Beam Bridge. — These are designed in the same manner as any other concrete floor. Spans up to 20 to 30 feet may well be made as a simple slab or uniform thickness spanning the opening. For railroad structures the loads are relatively so large that shearing stresses will usually require careful attention. For longer spans a gain in economy will result by the use of main horizontal girders of relatively great depth, with a floor supported by the girders and reinforced transversely. The bridge may be made either a "through" or "deck" girder, according to the requirements of the case, the latter being the more economical. Floors of reinforced concrete are also used for steel truss and girder bridges to a considerable extent where a solid floor is desired. The details are arranged in a variety of ways, but the calculation and design of the reinforcement to meet the given conditions require no special consideration. The proper allowance for impact is an important point in this connection. Durability is an important factor favorable to the use of reinforced concrete for bridge floors.

197. Concrete Trestles. — Where several short spans are required and concrete is used for both the girders and the piers, the latter may usually be made of comparatively small cross-section, — much smaller than possible if ordinary masonry be used. The structure then approaches the ordinary floor and column construction in the relations of its parts. The piers, if lightly loaded, may consist merely of two or more columns connected by a suitable portal. In some extreme cases designs have been carried out in which the supporting piers or towers have been arranged in a manner similar to a steel trestle, even to the diagonal bracing. It would seem, however, that the treatment of concrete should be on somewhat different lines than is best suited to such a material as steel, and that structural forms in concrete should be somewhat massive and limited in general to the beam and the compression member.

Where the piers are made small, as here assumed, they must be built rigidly in connection with the girders of one or more spans, as are the columns in a building. The girders must be designed with proper reference to their continuity, and the piers must be able to resist a certain amount of bending moment. This moment can be estimated in the manner suggested in Chapter VII, Art. 167.

As an example, let Fig. 121 represent a concrete trestle of monolithic construction. The girders are continuous and the

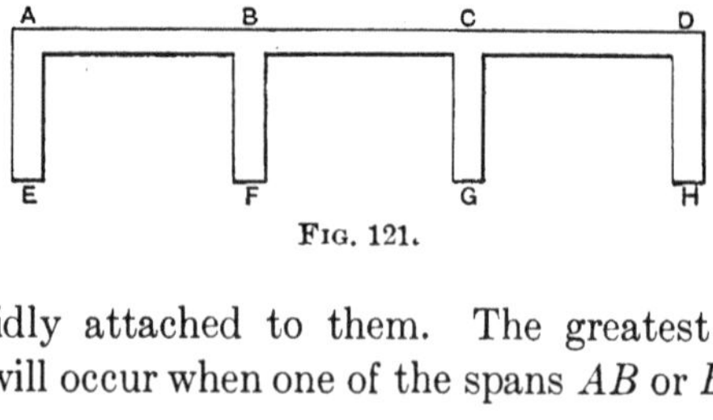

FIG. 121.

piers are rigidly attached to them. The greatest moment in the pier BF will occur when one of the spans AB or BC is loaded.

Suppose BC be loaded. Then calculate the negative moment at B, assuming BC to be fixed at the ends. This moment will be equal to $-\frac{1}{12}\,pl^2$, where p = load per foot and l = span length. Now this moment is distributed at the joint B among the three members AB, BF, and BC in proportion to the value of I/l for the three members, the length l being taken as the estimated length to the point of inflection in each case (the full length of BF). This will determine approximately the moment in BF. The *maximum* negative moment in BC and AB will occur when both spans are loaded and will be approximately equal to $\frac{1}{10}\,pl^2$. (See Chapter VII, Art. 155.) The end piers or abutments must be designed also as retaining walls.

198. Pipe and Box Culverts. — For small openings the monolithic pipe or box form is very advantageous. This form of structure is a complete opening in itself and so long as intact will do good service. Considerable settlement, as a whole, may be permissible, and hence solid foundations may not be needed.

The cross-section may be circular, elliptical or rectangular. Theoretically, the elliptical form is the best as corresponding more nearly to the requirements for resisting the earth pressure. The circular is practically as good for small openings, while for large openings the rectangular form will often be the best on account of its simplicity and the lesser head room required. Where the culvert is manufactured at a shop and transported to the site, the circular or elliptical forms will usually be the most advantageous. As the loads coming upon such structures are not accurately known an exact analysis of the stresses is impossible, but the results obtained for certain simple cases will be useful as a guide to the judgment. The general method of analysis employed in Chapter VIII has been used. The details of the analysis will be omitted.

199. *The Circular Culvert.* — Two cases have been analyzed; (1) for a uniform load, and (2) for a concentrated load.

Case I; Uniform load. (Fig. 122.) It is assumed that the pressure on the pipe is exerted in parallel lines (as downward and upward) and is uniformly distributed with respect to a plane perpendicular to the direction of the pressure.

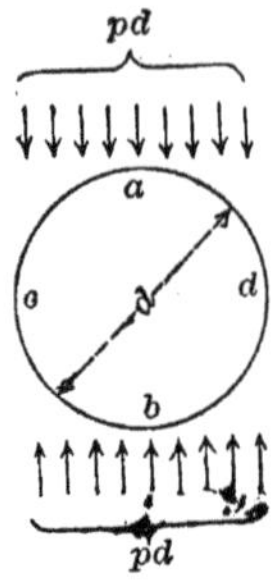

FIG. 122.

Let d = diameter of pipe;
p = pressure per unit area as measured perpendicularly to the pressure;
M = bending moment in pipe in a length of one unit;

Then the following equations result.

$$M_a = M_b = \tfrac{1}{16}\,pd^2 \quad . \quad . \quad . \quad . \quad . \quad . \quad . \quad . \quad (1)$$
$$M_c = M_d = -\tfrac{1}{16}\,pd^2 \quad . \quad . \quad . \quad . \quad . \quad . \quad . \quad (2)$$

If the lateral pressure, measured in a similar way, be p' per unit area, then the moments due to this pressure will be

$$M_a = M_b = -\tfrac{1}{16}\,p'd^2 \quad . \quad . \quad . \quad . \quad . \quad . \quad . \quad (3)$$
and
$$M_c = M_d = \tfrac{1}{16}\,p'd^2 \quad . \quad . \quad . \quad . \quad . \quad . \quad . \quad . \quad (4)$$

For equal horizontal and vertical forces (equivalent to a uniform radial pressure), the moments at all points are zero. Usually the lateral pressure will be much less than the vertical pressure; probably not more than one-fourth or one-fifth as much. Assuming a ratio of one-fourth, the resulting total bending moments at the points a, b, c, d, will be $\tfrac{3}{64}\,pd^2$, positive at the top and bottom and negative at the sides.

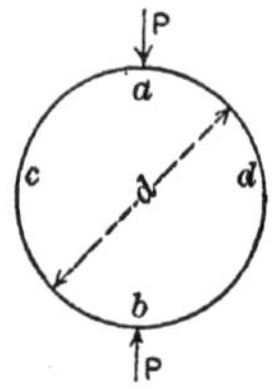

FIG. 123.

Case II; Concentrated loads at opposite points (Fig. 123).

In this case the moments are

$$M_a = M_b = .16\,Pd \quad . \quad . \quad . \quad . \quad . \quad . \quad . \quad (5)$$
$$M_c = M_d = -.09\,Pd \quad . \quad . \quad . \quad . \quad . \quad . \quad (6)$$

200. *The Rectangular Culvert.* — Case I; Uniform loads (Fig. 124).

Let l_1 = width of culvert;
l_2 = height of culvert;
I_1 = moment of inertia of top and bottom, assumed as equal;
I_2 = moment of inertia of sides;
p = vertical load and foundation reaction per unit area.

Then

$$M_a = M_b = \frac{pl_1^2}{8} \cdot \frac{\frac{1}{3}l_1/I_1 + l_2/I_2}{l_1/I_1 + l_2/I_2} \quad \ldots \quad (7)$$

$$M_c = M_d = M_a - \tfrac{1}{8}pl_1^2 \quad \ldots \quad (8)$$

The moments at e and f are equal to M_c.

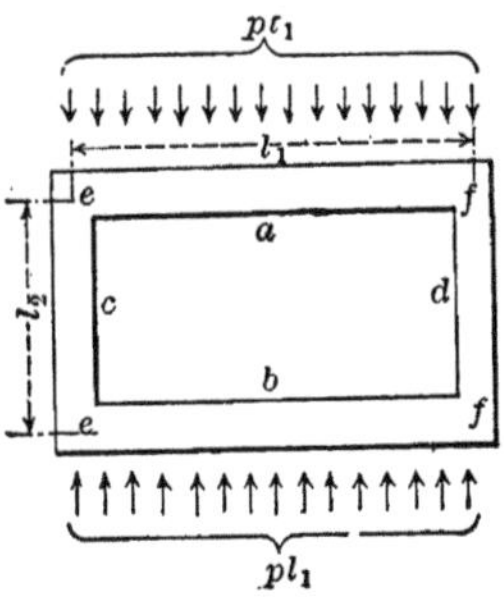

FIG. 124.

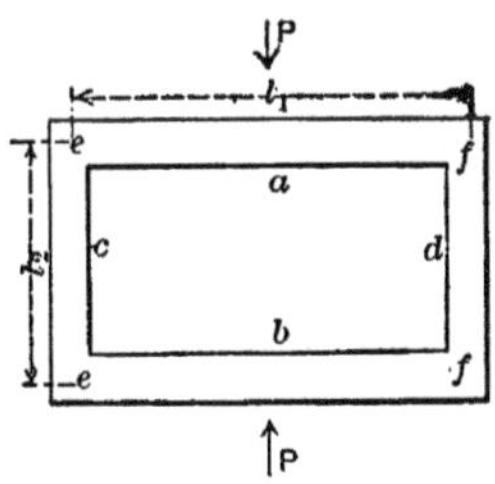

FIG. 125.

For a square culvert with uniform section $M_a = \frac{1}{12}pl^2$ and $M_c = -\frac{1}{24}pl^2$.

For equal vertical and lateral loads the moments in the square culvert become $M_a = M_c = +\frac{1}{24}pl^2$ and $M_e = -\frac{1}{12}pl^2$ as in a beam with fixed ends.

Case II; Concentrated loads. (Fig. 125.)

For vertical loads applied centrally,

$$M_a = M_b = \frac{Pl_1}{4} \cdot \frac{\frac{1}{2}l_1/I_1 + l_2/I_2}{l_1/I_1 + l_2/I_2} \quad \ldots \quad (9)$$

$$M_c = M_d = M_a - \tfrac{1}{4}Pl_1 \quad \ldots \quad (10)$$

For the square form, $M_a = \frac{3}{16} Pl_1$ and $M_c = -\frac{1}{16} Pl_1$; and for equal lateral and vertical forces $M_a = M_c = \frac{1}{8} Pl_1$ and $M_e = -\frac{1}{8} Pl_1$ as for fixed beams.

201. Arrangement of Reinforcement.—The bending moments here determined are based on the assumption that the entire section is reinforced so as to act as a monolithic structure. This of course requires proper reinforcement for negative as well as positive moments.

In the circular form a wire mesh is convenient, especially for small diameters. A single mesh will be sufficient, placed near the intrados at top and bottom and near the extrados at the sides, crossing the central axis at about the quarter point.

In the rectangular form, if reinforcement for negative moments at the corners is omitted, then the four sides will act as simple beams, the concrete cracking more or less on the outside near the corners.

Longitudinal reinforcement should be provided to some extent. Where foundations are good a very small amount will be sufficient, but if settlement is likely to occur the longitudinal reinforcement becomes of much importance. The entire culvert will act as a beam subjected in the main to positive bending moments. Most of the reinforcement should therefore be placed along the bottom of the culvert.

201*a*. Tests of Reinforced Concrete Rings and Culvert Pipe.—Large reinforced concrete rings and pipe have been tested by Professor Talbot, with results agreeing closely with the theoretical analysis of Art. 199. Loads were applied in two ways, (*a*) as concentrated loads, in which the pipe was supported along an element at the bottom and the load was applied along an element at the top; and (*b*) as distributed loads, in which the pressures at bottom and top were distributed as uniformly as possible over the entire horizontal projection of the pipe by means of a carefully constructed sand box.* These

* For full details see Bulletin No. 22, Eng. Exp. Sta., University of Illinois, 1908.

methods of loading correspond to the cases of concentrated and distributed loads discussed in Art. 199. All rings and pipe were 48-in. internal diameter and 4-in. thick, the rings being 24 in. long, and the pipe sections 102–104 in. long. The latter were made with the usual bell end. The reinforcement consisted of ¼-in. corrugated bars except in No. 982, in which ½-in. bars were used, and in No. 988, in which No. 3 Clinton wire mesh was used. The rings were made circular and the reinforcement placed near the intrados at top and bottom and near the extrados at the sides. In the pipe, the reinforcement was made circular and the concrete cast with a vertical diameter 4 in. greater than the horizontal, giving a similar relative position for the reinforcement. The results are given in Tables 23 and 24. In these tables the value of t is the net or effective thickness of the pipe as measured to the center of the steel. In the last column of Table 23 the theoretical strength is the strength calculated by means of eq. (5), Art. 199, assuming the bending resistance of the pipe to be .87 Af_st per lineal foot, using the yield-point of the steel for the value

TABLE No. 23.

RESULTS OF TESTS ON REINFORCED CONCRETE RINGS.

(TALBOT.)

(Concentrated Loads.)

Diam. of rings = 48 ins.; thickness = 4 ins.; age 1–3 mos.

No.	Reinforcement, Per Cent.	Load at First Crack, Lbs/lin.ft.	Maximum Load, Lbs/lin.ft.	t inches.	Ratio of Theoretical to Actual Strength.
926	0.73	1500	2850	2.75	1.12
928	0.80	1400	3550	2.5	0.82
931	0.73	2150	2500	2.75	1.28
932	0.66	1500	3000	3.0	1.16
933	1.00	1170	3170	2.0	0.73
934	0.80	1300	3150	2.5	0.92
952	1.00	1000	2350	2.0	0.99
953	0.89	1200	3600	2.25	0.73
971	0.73	1500	4120	2.75	0.77

TABLE No. 24.

RESULT OF TESTS ON REINFORCED CONCRETE RINGS AND PIPES.

(TALBOT.)

(Distributed Loads.)

Diam. of rings and pipe = 48 ins.; thickness = 4 ins.; age of rings 1–3 mos.; age of pipe 4–6 months.

REINFORCED CONCRETE RINGS.

No.	Reinforcement, Per Cent.	Load at First Crack, Lbs/lin.ft.	Critical Load, Lbs/lin.ft.	Maximum Load, Lbs/lin.ft.	t inches.	Ratio of Theoretical to Actual Strength at Critical Load.
923*	0.80	2250	7000	10500	2.5	1.06
921	0.80	3500	10000	23500	2.5	0.74
922	0.80	3250	10000	18500	2.5	0.74
927	0.80	3250	8000	26000	2.5	0.93
951	0.80	3200	9000	25000	2.5	0.83
972	0.73	4500	8000	17500	2.75	1.03
976	0.66	4000	9000	19000	3.0	0.99
977	0.66	4000	10000	21000	3.0	0.89
			REINFORCED CONCRETE PIPE.			
981	0.66	8360	19500	31500	3.0	0.55
982	1.39	10960	15000	24800	3.0	1.49
983	0.66	4950	12500	23800	3.0	0.86
988	0.88	6700	9000	31400	3.0	

* No lateral restraint.

of f_s. This was taken at 46,400 lbs/in² for the rings and 55,000 lbs/in² for the pipe. The tests showed that the primary cause of failure was the failure of the steel in tension. In Table 24, similarly, the theoretical strength is determined on the basis of eq. (1) of Art. 199. In the case of the distributed load tests the "critical load" was estimated as the load beyond which the increased resistance was primarily due to increased lateral resistance of the sand filling and not of the ring or pipe itself. The ultimate resistance would obviously be chiefly dependent upon the character of the filling about the pipe.

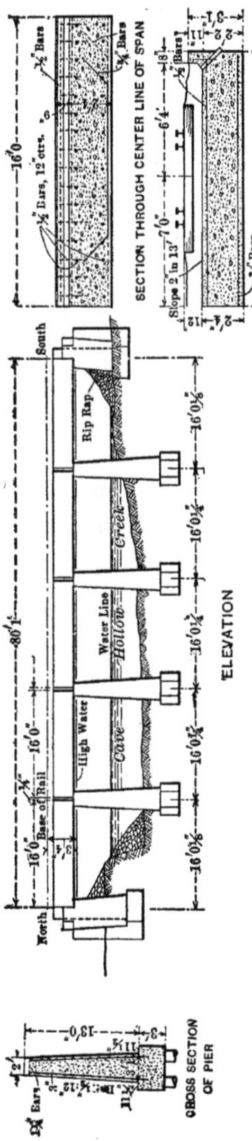

FIG. 126.

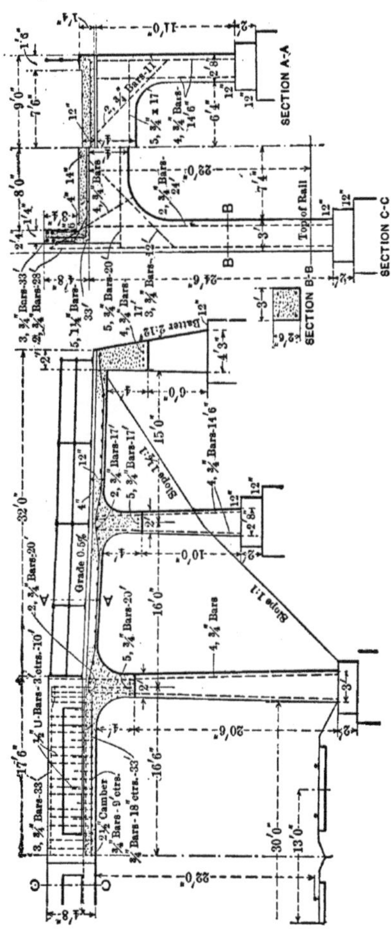

FIG. 127.

The results of the concentrated load tests show close agreement between the theoretical and experimental values. This means merely that the resisting moment of a pipe may be taken the same as that of a straight beam of the same thickness and reinforcement. The tests under distributed loads show that the theoretical strength can be secured by very careful bedding and testing. It would seem that in practice nothing should be allowed for lateral support and that the theoretical moment of $1/16\, pd^2$ may be used if the filling and bedding are carefully done. If poorly done a larger bending moment will exist and will need to be considered.

202. Illustrative Examples.—Fig. 126 illustrates a simple beam bridge or "trestle" on the Chicago, Burlington and Quincy R.R.* The girder consists of a slab twenty-four inches in thickness, reinforced as shown in the illustration. The piers are separate structures.

Fig. 127 represents a concrete highway bridge as an overhead crossing of the Big Four R.R. This design illustrates the deep girder with floor-slab reinforced transversely, and also the "trestle" in which the piers are columns built as one piece with the girders.†

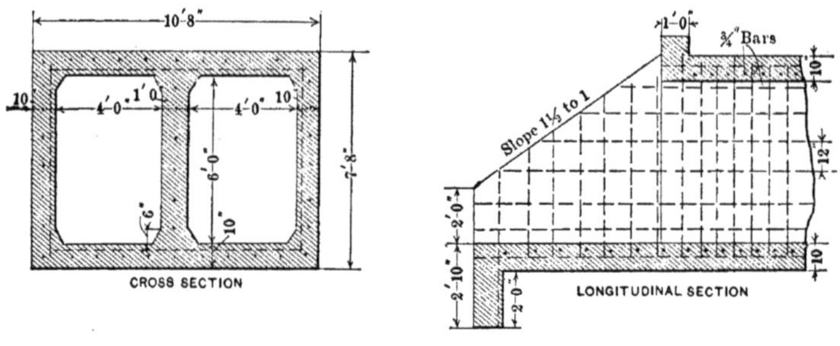

FIG. 128.

Fig. 128 illustrates a standard design for a monolithic box culvert. It is not reinforced for negative moment at the

* R. R. Gaz., Vol. XL, 1906, p. 713.
† R. R. Gaz., Vol. XL, 1906, p. 497.

corners. This form of construction is applicable to many other structures as subways, tunnel linings, etc. No special consideration of these various applications of the reinforced beam is required in this place. A clear understanding of the general principles of reinforced concrete design will enable the details to be suitably modified to meet the conditions of the case.

Conduits and Pipe Lines.

203. For conduits not under pressure, large sewers and the like, reinforced concrete lends itself to convenient and economi-

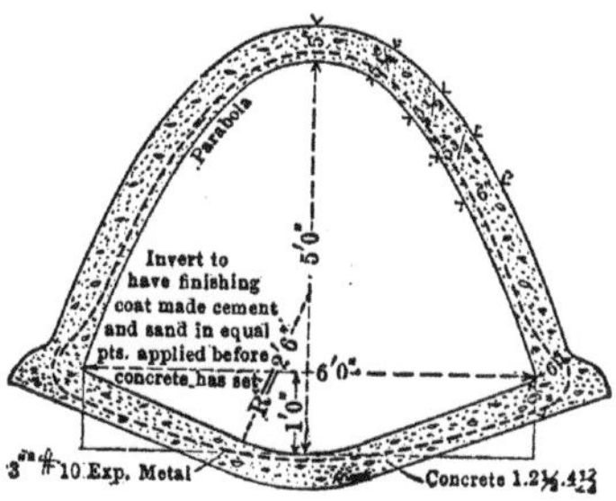

Fig. 129.

cal construction. As to the analysis and design, these structures are only special cases of the monolithic pipe or box discussed in preceding articles. The character of the foundation

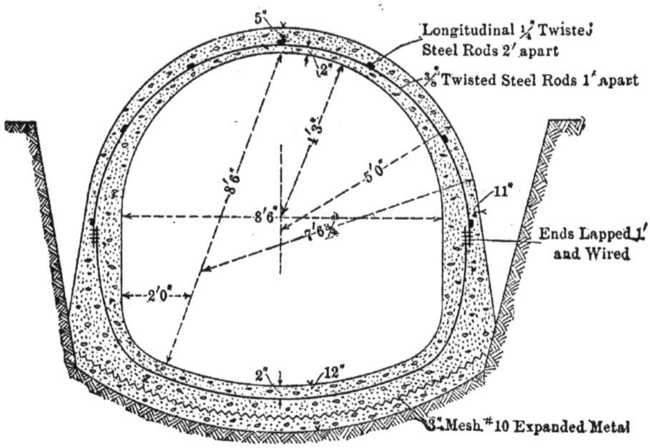

Fig 130.

and convenience in construction will lead to various modifications of design.

Fig. 129 is a typical cross-section of a large sewer for Harrisburg, Pa. A mesh of expanded metal is used for reinforcement, arranged to resist positive moments excepting at bottom and corners.

Fig. 130 illustrates a large conduit of the Jersey City Water Supply. This section is employed where the bottom is soft, special reinforcement being used in the invert. The position of the reinforcement to carry positive moments at crown and negative moments at sides should be noted.

Reinforced concrete has also been used to some extent for pipes under pressure, but it is very difficult to secure imperviousness under heads of considerable magnitude. In pressure pipes the tensile stress is entirely taken by the steel, the concrete furnishing merely the impervious layer and resisting bending due to earth loading.

Tanks, Reservoirs, Bins, etc.

204. For covered reservoirs reinforced concrete is very well adapted. The rectangular form with flat cover is usually the most convenient; its design involves the same features as building design with the additional one of imperviousness. Elevated towers and tanks may also be made of concrete, but high pressures are difficult to deal with.

Bins and coal pockets are structures for which concrete is well adapted. For the storage of coal unprotected steel is not durable, but reinforced concrete furnishes an almost ideal material, lending itself readily to the necessary form for strength and furnishing the desired durability.

Reinforced concrete is advantageously used in other minor forms of structures and structural elements. Noteworthy among such uses are its employment for piles, fence posts, and poles for various purposes. For piles it is especially advantageous in situations where continual submergence is not certain. For further illustrations the reader is referred to the larger works on the subject and especially to the American works of Messrs. Buell and Hill and Mr. Reid.

CHAPTER XI.

REINFORCED-CONCRETE CHIMNEYS.

205. General Description.—Over 400 reinforced-concrete chimneys have already been built in America, one at least 350 ft. high, and 18 ft. in diameter. Fig. 131 represents the general features of such chimneys as usually built; they are the shaft or outer shell, the base, and the lining or inner shell.

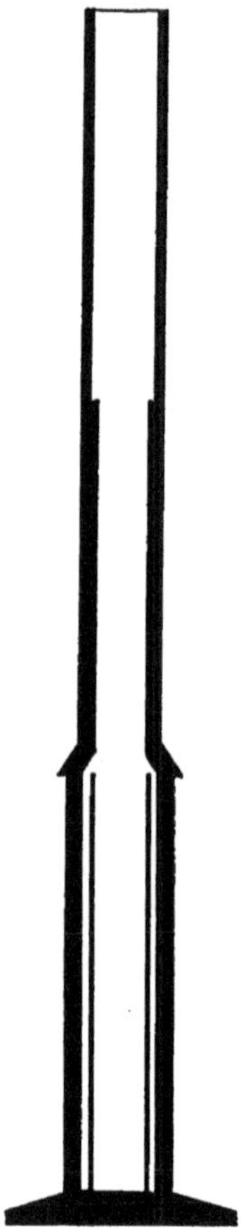

FIG. 131.

The outer shell is usually made of two or more thicknesses, the maximum thickness being about 12 in. in a large, tall chimney, and the minimum is as small as 4 in. in some. Outer shells are usually built with an offset at the top of the lining to avo d a large change in internal diameter, and for architectural effect, but as an offset is a weak place unless very well reinforced, some chimneys have been built with uniform outer diameter. Shells are generally made cylindrical, but at least one firm of builders advertises taper or conical chimneys.

Bases are generally square or octagonal in plan, and thinner at the perimeter than at the shells; but small bases have been built uniform in thickness for simplicity, and large ones have been built thin near the center as well as at the perimeter to effect a saving of concrete.

Linings for the protection of the outer shell from excessive heat, were built to the top in some of the earlier chimneys.

Now they are generally built to about one-third the height; but there are some ch mneys in use without any l uing. There s no rigid connection between the shells except at the base, hence the lining is free to expand and contract and is not subjected to wind stresses. The thickness is generally 4 or 5 in. and the air space between the shells is made as small as the forming will allow, about 4 in. The air space communicates with the outer air through vent holes.

206. Design of a Chimney.—The inner diameter and the height of the chimney are determined by the requirements of draft and capacity—matters which do not fall within the scope of this work.

The outer shell may be made 4 to 6 in. at the top, depending on the diameter, the thickness to be increased one or more times in the height as required. While many changes in thickness would result in a saving of concrete, it must be noted that each change means an alteration in forms and hence an expense. Thicknesses are chosen tentatively, also amounts of reinforcements, both vertical and circumferential, and then various sections are investigated for stress. Methods for computing wind and temperature stresses are explained in Arts. 207–209 and 211–214.

A base should be made with such an extent of bottom that its greatest pressure on the earth due to weight of chimney, weight of earth filling over the base, and wind pressure, will not exceed the permissible limit; and the base itself must be strong enough to withstand the pressures on its top and bottom. Methods for investigating these points for a given base are explained in Arts. 215–217.

Linings are designed, as yet, by precedent. In the report on his investigations of reinforced-concrete chimneys,* Sanford E. Thompson recommends that for temperatures above 750° F. fire-brick be used for linings, but for lower, suitable cement mortar may be safely used. Inasmuch as a number of outer

---

* For abstract, see Eng. News, Jan. 9, 1908.

shells have cracked badly near the top of the lining, he suggests that linings be built higher than one-third the height or else that the outer shell be extra-reinforced near the top of the lining. No reports of injured linings were received by him. The usual thickness, 4 in. for moderate heights, and the reinforcements used would seem adequate but it should be noted that comparatively few owners have examined the linings of their chimneys. Published descriptions of reinforcements are meager. In one lining, 90 ft. high, the percentage of hoop steel is $\frac{1}{4}$ and the hoops are spaced 20 in.; in another 60 ft. high, it is $\frac{1}{6}$% and the spacing is 36 in. In the latter the percentage of vertical steel is 1.8.

207. **Wind Stresses in the Outer Shell.**—On a horizontal section of a chimney sustaining no wind pressure, the "fiber stress" in the concrete is a uniform compression. Wind pressure changes this uniform stress, increasing the intensity of the compression on the lee side and decreasing it on the windward. The decrease may be larger than the preexistent intensity, the net result being a tensile stress. Two cases will be distinguished; in both it is assumed, just as in the most widely used flexure formulas for the working strength of an ordinary reinforced-concrete beam, that the fiber stress is a uniformly varying one.

Notation.—In this connection see Figs. 132 to 136. Also let

A = area of chimney section under consideration;
A_s = total area of all steel sections there;
W = weight of superincumbent portion of chimney;
P = wind pressure on that portion;
M = bending moment at the section;
e = distance from the center of the section to where the resultant of the weight and wind pressure cuts the section, "eccentric distance";
f_c = unit stress in concrete adjacent to the steel at lee side;
f_c' = unit stress in concrete adjacent to steel at windward side;
f = unit stress on concrete at the lee side;
f' = unit stress on concrete at the windward side;

f_s = unit stress on steel at the windward side;
m = a coefficient such that $f_c = mW/A$;
m' = a coefficient such that $f_c' = m'W/A$;
p = steel ratio, i.e., A_s/A; and
n = ratio of modulus of elasticity of steel to that of concrete (taken as 15 in all numerical work following).

208.—*Case I.* The stress at the windward side is compressive or a tension of low intensity, say, 50 lbs/in.² This case obtains in all sections where the resultant of W and P falls within or not far without the kern* of the section. (The kern is a circle concentric with the hollow circle, its radius being $\frac{1}{4}r_2[1+(r_1/r_2)^2]$. Since in chimneys r_1/r_2 is nearly 1, the kern radius is nearly $\frac{1}{2}r_2$; it may be taken as $\frac{1}{2}r$). Fig. 133*a* represents the variation in the concrete stress (wholly compressive) when the resultant falls well within the kern, and Fig. 133*b* represents it (part tensile) when the resultant falls outside.

First Method.—This is carried out graphically by means of a diagram (Fig. 132); it will be described by examples, and then the analysis on which the diagram is based will be given. At the base of the diagram there are given values of the eccentricity, e/r from 0 to 0.8; at the left side values of the coefficient m, and at the right values of the coefficient m'.

---

* Imagine all the forces acting on either side of a section of a beam, or column, etc., to be compounded into a resultant R, or if that is impossible, into two forces: one, N, perpendicular to the section and one in the section (always possible). If R or N cuts the section at its centroid, then the normal stress at the section is all one kind, tension or compression, and it is a uniform stress. If R or N cuts the section and not very eccentrically, then the normal stress will still be of one kind but it will not be a uniform one. That part of the cross-section within which R or N must cut the section in order that the normal stress may be of one kind, is called the *kern* of the section.

When the normal stress varies uniformly, then the kern for any section can be determined easily. For this kind of normal stress the kern for a hollow circle is described above and for a square an octagon and a circle in Art. 216.

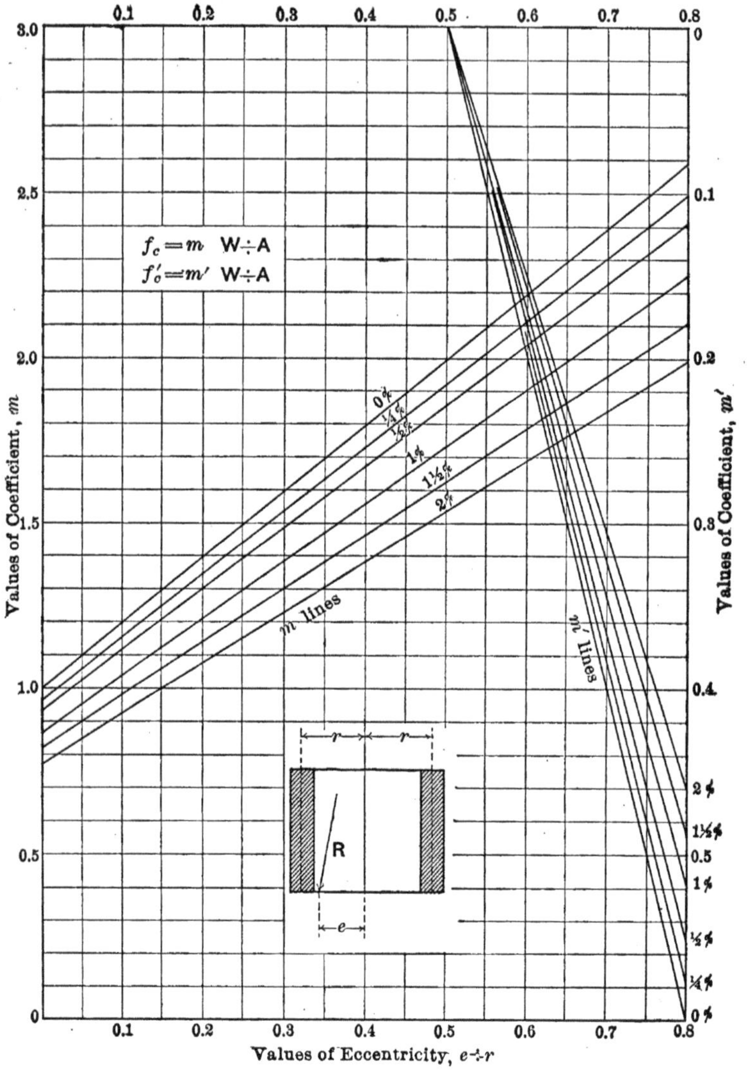

FIG. 132.—Wind Stresses in Chimneys.

The oblique lines relate to various percentages of steel from 0 to 2, and may be called "percentage lines." It will be noticed that there are no m' percentage lines for eccentricities less than 0.5; for such small values, the stress f_c' is com-

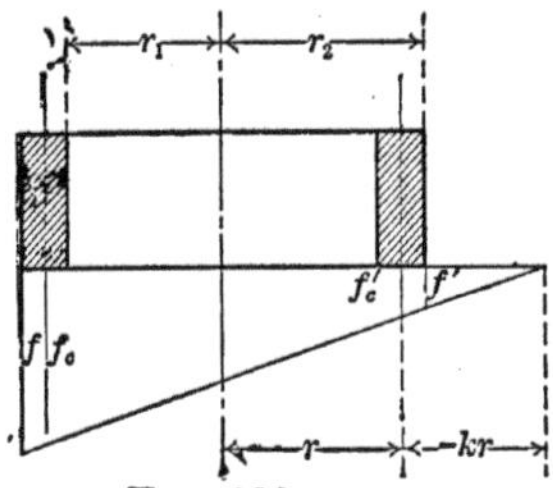

FIG. 133*a*.

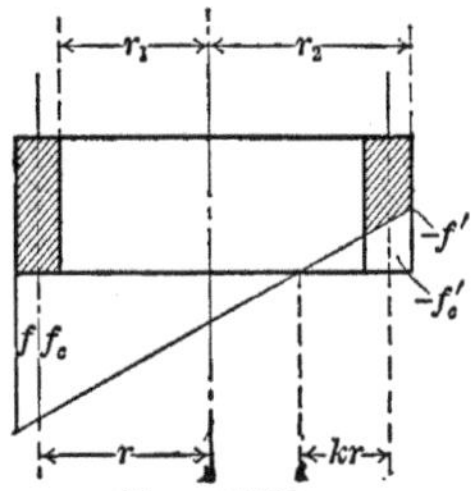

FIG. 133*b*.

pressive (and less than f_c) and there is no need ever to determine these small values of f_c'.

Examples.—(1) A reinforced-concrete chimney is 150 ft. high; its outer diameter is constant and equal to 12 ft. 2 in.; the upper 100 ft. of the shell is 6 in. thick and the lower 50 ft. is 8 in. thick. At a section 50 ft. from the top the vertical reinforcement consists of sixteen $\frac{3}{4}$-in. rods. The extreme unit stresses at this section are required, the chimney being under wind pressure assumed to be equivalent to 30 lbs/ft² of "projected area."

Taking the weight of concrete as 150 lbs/ft³, W is about 137,500 lbs., and since M is about 5,475,000 in-lbs., $e=M/W=39.8$ in., and $e/r=0.57$. Since $A_s=7.07$ in², and $A=2636$ in², $p=0.0027=0.27\%$. With these values of p and e/r, the diagram (Fig. 132) gives $m=2.06$ and $m'=0.135$; hence $f=2.06\ W/A=107$ lbs/in², and $f'=0.135\ W/A=7$ lbs/in². This f' being a small tension, the stress condition at the section under consideration does fall under Case I.

(2) At the section 100 ft. below the top of the chimney the vertical reinforcement consists of forty-eight $\frac{3}{4}$-in. rods. It is required to determine the extreme unit stresses there.

Here W is about 275,000 lbs., and M about 21,900,000 in-lbs.; hence $e=M/W=79.6$ in. and $e/r=1.137$. Since $A_s=21.2$ in², and $A=2636$ in², $p=0.008=0.8\%$. These values of e/r, and p fall beyond the limits of the diagram; hence the stress condition does not fall under Case I probably. Substituting in eq. (6), it will be found that $m'=1.15$; and hence $f'=1.15\ W/A=120$ lbs/in² approximately; this being a high tension, the stress condition does not fall under Case I and the methods of Case II should be applied (see ex. 1, Art. 209).

Analysis for the Diagram.—In the case of a uniformly varying stress, the average unit stress for any portion of the section equals the actual unit stress at the centroid of that portion (or at any point of the section whose distance from the neutral axis equals that of the centroid). Hence the average unit stress in the concrete is $\frac{1}{2}(f_c+f_c')$; also the average unit stress in the steel is $\frac{1}{2}(f_c+f_c')n$ (see Fig. 133). (It is supposed that the vertical steel is securely tied to the circumferential so that the former will not buckle.)

And since the total stress on the section equals W,

$$\tfrac{1}{2}(f_c+f_c')A+\tfrac{1}{2}(f_c+f_c')npA=W. \quad . \quad . \quad . \quad . \quad (1)$$

In the case of a uniformly varying stress, the point of application of the resultant of the stress on any portion of the section lies at a distance from the neutral axis equal to the ratio between the square of the radius of gyration of the portion with respect to the neutral axis and the distance of the centroid of that portion from the same axis. Now the radius of gyration of the concrete section is nearly the same as that of the steel circle (radius r), and hence the resultants of the concrete and steel stresses practically coincide. The square of the radius of gyration of this circle with respect to the neutral axis is $\frac{1}{2}r^2+(1-k)^2r^2$ (see Fig. 133); hence the arm of the resultants with respect to the neutral axis is

$$[\tfrac{1}{2}r^2+(1-k)^2r^2]/(1-k)r,$$

and the arm with respect to the center of the section is $r/2(1-k)$. Since the sum of the moments of these two resultants with respect to the center equals the bending moment,

$$\tfrac{1}{2}(f_c+f_c')A\frac{r}{2(1-k)}+n\tfrac{1}{2}(f_c+f_c')pA\frac{r}{2(1-k)}=We. \quad . \quad (2)$$

From eqs. (1) and (2), it follows that

$$k=1-\frac{0.5}{e/r}. \quad . \quad . \quad . \quad . \quad . \quad . \quad . \quad . \quad (3)$$

From the similar triangles in Figs. 133*a* and 133*b*, and eq. (3), it follows that

$$f_c' = -f_c k/(2-k) = f_c(1-2e/r)/(1+2e/r),$$

and if this value be substituted in (1), then the equation gives

$$f_c = \frac{1+2e/r}{1+np}\frac{W}{A}; \quad . \quad . \quad . \quad . \quad . \quad . \quad . \quad . \quad (4)$$

hence, also,

$$f_c' = \frac{1-2e/r}{1+np}\frac{W}{A}. \quad . \quad . \quad . \quad . \quad . \quad . \quad (5)$$

Since by definition, $f_c = mW/A$ and $f_c' = m'W/A$,

$$m = \frac{1+2e/r}{1+np}, \quad \text{and} \quad m' = \frac{1-2e/r}{1+np}. \quad . \quad . \quad . \quad (6)$$

The straight line in Fig. 134 was plotted from eq. (3), and all lines in Fig. 132 from eqs. (6).

It should be noticed that f_c and f_c' are not the unit stresses for the extreme fiber; these latter might be obtained from the former and k by proportion (see Figs. 133*a* and 133*b*), or by the

Second Method.—This is the ordinary method for combining "direct" and flexural stress; it gives the unit stresses for the extreme fibers. Thus, I denoting the moment of inertia of the concrete-steel section about a diameter, computed as explained below, then

$$f = \frac{W}{A} + \frac{Mr_2}{I},$$

and

$$f' = \frac{W}{A} - \frac{Mr_2}{I}.$$

If, in a given instance, f' comes out negative, then the stress at the windward side of the section under consideration is tensile. The greatest compressive unit stress in the steel is less than nf, and, if some of the steel is under tension, its

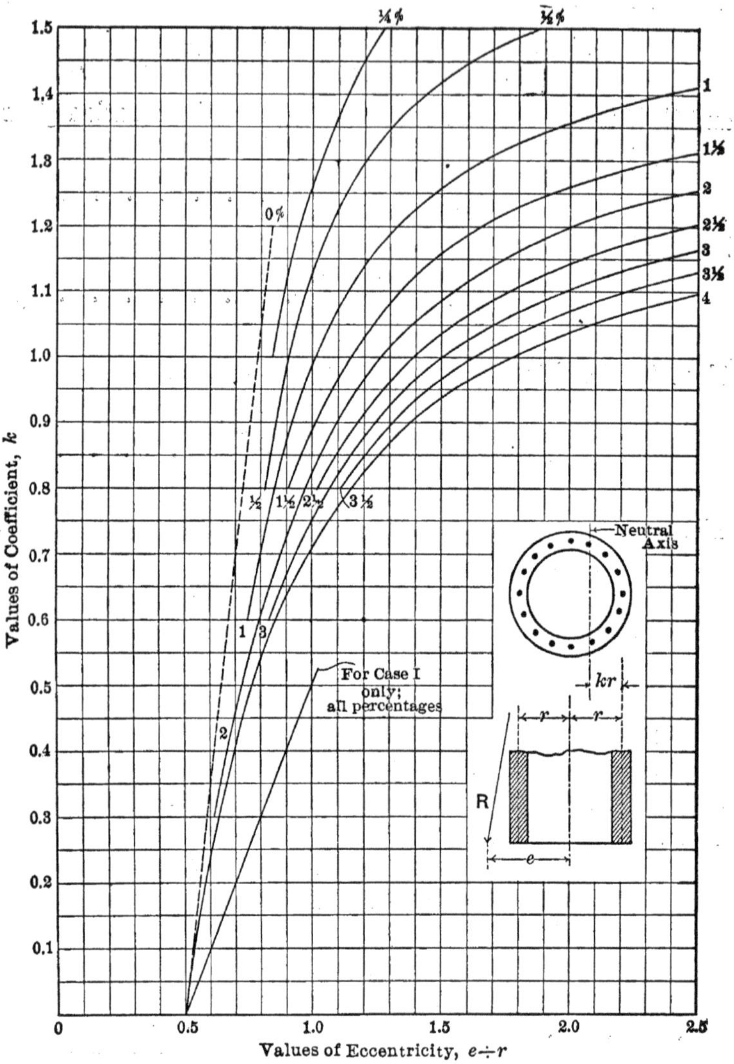

FIG. 134.—Neutral Axes in Chimneys.

greatest tensile unit stress is less than nf'. The first of these maxima is safe if f is safe, and the second is insignificant.

In computing the moment of inertia I, the steel sections must be weighted n-fold; thus let I_c=moment of inertia of the concrete section with respect to a diameter, and I_s=moment of inertia of all the steel sections with respect to the same line, then $I=nI_s+I_c$. Instead of using the actual steel sections to compute I, one may substitute with sufficient accuracy the section of a cylindrical shell rolled from the steel, the mean radius of the shell being r. Then, as the concrete sectional area is practically the same as the total,

$$I=nA_s\tfrac{1}{2}r^2+A\tfrac{1}{4}(r_1^2+r_2^2)=\tfrac{1}{4}A(2npr^2+r_1^2+r_2^2).$$

For Ex. (1), $I=\frac{1}{4}\ 2636(30\times 0.0027\times 70^2+67^2+73^2)=$ 6,731,685 in⁴; hence $f=111$ and $f'=-7$ lbs/in², the negative sign indicating tensile stress at the windward side. For Ex. (2), $I=6{,}970{,}000$ in⁴, $f=334$ and $f'=-126$ lbs/in².

209. *Case II.*—The eccentricity is so great that the resultant stress at the windward side is a tension whose intensity is so high that the concrete has been cracked or is near the cracking stage; in other words, the tensile stress condition resembles somewhat that at the section of maximum moment in an ordinary reinforced-concrete beam under full safe load. In the computation for this case the tensile strength of the concrete will be entirely neglected, as is almost universally done nowadays for concrete beams.

Practical formulas for unit stresses based on this "common theory" cannot be deduced for this case. But a diagram can be constructed by means of which unit stresses can be easily determined for any section of a given chimney; also the amount of vertical reinforcement required at any section of a given concrete shell can be readily determined by it. Such a diagram will now be described by example, and then the analysis on which its construction is based will be given. At the base (Fig. 135) there are given values of "eccentricity"

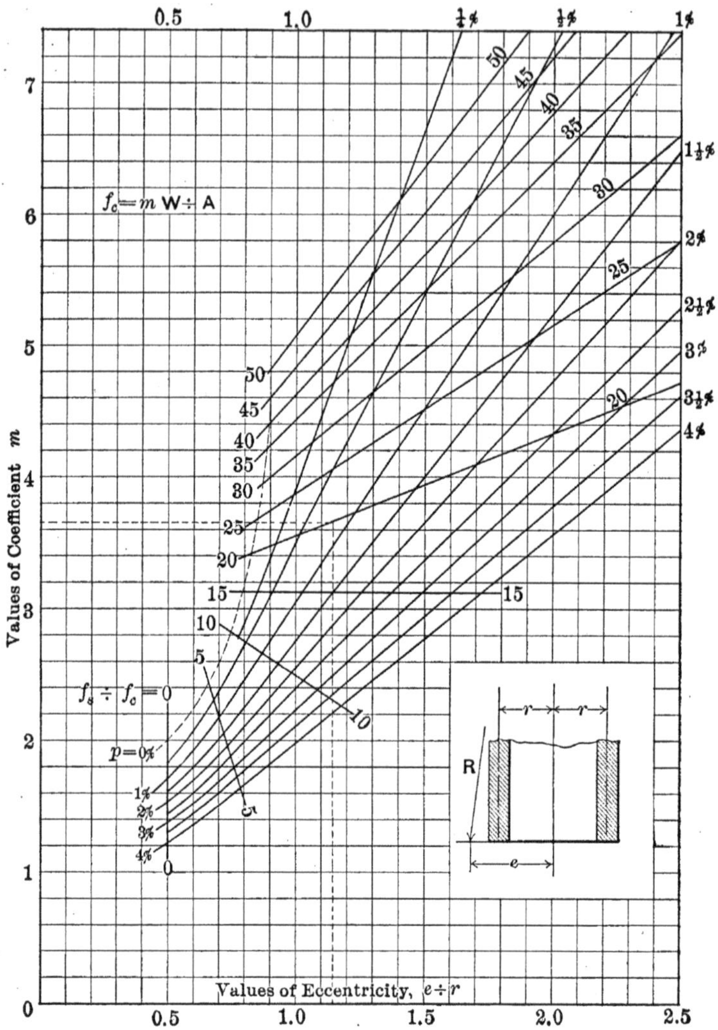

FIG. 135.—Wind Stresses in Chimneys

e/r, and at the left side values of m. The curves relate to various percentages of steel, from 0 to 4%, and will be called percentage ines; and the straight lines relate to various ratios of f_s/f_c and will be called ratio lines.

Examples.—(1) Ex. (2), Art. 208, will be used as illustration. From its solution, $W=275,000$ lbs., $M=21,900,000$ in-lbs., $p=0.8\%$, and $A=2.636$ in^2 ; hence $e=21,900,000/275,000=79.6$ in., and $e/r=79.6/70=1.14$; also $W/A=104$ lbs/in^2. Now entering the diagram at $e/r=1.14$, we trace vertically upward to a point corresponding to an 0.8 percentage-line, and then horizontally to the left side, taking out the value $m=3.65$. We also note that the turning point is practically at the 20 ratio-line. Hence $f_c=3.65\times104=380$ lbs/in^2, and $f_s=20\times380=7600$ lbs/in^2. It may be noticed that f_c is not the unit stress at the remotest fiber, and hence not the maximum compressive unit stress in the section. The maximum can be readily computed from

$$f=f_c+(f_c+f_s/n)t/4r,$$

in which t denotes thickness of the concrete shell. (The formula may be deduced from similar triangles in the lower part of Fig. 136.) Here

$$f=380+(380+500)\ 6/280=399 \text{ lbs/in}^2.$$

(2) How much vertical reinforcement is needed at the base of the chimney, the working strengths of concrete and steel being limited to 500 and 15,000 lbs/in^2 respectively?

W is about 456,000 lbs., the wind pressure about 54,750 lbs., and M about 4,106,000 ft-lbs.; hence $e=4,106,000/456,000=9$ ft. $=108$ in., and $e/r=108/69=1.56$. The section area A is 3490 in^2, and $W/A=130$ lbs/in^2; hence if the amount of steel is just sufficient to make $f_c=500$, then $m=500/130=3.85$. Now entering the diagram at $e/r=1.56$ and $m=3.85$, we trace vertically and horizontally from these places respectively, and note the intersection at about $p=1.9\%$ and $f_s/f_c=19$. With this percentage of steel, $f_s=19\times500=9500$ lbs/in^2. (This is a low working stress; use of a thicker shell will make higher values possible without increase of amount of steel. Several trial sections with different thicknesses may be quickly analyzed by means of the diagram, and an economical size determined.)

Analysis for the Diagram.—The ring $NPNQ$ (Fig. 136) represents a section of a chimney, NN the neutral axis, and

NQN the compression area, the wind blowing from the right. In addition to the foregoing notation, let

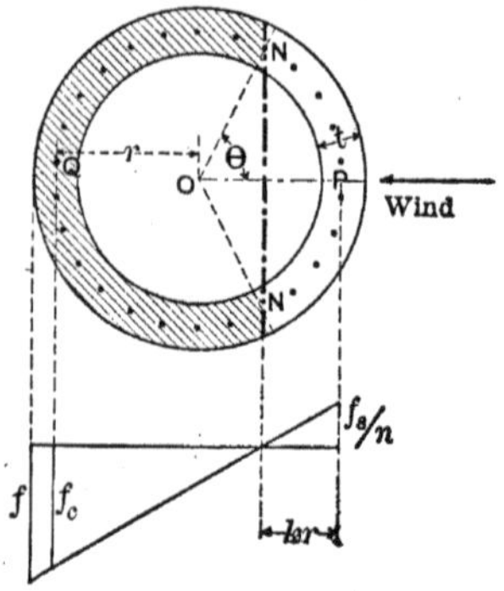

FIG. 136.

C_c=resultant compresive stress in the concrete;

C_s=resultant compresive stress in the steel;

T=resultant tensile stress in the steel;

a_c=arm of the resultant compression C_c+C_s with respect to the center O; and

a_t=arm of the resultant tension T with respect to the center O.

The resultant normal stress on the section equals the bending moment, that is,

$$C_c+C\ -T_s=W \quad . \quad . \quad . \quad . \quad . \quad . \quad . \quad (1)$$

and

$$(C_c+C_s)a_c+T_sa_t=M. \quad . \quad . \quad . \quad . \quad . \quad . \quad (2)$$

These two equations constitute the basis of the solution; however, they must be modified considerably, and this will be done presently. Let

x_1=distance from *NN* to the centroid of the arc *NPN*;
x_2=distance from *NN* to the centroid of the arc *NQN*;
y_1=radius of gyration with respect to *NN* of arc *NPN*;
y_2=radius of gyration with respect to *NN* of arc *NQN*; and
θ=angle *NOP*.

Now the average unit compressive stress in the concrete is $f_cx_2/(1+\cos\theta)r$, that in the compressive steel is $nf_cx_2/(1+\cos\theta)r$, and that in the tensile steel is $nf_cx_1/(1+\cos\theta)r$. And since the area of the section of the compressive concrete is practically $A(1-\theta/\pi)$, that of the section of the compressive steel $pA(1-\theta/\pi)$, and that of the tensile steel $pA\theta/\pi$, it follows that

$$C_c=A(1-\theta/\pi)f_cx_2/(1+\cos\theta)r,$$

$$C_s=pA(1-\theta/\pi)nf_cx_2/(1+\cos\theta)r,$$

and $$T_s = pA(\theta/\pi)nf_c x_1/(1+\cos\theta)r.$$

These values of C_c, C_s, and T_s substituted in eq. (1) give

$$(1-\theta/\pi)(1+np)x_2/r-(\theta/\pi)np\,x_1/r=(1+\cos\theta)W/Af_c. \quad (3)$$

From any source of information on the centroid of a circular arc, it can be shown that

$$x_1=r\left(\frac{\sin\theta}{\theta}-\cos\theta\right) \quad \text{and} \quad x_2=r\left(\frac{\sin\theta}{\pi-\theta}+\cos\theta\right).$$

Imagining these values substituted in eq. (3), it will be seen that the left-hand member is a function of θ, n, and p only. Denoting this function by $F_1(\theta, n, p)$, the equation can be written thus,

$$f_c=\frac{1+\cos\theta}{F_1(\theta, n, p)}\frac{W}{A}; \quad . \quad . \quad . \quad . \quad . \quad . \quad . \quad (4)$$

hence, also,

$$m=\frac{1+\cos\theta}{F_1(\theta, n, p)}, \quad . \quad . \quad . \quad . \quad . \quad . \quad . \quad (5)$$

that is, m depends on θ, n, and p only.

Referring to statement in the preceding article about the point of application of resultant stress and to Fig. 136, it will be seen that

$$a_t=y_1^2/x_1+r\cos\theta.$$

And since the resultants C_c and C_s are practically equally distant from the neutral axis, the arms of C_c+C_s and C_s are practically equal; that is, approximately,

$$a_c=y_2^2/x_2-r\cos\theta.$$

If now these values of a_t and a_c and those for C_c+C_s and T_s be substituted in eq. (2), it will reduce to

$$\left(1-\frac{\theta}{\pi}\right)(1+np)\frac{x_2}{r}\left(\frac{y_2^2}{x_2r}-\cos\theta\right)+\frac{\theta}{\pi}np\frac{x_1}{r}\left(\frac{y_1^2}{x_1r}+\cos\theta\right)$$
$$=(1+\cos\theta)\frac{W}{Af_c}\frac{e}{r}. \quad . \quad . \quad . \quad . \quad . \quad (6)$$

It can be shown that $y_1^2=r^2[1+\frac{1}{2}\cos 2\theta-\frac{3}{4}(\sin 2\theta)/\theta]$ and $y_2^2=r^2[1+\frac{1}{2}\cos 2\theta+\frac{3}{4}(\sin 2\theta)/(\pi-\theta)]$. Imagining these values of y_1 and y_2 substituted in eq. (6), it will be seen that the

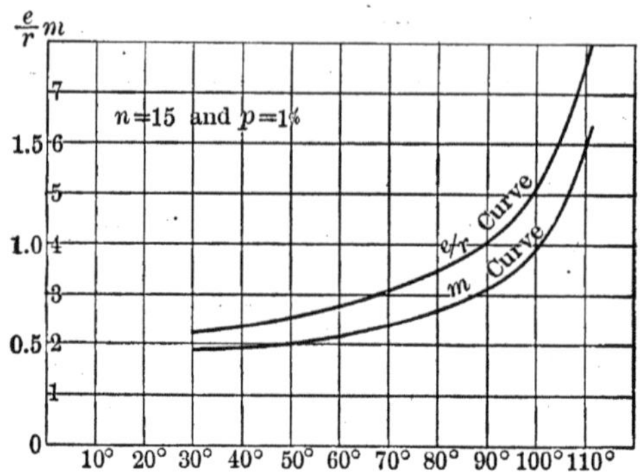

FIG. 137.

left-hand member is a function of θ, n, and p only. Denoting this function by $F_2(\theta, n, p)$, the equation can be written,

$$(1+\cos\theta)(e/r)W/Af_c=F_2(\theta, n, p). \quad . \quad . \quad . \quad (7)$$

Division of eq. (7) by eq. (4) gives

$$\frac{e}{r}=\frac{F_2(\theta, n, p)}{F_1(\theta, n, p)}, \quad . \quad . \quad . \quad . \quad . \quad . \quad . \quad . \quad (8)$$

that is, e/r depends on θ, n, and p only.

Equations (5) and (8) are the desired modifications of eqs. (1) and (2). If both be plotted on a θ base for a given set of values of n and p ($n=15$ and $p=0.01$, say), a pair of curves results as sketched in Fig. 137, from which may be taken the value of m for any value of e/r. Finally, if such simultaneous values of m and e/r (corresponding to one and the same value of θ) be taken off from this pair of curves and these values be plotted on an e/r base, the resulting curve is the 1% line of the diagram, Fig. 135.. In a similar manner the other percentage lines can be determined.

The ratio lines may be determined as follows: From similar triangles in the lower part of Fig. 136 it is plain that

$$f_s/nf_c=(1-\cos\theta)/(1+\cos\theta), \quad \text{or} \quad f_s/f_c=n\tan^2\theta/2.$$

From this equation, it appears that for $f_s/f_c=5$, say, and $n=15$, $\theta=60°$, irrespective of the value of p. Now the values of e/r for $\theta=60°$, and $p=0$, $\frac{1}{2}$, 1, $1\frac{1}{2}$, etc., may be read off from the corresponding pairs of curves (like Fig. 137), and these values of e/r may be marked off on the corresponding percentage curves in the diagram; the points so marked off fix the $f_s/f_c=5$ line. In a similar manner the other ratio lines can be determined.

Since $kr=r-r\cos\theta$ (see Fig. 136), $k=1(1-\cos\theta)$. From this formula and Fig. 137 the value of k may be obtained for any value of e/r and $p=1\%$. In this way the 1% curve in Fig. 134 was obtained; and in a similar way the others.

210. Wind Pressure.—Recent experiments made on the Eiffel Tower and at the National Physical Laboratory of England show that the pressure per square foot on square flat surfaces from 10 to 100 square feet in extent is 0.0032 times the square of the wind velocity in miles per hour. There is some evidence that the pressure on a cylindrical surface is about two-thirds that which would exist on an axial section of the cylinder ("projected area"). On the basis of the above, 20 pounds per square foot of projected area is a safe value for chimneys. The Prussian Regulations permit use of 17 pounds per square foot; in American practice considerably higher values are used.

211. Temperature Stresses.—Fig. 138, which is from a photograph, shows plainly some large vertical and horizontal cracks in the outer shell of a chimney. The vertical cracks are doubtless due to temperature and probably the horizontal ones also. For since the inner part of the shell is hotter than the outer, the inner tends to expand more circumferentially and vertically than the outer, so that it stretches the outer part and is itself compressed vertically and circumferentially by the outer part. If the circumferential or vertical tensions in the outer part are excessive, the concrete will crack on vertical or horizontal planes respectively.

212. *Circumferential Temperature Stress.*—The following are formulas for the greatest unit compression in the concrete and the unit tension in the steel at any place in a chimney:

$$f_c = \tau K E_c m_c \quad \text{and} \quad f_s = \tau K E_s m_s,$$

m_c and m_s being certain multipliers which depend on the percentage of hoop reinforcement, the position of the hoops

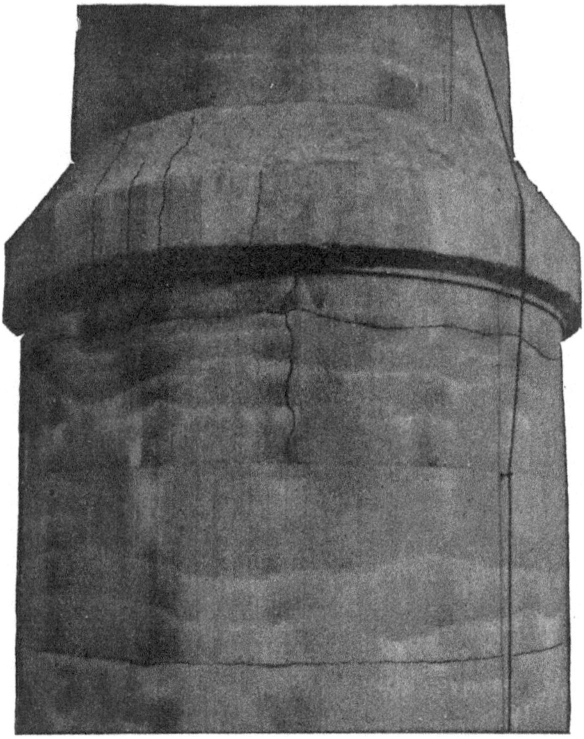

FIG. 138.

relative to outer and inner surfaces of the shell, and the ratio of outer to inner radii. For meaning of the other symbols see page 417. Formulas for the multipliers are complicated (page 419) but can be made available for practical use by graphical means. They have been plotted in a diagram (Fig.

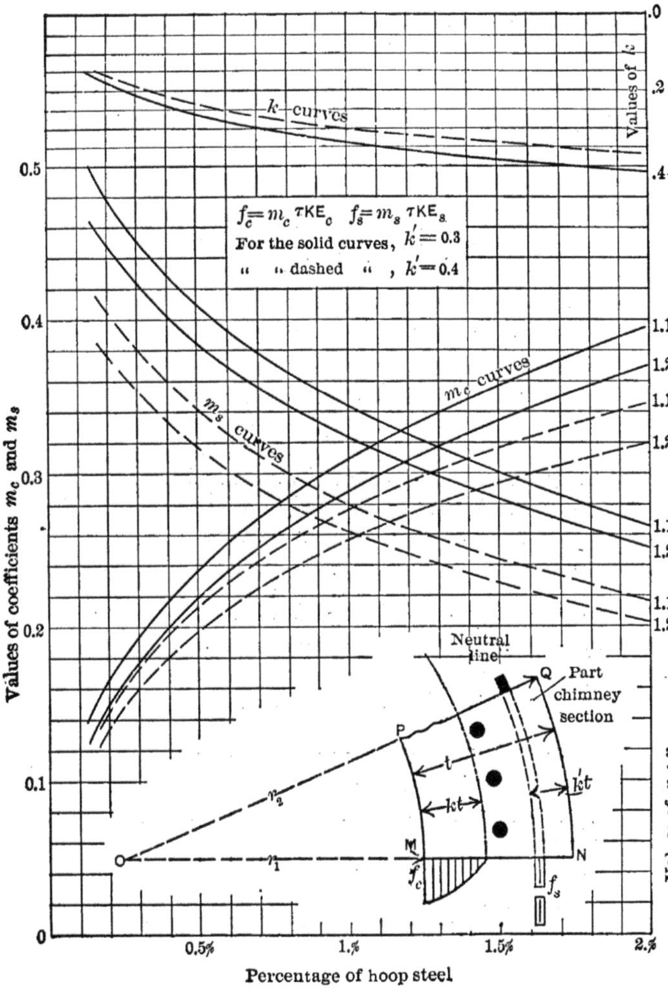

k-curves
Values of k
.0
.2
.4
$f_c = m_c \tau KE_c$ $f_s = m_s \tau KE_s$
For the solid curves, $k' = 0.3$
" " dashed " , $k' = 0.4$
m_c curves
m_s curves
1.1
1.2
1.1
1.2
1.1
1.2
1.1
1.2
Values of coefficients m_c and m_s
0.5
0.4
0.3
0.2
0.1
0
Neutral line
Q
Part chimney section
P
t
kt
k't
r_2
r_1
O
M
N
f_c
f_s
Values of $r_2 \div r_1$
0.5%
1.%
1.5%
2.%
Percentage of hoop steel

FIG. 139.

139) for percentages of steel from 0.2 to 2, for proportionate depths of steel from 0.3 to 0.4, and for ratios of outer to inner radii from 1.1 to 1.2. Inspection of the diagram shows:

(1) The higher the percentage of steel, the lower is the unit stress in the steel f_s and the higher is that in the concrete f_c.

(2) Increasing the thickness of the shell within practical limits decreases the unit stresses in concrete and steel but not materially, less than 10%.

(3) Moving the steel inward from $0.3t$ to $0.4t$ decreases both unit stresses, from 10 to 20% depending upon the amount of reinforcement.

Example.—The internal diameter of a chimney shell is 12 ft., the thickness of its walls is 6 in., the hoops are ½-in. rounds 10 in. apart, and are placed so that the center of the steel is 2 in. from the outer surface of the shell. What are the temperature stresses in steel and concrete due to a temperature difference of 200° F.?

The percentage of steel is $0.1964/(6\times10)=0.0033=1.3\%$, $r_2/r_1=1.1$, and $k'=2/6=1/3$. For $k'=0.3$, $p=1\%$ and $r_2/r_1=1.1$, the diagram gives $m_c=0.18$ and $m_s=0.375$; for $k'=1/3$, $m_c=0.19$, and $m_s=0.40$ about. Hence if $K=0.000006$, $E_c=2{,}000{,}000$ and $E_s=30{,}000{,}000$ lbs/in²,

$$f_c=200\times0.000006\times2{,}000{,}000\times0.19=456$$

and

$$f_s=200\times0.000006\times30{,}000{,}000\times.40=14{,}400 \text{ lbs/in}^2.$$

Analysis for the Diagram.—In Fig. 139, $MNPQ$ represents a portion of a horizontal section of a chimney and O the center of the section. The "neutral line" represents the neutral (cylindrical) surface which is not stressed, within and without which the concrete is under compression and tension respectively. The distance of the circumferential steel from the outer surface of the shell is called $k't$; r denotes the radius of any circumferential "fiber," C the total compressive stress on the vertical section MN per foot of height, and T the total tension in the circumferential steel per foot of height. Also let

t = thickness of concrete shell at the section under consideration;
r_1 = inner radius of shell;
r_2 = outer radius of shell;
τ = difference of temperatures of concrete at outer and inner faces;
K = coefficient of expansion of concrete and steel;
E_c = modulus of elasticity of concrete in compression;
E_s = modulus for steel;
f_c = temperature unit stress in concrete (circumferential) at inner face; and
f_s = temperature unit stress in circumferential steel.

In this analysis the tensile value of the concrete is neglected and an average modulus of elasticity for concrete in compression is assumed for all unit stresses, just as in computations on the strength of reinforced-concrete beams. The temperature gradient is assumed to be straight and the coefficients of linear expansion for concrete and steel are taken as equal and constant for all temperatures involved (see Art. 28). Furthermore, it is assumed that the thickness of the shell remains unchanged and that the radii of all circumferential fibers are increased equally. (Strictly this is not true for there are radial expansions and contractions accompanying the circumferential stresses, and then there is radial shortening accompanying the radial compressive stress; but these are small and their observance is out of place in this analysis involving as it does, approximation of a larger order. On account of the unequal vertical expansions there will be circumferential expansion at the top, the chimney "belling" out there, and some contraction lower down. In the following analysis, these are neglected; the resulting error is probably small except for stresses near the top. The assumption that the modulus of elasticity is constant for the range in temperature may be quite erroneous, and if so, the analysis following is in error; if the modulus is lower for the higher temperatures then the actual unit stresses are lower than those here found.)

If Δr denotes the radial increase mentioned, then the actual elongation of each circumference is $2\pi\Delta r$, and the actual unit elongation is $\Delta r/r$. The temperature at the distance r from the center (reckoned from the temperature at the outer surface as zero) is $\tau(r_2-r)/t$, and hence the free elongation of the circle of radius r would be $2\pi rK\tau(r_2-r)/t$ and the free unit elongation $K\tau(r_2-r)/t$. The difference between the free and the actual unit elongations is the prevented unit elongation, and hence the corresponding preventing unit stress in the concrete is

$$f=\left[\frac{\tau}{t}(r_2-r)K-\frac{\Delta r}{r}\right]E_c; \quad . \quad . \quad . \quad . \quad . \quad (1)$$

for the steel, r becomes equal to $r_2-k't$ and

$$f_s=\left[\frac{\Delta r}{r_2-k't}-\tau k'K\right]E_s. \quad . \quad . \quad . \quad . \quad . \quad (2)$$

At the neutral surface $f=0$, and $r=r_1+kt$; hence on substituting these in eq. (1), it will be found that

$$\Delta r=\tau K(1-k)(r_1+kt). \quad . \quad . \quad . \quad . \quad . \quad . \quad . \quad (3)$$

A value of the compression at the inner face may be obtained from (1) by substituting for r its value there and for Δr its value from eq. (3); the expression will reduce to

$$f_c=\tau KE_c[1-(1-k)t/r_1]k. \quad . \quad . \quad . \quad . \quad . \quad . \quad (1)'$$

A new value of f_s can be obtained from (2) by substituting for Δr its value from eq. (3); it will reduce to

$$f_s=\tau KE_s\left[\frac{(1-k)(1+kt/r_1)}{(r_2/r_1)-k't/r_1}-k'\right]. \quad . \quad . \quad . \quad (2)'$$

These expressions for f_c and f_s contain only quantities ordinarily known and k; it remains to determine k. This is done by means of the condition that C and T are equal. Now,

$C=\int_{r_1}^{r_1+kt} f\,12dr$ and $T=f_s p\,12t$; and if in these, the values of f and f_s from eqs. (1) and (2) be inserted, and the resulting expressions for C and T be equated, and then if the operations indicated in the equation be performed, the following may be arrived at:

$$\Delta r = K\tau t(k-\tfrac{1}{2}k^2+npk')(r_2/t-k') \div [np+(r_2/t-k')\log(1+k't/r_1)]$$

The logarithm is Naperian, or natural. Equating this value of Δr to that given by eq. (3) and simplifying, one may arrive at

$$\frac{k-\frac{1}{2}k^2+npk'}{(1-k)(k+r_1/t)}=\log(1+kt/r_1)+\frac{np}{r_2/t-k'}. \quad . \quad . \quad (5)$$

Now this equation determines k for given values of k', n, and r_2/r_1, and yet it cannot be solved for k on account of the logarithmic term. However, values of p can be determined for given values of k, k', n, and r_2/r_1, and thus a diagram can be made like the group of k curves (Fig. 139) but with much larger vertical scale, from which k may be taken off for any given case (p, k', n, r_2/r_1). Then this value may be used in eqs. (1)′ and (2)′ to determine the bracketed coefficients, that is, m_c and m_s.

213. *Vertical Temperature Stress.*—A satisfactory analysis is not available; the following approximation indicates the "order of magnitude" of these stresses. Horizontal sections through the unheated chimney are still horizontal after heating just as plane sections of a beam remain plane during bending. Hence the inner part of the chimney shell when hot is compressed while the outer part is stretched, and somewhere between there will be a neutral surface whose distance from the inner surface is here called kt (see Fig. 140). If the tensile strength of the concrete is neglected, then the entire vertical tension must be ascribed to the steel, and the neutral surface located between the steel and inner surface as shown.

The temperature difference between the inner surface and the neutral surface is τk, and the temperature difference between the neutral surface and the steel is $\tau(\frac{1}{2}-k)$; hence the unit stress at the inner surface f_c, and the unit stress in the steel f_s are given by

$$f_c = k\tau KE_c, \quad \text{. . . . . (1)}$$

and

$$f_s = (\tfrac{1}{2}-k)\tau KE_s. \quad \text{. . . . . . . (2)}$$

To determine k equate the total compression and the tota tension per unit of circumference; thus p denoting the (vertical) "steel ratio," or total vertical steel area divided by

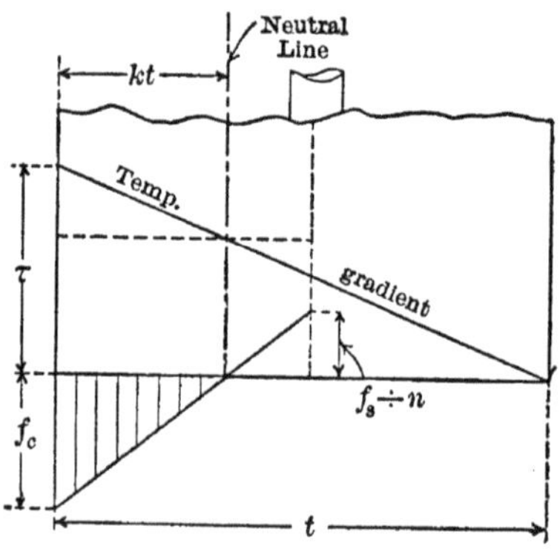

FIG. 140.

total area of cross-section of the shell, then $\frac{1}{2}k\tau KE_c kt = (\frac{1}{2}-k)\tau KE_s pt$, which simplified and solved for k gives

$$k = np[\sqrt{(1+1/np)}-1]. \quad \text{. . . . . . (3)}$$

From this equation the following table was computed, n taken as 15:

$p=$	0.05	0.10	0.15	0.20	0.25	0.30	0.35	0.40
$k=$	0.21	0.26	0.30	0.32	0.34	0.36	0.37	0.38

For 2% of steel, $\tau = 200°$, and $K = 0.000006$, formulas (1) and (2) give $f_c = 770$ and $f_s = 6500$ lbs/in^2.

214. Chimney Temperatures.—Sanford, in the report alluded to in Art. 206, states that "the temperature in an ordinary chimney seldom exceeds 700° F. at the base and 400 to 500 is more usual."

Whatever the difference between the temperatures of the chimney gas and the outer air, the difference τ between the temperatures of the concrete at the inner and outer face of the chimney shell (on which the temperature stresses depend) is less, for it is known that at the surfaces of a heat barrier there is a drop in temperature in the direction of the heat flow. And at the outer surface of a chimney the drop is considerable, as is known to any one who has placed his hand upon the surface on a cold day; to him it felt warm while the air temperature may have been zero or less. Lange* has computed the temperature drops at the surfaces of some brick chimneys (various diameters, thickness of walls, gas, and air temperatures) and while the computations seem to be based on uncertain values of thermal conductivity, emissivity, and absorption of the chimney material, still they are doubtless reliable enough to indicate that the temperature difference for a concrete shell may be as little as 50% of the temperature difference for the chimney gases and air.

215. Bases.—In the two succeding articles there are explained methods for computing the maximum pressures between the base of a given chimney and its earth or stone foundation, and the strength of the base to withstand those pressures and the forces on its top. The notation is as follows (see also Figs. 141 and 142):

W = total weight of chimney and earth filling over the base;
M = wind moment at the bottom of the base;
A = area of the bottom of the base;
p_2 = maximum unit pressure on bottom;
p_1 = minimum unit pressure on bottom;
r = kern radius of bottom in direction of wind;
e = eccentricity at bottom of resultant of the wind pressure and W, $e = M/W$.

* Der Schornsteinbau.

216. *Earth Pressures.*—It is assumed that the pressure is a uniform y varying one. If the resultant of the wind pressure on the chimney, the weight of the chimney and earth filling over the base cuts the kern of the bottom of the base, then there will be no tendency for the windward toe to lift, or, in other words, there will be no tendency to tension between the earth or rock foundations and the windward toe. Fig. 141 shows the kerns for square, octagonal, and circular bottoms. It is good practice to make the bottom, in a given case, large enough so that its kern will intercept the line of action of

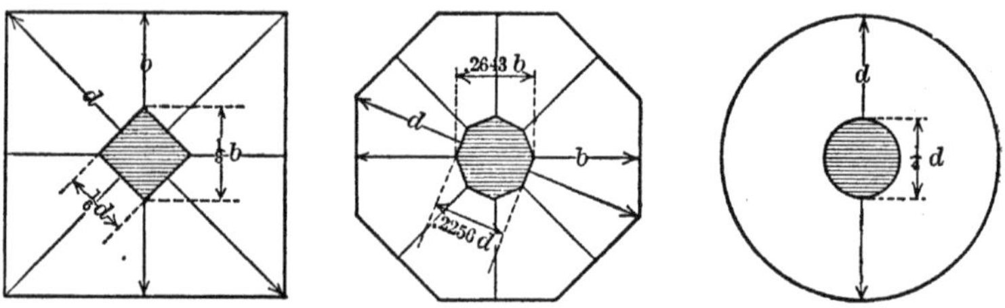

Fig. 141.

the resultant R; however, this is not an absolutely essential requisite for stability and safety.

If the resultant R cuts the kern then the computation for the greatest and least unit pressure is comparatively simple. These are the formulas:

$$p_1=\frac{W}{A}+\frac{M}{Ar} \quad \text{and} \quad p_2=\frac{W}{A}-\frac{M}{Ar}. \quad . \quad . \quad . \quad (1)$$

Evidently the greater pressure p_1 is a maximum when the kern radius is minimum; hence p_1 is maximum when the wind pressure is parallel to the longest diameter of the base.

If the resultant R does not cut the kern then there is a neutral axis as it were, that is, only a part of the bottom is under pressure. This neutral axis is perpendicular to the direction of the wind pressure. Three cases are noted:

(*a*) *Square Bases.*—If the direction of the wind is parallel to a side of the square, then the distance x of the neutral axis from the windward of the square is

$$x=3(\tfrac{1}{2}b-e), \quad . \quad . \quad . \quad . \quad . \quad . \quad . \quad (1)$$

and the greatest unit pressure is

$$p_1=\frac{4}{3(1-2e/b)}\frac{W}{b^2}. \quad . \quad . \quad . \quad . \quad . \quad . \quad . \quad (2)$$

These formulas follow from two equations obtained by equating the total pressure on the bottom to W, and the moment of that pressure about the diameter parallel to the neutral axis to $M=We$.

If the direction of the wind is parallel to a long diameter of the square, then the position of the neutral axis (see Fig. 142) is given by

$$\frac{1-\frac{1}{2}k^3(2-k)}{6(1-k)+k^3}=\frac{e}{r}, \quad . \quad . \quad . \quad . \quad . \quad . \quad . \quad (3)$$

and the value of the greatest unit pressure by

$$p_1=\frac{2-k}{1-k+\frac{1}{6}k^3}\frac{W}{b^2}. \quad . \quad . \quad . \quad . \quad . \quad (4)$$

In Fig. 142 there are two curves marked "square;" one gives values of k and the other values of m for use in $p_1=mW/b^2$. Eqs. (3) and (4) were deduced from the same principles employed to deduce (1) and (2).

(*b*) *Octagonal Bases.*—Exact formulas are not practical. Since an octagon does not differ much from a co-centric circle whose diameter equals the mean of the greatest and least diameters of the octagon, it must be that the neutral axis and the greatest unit pressure for an octagonal bottom do not differ materially from those for such

(*c*) *Circular Bases.*—In Fig. 142 there are two curves marked "circle;" one of these gives values of k and the other values of m for use in $p_1 = mW/A$, A denoting area of the circle and p_1 the greatest unit pressure on it.

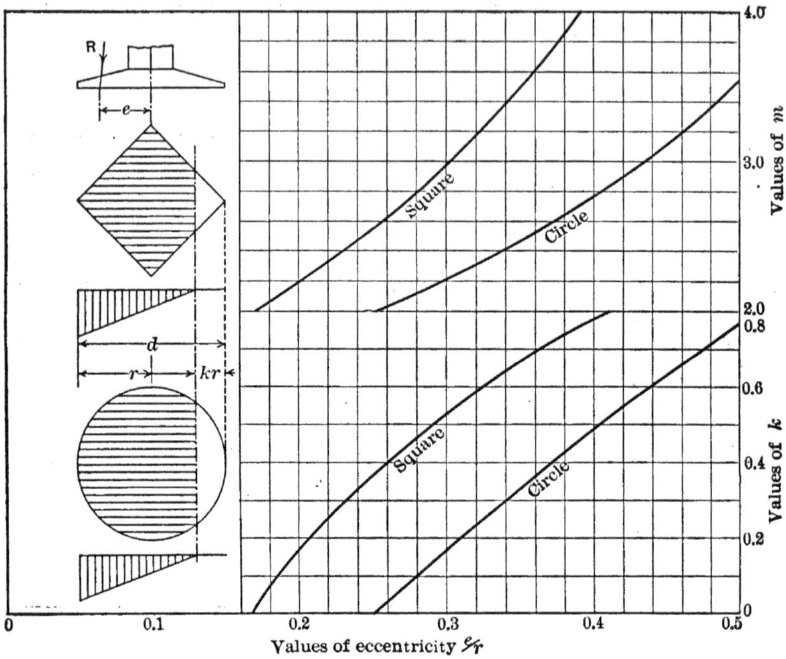

FIG. 142.

217. *Design of Bases.*—Having determined the diamete of the base from a consideration of the earth pressures as explained in the preceding article, it rema'ns to determine thicknesses of the bases and the reinforcement. Only rough approximate methods are available. A chimney base is essentially a column footing for methods of the design of which see Art. 169. But while column footings are regarded as always subjected to a uniform earth pressure, a chimney base should not be so regarded; and the outstanding cantilever part of a base should be figured for the earth pressure, distribution obtaining with the maximum wind pressure as explained in the preceding article. Also while the column

is solid, the chimney is not; and in the latter case tensile stresses may arise at the top side of the base at its center. Such tension will probably obtain when the inner diameter of the chimney is greater than twice the length of the outstanding cantilever and no wind blowing.

INDEX.

www.ingramcontent.com/pod-product-compliance
Ingram Content Group UK Ltd.
Pitfield, Milton Keynes, MK11 3LW, UK
UKHW032050260826
14628UKWH00008B/3292